Contemporary Chemistry Science, Energy, and Environmental Change

Contemporary Chemistry Science, Energy, and Environmental Change

Charles G. Wade

Macmillan Publishing Co., Inc.
New York
Collier Macmillan Publishers
London

Macmillan Publishing Co., Inc.
866 Third Avenue, New York, New York 10022

Collier-Macmillan Canada, Ltd.

Library of Congress Cataloging in Publication Data

Wade, Charles Gordon, (date)
  Contemporary chemistry.

  Includes bibliographies and index.
  1. Environmental chemistry.   2. Pollution.
3. Power resources.   I. Title.
QD31.2.W2   628.5   75–9579
ISBN 0–02–423650–0

Printing: 1 2 3 4 5 6 7 8   Year: 6 7 8 9 0 1 2

To my parents and my family

Unit V
p. 239-271   8, 10-13
331 -365   2,5-7,10-12,17
371-412   5-7,14,19,22

Preface

107-12 279-321
325-327
126-132
144-149

IV 5-11, 32
155-218
225-238

This book has developed from a course I have taught for several years to non-science majors, and it has the same goal as that course: to present to the non-scientist the scientific and technical aspects of environmental and energy problems. It attempts to give the reader a sound basis for considering the essential factors in the decisions that must be made in these areas in the next decades. The approach considers not only our present policies in these areas but also explores what science can predict about the effects of the many proposed alternatives to our present policies. Within the very limited scientific background assumed for the reader, the text stresses the limitations of our scientific knowledge of the environment but emphasizes the importance of scientific input when attempting to formulate policies relating to the topics in this book. In the discussions of environmental impact, I have attempted to make clear where solid scientific evidence exists, so that firm predictions can be made; where scientific information is meager, so that predictions must be based on assumptions; and where scientific information is virtually nonexistent, so that decisions must be made on intuition.

Since science seldom proposes an obvious choice between alternative solutions to an environmental or an energy problem, I have emphasized that value judgments must usually be applied to the existing scientific evidence. Because of my personal observations that political action is almost always involved in achieving positive change in environmental problems regardless of the state of scientific knowledge, I have included where appropriate some relevant economic and political realities. In several instances I have applied my own value judgments and argued for a particular solution or a particular policy. I have found that students in my classes, who have had majors in the arts, architecture, business, education, and social sciences, appreciate such an approach much more than one that merely presents facts.

Many of the examples used in the text are the result of trial and error in attempting to convey science to a large class. My approach is to focus on relevant problems, then to develop the science needed to discuss these problems. This is, therefore, not a science text that uses energy and environmental problems as examples, but rather a text that looks at environmental problems and the science relevant to them.

I have attempted to make the text useful to both the student and the teacher. One feature is a section in each chapter of term paper and research paper topics on material related to, but not discussed explicitly in, the chapter. Appropriate references for each of these "Questions Requiring Outside Reading" are found in the individual questions. Each chapter concludes with an extensive list of annotated references; most of these are in magazines, newspapers, or readily available general scientific publications.

Because of the rapid changes that were occurring relevant to many of the topics discussed herein as the book was being completed during 1975, the reader should expect to find that the thrust of the discussions relating to such topics as energy sources, automobile emission controls, and economics may deviate from the public emphasis which exists at the time of the reading.

I acknowledge the contributions of many scientific colleagues and other associates whose ideas and discussions have contributed to my own views. Very important also have been the hundreds of students who were sufficiently interested in these topics to challenge and to question and thus contribute to my own understanding. Drs. Cecil Dybowski and Richard Orwoll kindly read part or all of the manuscript and offered many constructive criticisms.

My wife, Kim, did much of the researching, proofreading and indexing.

Esther McCormick helped enormously with the proofreading and Bill McCormick supplied photographs for Chapter 7; both made many beneficial suggestions on various aspects of the manuscript.

Mr. James Smith and Mr. Leo Malek of Macmillan Publishing Company, Inc., were especially helpful and considerate. Their expertise made the entire project much easier.

Finally, special thanks are due to Rayna Kolb without whose organizational skills, expert typing, and unique ability to do things correctly this book (and many other projects) would have been an almost unendurable burden.

C. G. Wade
Austin, Texas

Contents

ix

3 Air Pollution   91

4   **Energy   155**

5 Energy Options: Fuels, Sources, Conservation and Policy   225

6 Water Pollution   279

7 Solid Waste and Recycling   330

8 Pesticides   371

9 Legalities, Politics, and the Future 421

Contemporary Chemistry Science, Energy, and Environmental Change

1

An Introduction to the Environment, Energy and Chemistry

The environment is similar to a ball of snakes.
If you pull on one snake, they all move.
John Muir

Complexity of the Environment

In this book we will attempt to analyze energy, resources, and environmental problems. We shall make an effort to be objective, pointing out limits of knowledge in given areas and contrasting the possible effects of various proposed solutions. In this initial chapter we consider some fundamentals that will form the basis for many discussions in subsequent chapters.

The dominant aspect of the environment is its complexity. The first venture of the layman or the scientist into environmental problems invariably leads to a keen awareness of the limits of man's knowledge in this area. Nearly everyone has an appreciation of the vast gaps in knowledge that exist in just one facet of the environment, human physiology. A perfect treatment of environmental problems would require a complete knowledge of the physiology of *all* organisms including, for example, how they react to a change in air pollutants. A *complete* solution to environmental problems is obviously beyond man's intellectual capacities. However, our knowledge in small areas is sufficient to permit very reliable evaluations. It is on these that we shall concentrate.

The Role of Science in Environmental Problems

Science is a systematic structure used to describe nature. Because our approach in this book focuses on the scientific aspects of environmental problems, we should begin by considering the role science may fulfill in the delineation of and solutions to these problems. Science concentrates on quantitative aspects of problems. The usual evolution of the scientific approach is to acquire data, construct a model (or theory) that is consistent with the data, and make predictions from this model.

The popular concept of science is that it is "exact," that is, it provides precise answers that are consistent with processes observed in nature. However, what science *really* does is provide answers that are consistent with the input data and with the constructed model. A scientific prediction is no more exact than the data and the constructed model. If either or both of these are inadequate or inaccurate, the prediction can be no better. If the model is too

1

narrow to reflect nature accurately, the scientific predictions based on the model must be viewed very carefully.

In carefully defined, simple systems, science can be considered exact. That is, it provides answers consistent with observations of nature. When applied to problems as complex as those involving the environment, however, science provides exact answers in disappointingly few cases. The complexity of the environment makes the data inadequate, dooms the model to a simplified approximation, and leads to genuine disagreement among scientists about the validity of the scientific predictions. The disagreements are, in substance, reflecting the lack of knowledge about the model and about the errors introduced by the necessary simplifying approximations. For environmental problems science provides relatively few answers that are exact, considerably more answers that are approximately correct, and many answers that are little more than slightly refined guesses.

Thus it would be a mistake to assume that science has all the correct answers to the problems we shall discuss. It would also be a mistake to assume that scientists agree on which solutions are preferred or, in fact, agree on which problems are the most important. The scientific opinions will change with time; next year's model may result in a completely revised view of the particular problem. Part of our task will be to distinguish between those scientific predictions that are on solid ground, not likely to change, and those that are considerably less reliable because of data or modeling limitations.

The technical complexity of environmental problems arises in part because of the large number of variables involved compared to the typical laboratory research. For example, chemists have for years studied **photochemical reactions,** chemical reactions that occur only during illumination with intense light, such as that from a flashbulb or a strong lamp. In the laboratory, a study of a photochemical reaction might involve mixing, at most, three compounds in a vessel. The reaction is initiated by light from a strong source. The reaction proceeds and products are formed. Analysis of the reaction products would involve the isolation and identification of, perhaps, four or five compounds. In this experiment the initial concentrations, the pressure, the temperature, and the intensity of the light source would be carefully controlled. This is a carefully defined experiment and it represents about the maximum number of components with which chemists may confidently make accurate models and predictions.

Now, a very important photochemical reaction using light from the sun occurs in nature: the formation of smog. It provides a striking contrast to the simple laboratory experiment. Smog formation involves hundreds of compounds: over 1000 chemical species have been identified in the Los Angeles atmosphere. The sources of these chemicals are varied (automobiles, power plants, industrial plants) and have outputs that change with the time of day. Furthermore, smog formation occurs under a bewildering array of conditions:

varying pressure, temperature, wind velocity, humidity, and seasons, to mention just a few. Even the intensity of the light source, the sun, changes because of cloud formations and weather conditions. Identifying the key components of smog formation, forming a model, and proposing satisfactory cures is an enormous task.

Science does have an important role, however. It provides relevant data on the present status of environmental problems, and, even with its modeling limitations, science is still the best available means of indicating the problems and their possible solutions.

A rational solution to environmental problems must inevitably *begin* with the quantitative aspects, those supplied by science. In particular, the beginning should include where our present course has taken us, what solutions might be tried, and what the impact of those solutions might be. After the best available answers to these questions have been compiled, a second important aspect of rational decision making must occur: value judgments, the weighing of individual aspects of the problem. For this aspect of problem solving, considerably more than science is required. Social, economic, legal, and political decisions must be considered in many cases. The many aspects involved in a solution of environmental problems are best emphasized by consideration of a prime example, the automobile. In addition to being the source of over half of our air pollutants, the auto contributes to noise, to congestion, to highway proliferation, and to urban blight. It kills 50,000 persons per year in the United States, is an enormous user of resources, and has a lifetime of only a few years. Even when discarded it is a problem; the hulks are virtually valueless as raw materials. Yet when a recent national poll pointed all this out and asked if the automobile were worth it, 80% of the respondents said yes, though half of these indicated their answer would be different if an alternative to the auto were offered.

The automobile is obviously important in other ways: it transports 90% of us to work, to recreation, and to shopping. It is a dominant economic factor involving labor, the oil and gas industry, the steel industry, and the highway construction industry. In addition the auto provides convenience, a free lifestyle, a prominent indication of affluence, and, in some cases, a measure of sexuality. Any solutions to the environmental problems created by the car will affect all these factors. It is instructive to consider what might be involved in just three of the major proposals made to reduce the environmental impact of automobiles: limitation of vehicle emissions, conversion to electric cars, and limitation of auto access to the inner city.

The authority to establish emission limits is vested in a government agency, the Environmental Protection Agency (EPA). Presumably the EPA decision will involve a relative balance of health improvement benefits, present technology (how and by how much emissions can be reduced), and economic considerations (how much emission controls cost and who pays). In view of the broad impact of the auto and auto industry, one might expect other factors will be involved, especially political action through Congress by many industries, labor organizations, and consumer groups.

Conversion to electric cars also involves questions beyond mere technical aspects. A conversion from present autos to electric vehicles would require an increase in electrical power generation, thus adding an extra burden to an industry already in a crisis situation. The air pollution from transportation would be shifted from automobiles to power plants. Would that result in an overall improvement of United States air quality? Considerations of the total amount of fuel required in a gasoline engine versus an electric engine are also involved. If the electric cars are battery driven, provision must be made for recharging facilities at service stations. If present auto assembly plants are not utilized for making electric cars, what will happen to the workers in those areas?

Any decision to limit auto access to cities is fraught with political consequences. Such a limit would presumably be imposed with an eye toward development of alternative transportation (mass transit or car pools) for those dispossessed from cars. Mass transit development requires (as do highways) considerable public funding and therefore must compete with other worthy causes for its share of the federal budget; a political decision is thus required. Futhermore, once the mass transit exists, the public must be "reeducated" to its use because they are "conditioned" to relying on the automobile.

It must be clear, as a brief summary of the complexities of *one* source of an environmental problem indicates, that, although our approach will concentrate on technical aspects, we must keep in mind the important nontechnical factors such as economics and politics, for they are crucial to environmental improvement. Consequently, we shall give some attention to political and administrative ramifications in our discussions.

Energy and Thermodynamics

A theory is the more impressive the greater is the simplicity of its premises, the more different are the kinds of things it relates and the more extended is its range of applicability. Therefore, the deep impression which classical thermodynamics made upon me. It is the only physical theory of universal content which I am convinced that within the framework of applicability of its basic concepts, will never be overthrown.

Albert Einstein

Although environmental problems have a technical complexity, they are of course governed by the same physical laws one deduces from the study of less complex phenomena. As we proceed in our environmental study it will behoove us to consider what some of these laws can predict. The most fundamental of all physical laws are those relating to energy.

Energy is a fundamental ingredient to life on earth. Food, warmth, water, and weather cycles are a few of the benefits from just one form of energy, that of the sun, whose importance was recognized in antiquity. Engergy in the form of light was recorded in Genesis as the first gift to earth. The history of man is closely related to his understanding, development, and use of energy in its various forms. You undoubtedly have some intuitive feeling for the concept of energy. From an early age you have heard of the energy provided by food, of the energy "possessed" by an active child, and of the beneficial and harmful aspects of nuclear energy. Recently energy has become a popular topic for the media, largely because of the energy "crisis," fuel shortages, and increasing prices of fuels and electricity.

These few examples illustrate the variety of forms in which energy may exist. It is important that we carefully define three related terms: energy, work, and power. In common usage these three are often used interchangeably, but in scientific usage careful distinctions exist.

Energy

All matter and all things possess energy, though it may exist in a variety of forms. A rock possesses energy by virtue of its temperature, by virtue of the forces that hold it together, and perhaps by virtue of its altitude above the surface of the earth. A thrown rock has an additional form of energy. Energy is broadly classified into two types: kinetic energy and potential energy.

Kinetic energy is the energy of motion. The wind, a moving car, a light beam, a moving rifle bullet, and a falling object, all possess kinetic energy by virtue of their movement. Kinetic energy is difficult to store. Consequently it must be used at the time it occurs. **Potential energy**, on the other hand, is energy that can be stored for use at a later time. Examples of potential energy are a stone in a stretched slingshot, water in a tall tank (or behind a dam), and fuels such as coal and gasoline. All of these represent a quantity of energy that can be made available at an appropriate time.

Obtaining the services of potential energy usually requires its conversion into some other form of energy, usually kinetic energy. For example, the stone released from the slingshot loses potential energy but gains kinetic energy because of its motion. Releasing water from a tall tank or over a dam causes a similar conversion: from potential to kinetic energy. The potential energy in fuels is chemical in nature; it can be released by burning the fuel and changing the chemical nature of the constituents. In this conversion, the (chemical) potential energy is converted to kinetic energy, specifically into heat, which is a form of kinetic energy. Although conversion between the various forms of energy is possible, it rarely happens that one form of energy can be completely converted into another form. In most conversions some energy is inevitably lost. An understanding of energy conversion is fundamental to our view of energy in the environment.

Work

Work is a term usually associated with man's conversion of energy for some useful purpose. It provides a quantitative measure of that portion of the energy which accomplishes the useful purpose. That is, work denotes the special achievement of some portion of some form of energy. In the scientific definition work occurs or is done only when a force acts to *move* an object. Pushing on an immovable object such as a wall is *not* work. Only if the object moves in response to an exerted force is work done. An engine does work when it accelerates an automobile, a weight lifter does work when he lifts a barbell, and the wind does work when it rotates a windmill. In all of these cases, motion of a body is involved.

Power

In scientific use, power measures the amount of energy delivered or the amount of work done in a *time interval*.

$$\text{power} = \frac{\text{energy flow or work done}}{\text{time required}} \tag{1.1}$$

Power, since it depends on time, is thus quite a separate entity from energy or work. To illustrate, suppose a car, initially at rest, is accelerated until it reaches a speed of 60 miles per hour. It has a kinetic energy by virtue of its motion, and it takes a fixed amount of work to achieve 60 miles per hour for that car regardless of the time required to do it. However, it takes *twice* as much power to achieve 60 miles per hour in 15 seconds as it does to achieve it in 30 seconds. To reach 60 mph in 15 seconds will require the gasoline to be consumed twice as fast as it would be consumed at the slower acceleration.

Thermodynamics

Energy is of such fundamental importance that its study is one of the oldest and most thorough scientific interests. **Thermodynamics** is the study of energy, energy conversion, and work. The word "thermodynamics" is derived from "heat flow" and dates from the nineteenth century when the relationships between heat and energy were being investigated. With the possible exception of mathematics, no scientific discipline has wider use. The relationships that govern energy, its conversion to work, and its conversion to heat are among the most fundamental in science, and they are expressed elegantly by three concise, fundamental laws, **the laws of thermodynamics.** These three "laws," which are really scientific axioms, were slowly formulated, with extensive observation and experimentation, during the nineteenth and twentieth centuries. Although they are neither derived from a central theory nor written in stone by a lightning bolt, their demonstrated accuracy and validity, as applied to a multitude of problems, have led to universal acceptance.

The mathematical formulation of the laws of thermodynamics can be applied rigorously to a variety of topics. For example, thermodynamics yields the amount of energy available from fuels and chemical reactions and how this amount of energy will change with temperature and pressure. The maximum possible work available from engines and electrical power generators can be determined from thermodynamic laws as can the maximum amount of electrical energy that might be generated by a solar power scheme. If several chemical reactions might occur in a biological system, thermodynamics can predict the favored reaction. The three laws even demonstrate the impossibility of all perpetual motion machines. Adding to the beauty of thermodynamics is the fact that the laws can be stated concisely and logically in words as well as presented in a mathematical form for science and engineering applications. We will not consider the mathematical formulations, but we will attempt to develop an intuitive feeling for implications of the three laws.

To do this we must introduce the concept of the system and the surroundings. The **system** is a portion of the universe isolated from the remainder of the universe (the **surroundings**) by a conceptual boundary suitably chosen to fit the problem at hand. For example, to study the thermodynamics of home heating, one might choose a single room as the system; the surroundings are then the rest of the universe including the rest of the house. In another case, one might treat the heating unit as the system with the house and the remainder of the universe as the surroundings. Thermodynamics treats heat flow across the boundary, the energy changes in the system and the surroundings, and the work done.

First Law of Thermodynamics. The first law is the law of conservation of energy. It may be stated as:

> In all chemical and physical changes, energy is neither created nor destroyed, but merely transformed from one form to another.

or

> You can't get something for nothing.

The first law establishes the relationships between energy, heat flow, and work done. Energy can be neither created nor destroyed—merely transformed from one form to another, such as potential energy to kinetic energy. If the *system* gains or loses energy in any form, then an equal amount of energy in some form must be lost or gained, respectively, by the *surroundings*. The total energy, the sum of both, is constant for any process. From the first law, we

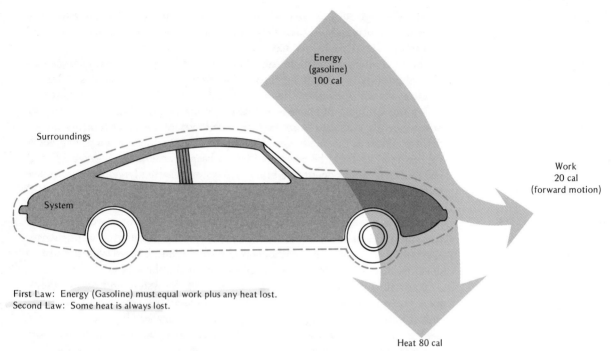

Energy
(gasoline)
100 cal

Surroundings

Work
20 cal
(forward motion)

System

First Law: Energy (Gasoline) must equal work plus any heat lost.
Second Law: Some heat is always lost.

Heat 80 cal

Figure 1.1.   *Thermodynamics of automobile. The dashed line is the boundary between the system (the car) and the surroundings (the rest of the universe).*

know that the energy of the universe has not changed since creation; the energy that was present then is numerically identical to the energy that is present today. However, the conversion of energy from one form to another has been extensive since creation.

The first law is universal and can be applied to any problem. As an example, consider the system of an automobile sitting at rest (Figure 1.1). A quantity of fuel, sufficient air to burn the fuel, and the rest of the universe constitute the surroundings. The potential (chemical) energy initially possessed by the fuel is transformed into two different forms of energy following its release upon combustion in the engine. Part of it is used to do work (make the car move), while a part (actually about 80 percent of the energy of the fuel) is returned as heat to the surroundings by the hot engine and exhaust. The work that is done moves the automobile and gives it kinetic energy. It is intuitively obvious that the resultant kinetic energy of the automobile could not exceed the chemical energy present in the fuel; one would not expect to get more energy transferred to the car than was supplied by the fuel. That is, you can't get something for nothing. The first law reflects this view and states furthermore that the energy available from the fuel is exactly equal to the sum of the work done on the automobile plus the heat lost to the environment.

__Second Law of Thermodynamics.__ The automobile illustrates another important concept of thermodynamics. In the conversion of energy that occurred, the heat lost from the car cannot be conveniently utilized for any "useful" purpose. It cannot be easily "reconverted" to gasoline or some other form of energy that might propel the car. Heat flowing to the environment is the least useful form of energy. It is irrevocably lost, a form of "heat tax" paid for the conversion of energy. The second law of thermodynamics relates to this "heat tax" and may be stated as:

> No device can be constructed that operates in a cycle (like an engine) and accomplishes nothing else except complete conversion of heat into work.

or

You can't break even.

The second law tells us that energy conversion is a one-way street; useful energy may undergo a 100% conversion into heat, but the reverse, the conversion of heat into work is always less than a 100% conversion (Figure 1.2). In the case of the automobile, even if some means could be derived to use the lost 80 calories of heat (Figure 1.1) to, say, drive a motorcycle, the second law guarantees that you could never impart 80 calories of kinetic energy to the motorcycle. Heat can never be converted completely into work.

The **efficiency** of a thermodynamic process is defined as

$$\text{efficiency (\%)} = \frac{\text{useful energy or work output}}{\text{total energy or work input}} \times 100 \qquad (1.2)$$

The efficiency tells what percentage of the available energy is converted into work. In its mathematical form, the second law gives the *maximum* **efficiency** to be expected for any process or machine. In the automobile, for example, the second law guarantees that the heat from the fuel cannot be transformed

Figure 1.2. *Second law forbids complete conversion of heat into work.*

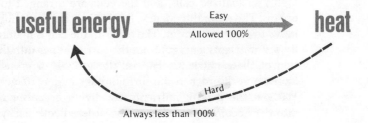

completely into work (energy that moves the car); some of the energy from the fuel *must* be returned to the surroundings as heat. When applied to electrical power plants, the second law shows that the maximum efficiency with which the plant can convert energy from the fuel into electricity is 47%. That is, at most 47% of the fuel's energy will appear as useful electricity; the remaining 53% is lost to the environment as heat. It is possible under the second law that the efficiency could be less than 47%; in fact it is around 33% in actual practice. The second law merely established an upper limit to efficiency.

The second law has other important implications. To investigate these we need some knowledge of spontaneous processes and of entropy. A **spontaneous process** is one that can occur naturally without an input of energy. Nonspontaneous processes, on the other hand, are those that can occur only when forced by means of an outside energy input usually due to man. Spontaneous processes abound: a ball will roll down an inclined plane or drop to the surface of the earth spontaneously, with no outside assistance; gas leaks spontaneously from the inside of a balloon (higher pressure) to the atmosphere (lower pressure); perfume diffuses spontaneously into all corners of a room upon opening the bottle. The reverse of each of these processes is nonspontaneous; to accomplish the reverse would require the expenditure of energy—you would have to *force* the reversal.

Closely associated with spontaneous processes is disorder or randomness. Thermodynamics uses the term **entropy** to measure disorder. The higher the degree of disorder or randomness, the greater the entropy. Casual observation provides an insight that natural processes tend toward higher disorder or increased entropy. For example, a drop of dye in water tends to diffuse throughout the container creating a mixture that is a highly disordered state compared to the starting conditions of pure dye plus pure water. Liquids tend to evaporate, entering a more highly disordered state by mixing with other compounds in air. Air leaks from a tire into the air, which is a more disordered state, because the lower pressure means the gas particles are farther apart on the average.

A careful analysis reveals that one can reverse the natural tendency toward disorder (that is, one can create order) in a *system* only at the expense of *increased* disorder in the surroundings. Consider, as an example, a living organism as a system. Such a system is highly ordered: molecules are arranged into highly specialized cells, and the cells are arranged to form membranes and organs. The natural tendency, however, is for the living organism to decay, to move to greater disorder. The order in the cell is maintained only through the flow of nutrients (energy) from the surroundings into the system. The production of these nutrients, such as the growth of crops, occurs *only* with an increase in disorder in the surroundings in accordance with the second law. Without the flow of nutrients, the living organism spontaneously dies and disorder becomes pronounced: the ordered cells decay and disintegrate.

Implications of Thermodynamics for the Environment

The first and second laws have broad implications for the environment. The first law, as noted previously, states that the energy present when the universe was formed is numerically identical to the energy present today. However, many energy conversions have occurred over the eons, and these have resulted in the inexorable change of energy from the more useful potential energy, such as that in fuels, into the less useful form of heat. The second law ensures that we can *never* regain the potential energy of the fuels from the heat. Thus any energy "crisis" that we face is not a shortage of energy, since it can never be destroyed, but rather a shortage of *useful* forms of energy. (This is to a certain extent of course a matter of semantics. Knowing that the energy once obtained from millions of barrels of gasoline still exists as heat in the environment does not alter the fact that you cannot buy gasoline!)

One implication of the second law is that whenever we attempt to order a part of nature, the disorder created in our environment (surroundings) *exceeds* the order created in the system. The life of man, particularly industrialized man, is characterized principally by efforts to increase the order of nature. Man converts nutrients, which are disordered, into crops to feed ever-growing numbers of human beings, whose bodies are highly ordered. He converts (disordered) ore into (ordered) iron and steel. He recovers disordered petroleum products and converts them into other energy forms and into highly ordered goods such as plastics, paints, and synthetic fabric. The environmental crisis might be considered an entropy crisis. Our efforts to order and conquer the earth must inevitably put greater and greater stress on the environment. The use of coal to generate electricity to heat homes, for example, creates increased disorder from strip mining, from the emission of pollutants into the air, and from an irreversible loss of heat during power generation.

Inasmuch as each organism's activities affect the disorder in the environment, the second law also means that all organisms are interconnected whether we like it or not.

Third Law; Fourth Law? We shall not specifically use the third law. It provides for a quantitative measure of entropy and is crucial to the logical application of thermodynamics.

It is worth noting that the obvious power and versatility of these three laws, coupled with the insight of decades of experience, have led scientists (and countless undergraduates who take laboratory courses) to speculate upon the possibility of the need for another law, the fourth, to explain nature completely, especially to explain nature's somewhat perverse qualities. Although never formally accepted, this fourth law (sometimes called Murphy's law) is widely known in scientific circles and is loosely stated as "if anything can go wrong, it will." Its applicability to environmental problems will be obvious.

Ecosystems and Cycles

The sun is the energy source that sustains life. The sun is a gigantic nuclear (fusion—see Chapter 4) reaction involving hydrogen and it liberates about 1×10^{26} calories of energy per second of which approximately 1×10^{17} calories per second reach our atmosphere. (In scientific notation 1×10^{26} is used to denote the number 1 followed by 26 zeroes, 80,000 is 8×10^4, 200 is 2×10^2, etc.) Of this amount, only about 2% is used to support life through photosynthesis in plants; the rest of the energy is reflected or absorbed by the atmosphere, land and oceans. All life is found within the biosphere—or ecosphere—a thin film of air, water, and soil approximately 9 miles thick. The atmosphere comprises 7 miles of this; the soil or crust extends only a few thousand feet down. This remarkably thin and intricate biofilm contains all the chemicals and produces all the nutrients of life.

An **ecosystem** is any group of organisms interacting with each other and their environment. The boundaries of an ecosystem are arbitrary. Thus a single drop of lake water, the lake itself, or the region in which the lake is located can be considered as ecosystems. The boundary surrounding an ecosystem is a conceptual one, however, because all ecosystems in the universe can (and do) interact.

Continuation of life requires the incessant recycling of atoms and a constant flow of energy within the biosphere; hence everything in this film is interconnected and interdependent. The cycles guarantee that nature will not face a natural resource shortage and, in addition, provide a natural purification scheme. The air helps purify the water; the water is used by the plants and animals; and the plants help renew the air. Vital chemicals such as carbon, nitrogen, oxygen, and water are recycled through the biosphere in cycles properly termed **biogeochemical cycles** because living organisms, geological formations, and chemical processes are involved. The cycles are extremely complex and only partially understood. Inasmuch as our life depends on these cycles, we shall consider one of these, the nitrogen cycle, in more detail.

The Nitrogen Cycle

Nitrogen is a key element in proteins, nucleic acids, enzymes, vitamins, and hormones. Nitrogen, as the gas N_2, comprises 79% (by volume) of our atmosphere. This gaseous form is rather inert, however, and to be useful to living organisms, nitrogen must be converted from the common atmospheric form, N_2, to other forms, and this is accomplished by the nitrogen cycle. The conversion process, which usually means a combination with hydrogen or oxygen, is referred to as **fixation.** Most of the nitrogen in living organisms, however, does not enter directly from the atmosphere. Instead, a group of bacteria in the soil and water can convert N_2 into the inorganic form **nitrate** (NO_3^-). This form is then utilized by the plants which convert it, via photosynthesis, into nucleic acids and proteins.

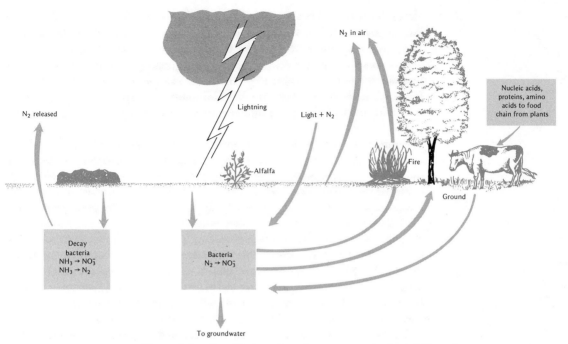

Figure 1.3. *Simplified version of nonaqueous portion of N_2 cycle.*

The distribution of nitrogen products up the food chain is begun by the herbivores, which eat the plants. Carnivores then eat the herbivores. Upon death, the nitrogen in plants and animals is bacterially converted into **ammonia** (NH_3), which another type of bacteria converts to either **nitrite** (NO_2^-) or N_2. The nitrite may be converted by further bacterial action into nitrate, which is then reused by the plants (Figure 1.3). This cycle coexists with an analogous cycle in the aqueous environment; the two are interconnected in many ways including nitrates and nitrites washed from land into the water.

This nitrogen cycle existed for thousands of years in equilibrium, and one could say with some confidence that the amount of nitrogen removed from the atmosphere by natural fixation processes was closely balanced by the amount returned to the atmosphere by organisms that convert nitrogen compounds into gaseous N_2. In this century man has manufactured nitrates and added them to the soil to improve production. These commercial fertilizers provide a nitrate source different from that of bacteria. Because of the increased commercial fertilizer tonnage, man now has the capability to interfere with this natural cycle.

Our knowledge of the biogeochemical cycles is incomplete; thus, we cannot accurately predict the effects of man's impact. In the case of nitrogen, our alteration of the cycle arises from the fact that, during the last century, heavy farming has resulted in the loss of nitrate from the soil to nearby water systems. We have compensated for this by the extensive planting of legumes and extensive use of commercial fertilizers. Photosynthesis requires nitrates, and their addition directly to the soil ("industrial fixation") in the form of chemical fertilizer accelerates plant growth. The use of fertilizer has increased fourteen-fold since 1945 and continued extensive use may have a long-range impact.

Commercial nitrate does not require bacterial action to be useful to plants, and its extensive use apparently reduces the bacterial capacity of the soil to fix nitrogen; this often requires the application of an increased amount of chemical fertilizer for the next crop. The agricultural use of legumes and fertilizer has caused the total nitrogen input to the soil to exceed (by perhaps 10%) the ecosystem input that existed before the advent of agriculture. This may mean that the soil (and water) input rate now exceeds the rate at which N_2 is returned to the atmosphere. Thus the nitrogen balance between the atmosphere and the soil-water system which existed before modern agriculture might be undergoing change. This could create an increased nitrogen buildup in the water, which would then increase the potential for eutrophication, or accelerated aging of water bodies, as discussed in a later chapter.

The biogeochemical cycles, and man's impact on them, will be of extreme importance in later chapters. Many (though not all) of the chemicals we term pollutants occur in nature and are integrated into the biogeochemical cycles which provide an air and water purification scheme. Man unfortunately tends to increase the concentration of the chemicals in a localized region—for example, a city—to the point where the natural cleansing scheme is overloaded; when this occurs we term these particular chemicals "pollutants." Man also produces chemicals *not* found in nature, DDT for example, and these tend to be *persistent* because they are not susceptible to the bacterial or chemical activity necessary for integration into the cycles.

Realistic Limits of Environmental Purity

As we consider pollution control schemes in later chapters, it will become relevant to inquire about what purity we should expect to achieve with respect to man's effluents and emissions. There are several reasons why it is unrealistic to expect to achieve *zero* pollution output from man's activities, that is, *no* pollutants released to the environment. As discussed earlier in the chapter, nature produces many of the chemicals we regard as pollutants, and to attempt to reduce man's output *drastically* below this natural level does not appear to be necessary. Furthermore, if complete purification were attempted, it would require an enormous effort. The second law of thermodynamics indicates that such efforts to increase order would only result in increased disorder elsewhere in the environment, for example, in the preparation of fuel to generate sufficient electricity to power the purification scheme.

Other difficulties would arise if a zero pollution output were attempted: no chemical analytical technique is available which is sufficiently sensitive to indicate whether zero pollution were achieved! For example, the smallest drop of water visible to the human eye is about one thousandth of an inch in diameter. Even this tiny drop contains an enormous number of molecules, approximately 500 trillion (5×10^{14}). To count the molecules in the drop would require 2 million years if we counted 10 per second continuously 24 hours per day! This drop could contain many impurities that are undetectable with even the most modern equipment. As many as 10,000,000 molecules of many of the common impurities could be present and not be detected. Radioactive chemical elements afford the most sensitive analytical approach; even a few thousand of these might go undetected if present in the drop.

A completely (chemically) **pure** drop of water, that is, one with *no* impurity molecules present, has probably never existed. We certainly have no analytical means to establish such purity. There is ample evidence that the purest water we can prepare in the laboratory has a wide variety of impurities present in very small concentrations. Water dissolves impurities from containers (even glass), from purification apparatus, and from the air. It is important to realize that "safe" drinking water, for example, is *not* free from impurities. Its safety derives from the fact that the inevitable impurities present are there in sufficiently small concentrations that they present no harm to the human system.

In the succeeding chapters, we will present considerable evidence that the reduction of environmental pollution requires urgent and vigorous attention. We should not, however, become so zealous as to press for a complete absence of man-made pollutant output. Rather, our aim should be the attainable and much more practical goal of attempting to reduce the total volume of pollutants with an eye toward not significantly increasing the naturally occurring levels. This will ensure that the natural purification schemes will not be overloaded and that man's impact will cause less change in the biogeochemical cycles than it has in recent years.

Some Elementary
Chemistry

Pollutants, fuels, and food are chemicals, and a knowledge of some very elementary chemical principles will bolster our discussions. Our look at chemistry will by no means be complete. Rather, it is intended only to be sufficient for the purposes of the succeeding chapters. In those we shall be interested primarily in the environmental importance of a compound and only secondarily in its reactions and geometric structure. In this section we will consider only nomenclature, molecular composition, and first principles of chemical reactions. The important chemical reactions of particular compounds will be discussed where relevant in later chapters.

All compounds are composed of combinations of **elements** of which slightly more than 100 are known; 88 elements occur naturally, the remainder have been created by man.

An **atom** is the smallest unit of an element which exhibits the chemical properties of that element. The atom is composed of a positively charged **nucleus** surrounded by a fuzzy cloud of negatively charged **electrons.** On a relative scale, the nucleus is dense and massive; it accounts for almost all of the mass of the atom but for almost none of the "size" of the atom. Because the fuzzy limits of the electron cloud cannot be accurately defined, the "size" of the atom is assumed to be the distance between the nuclei when the atoms combine to form a molecule as discussed in the following paragraphs. The electrons, on the other hand, fill most of the space occupied by the atom but are so light that they are almost negligible contributors to atomic mass. An electron weighs 2000 times less than the lightest nucleus (that of hydrogen). Thus the mass of an atom is concentrated in a tiny volume element at its center. If the nucleus were represented by the tiniest dot on this page, the size of the atom would be represented by a sphere 1 foot in diameter! The nucleus itself is composed of many particles. However, with the exception of nuclear reactions, which we discuss later in the text, the nucleus is sufficiently stable to be unaffected by most of man's activities and by most of nature's cycles.

Despite intensive research for more than 40 years, many details of nuclear structure are unknown and many questions about how the nucleus is held together (or "bound") remain unanswered. Sufficient information exists, however, to explain the significant characteristics of atomic structure. For our purposes we need consider only two nuclear particles: the **proton** and the **neutron.** These two particles suffice to explain the mass of any nucleus, the total positive charge on the nucleus, and the number of electrons present outside the nucleus.

A proton possesses a charge of $+1$ and a **mass number** (a rounded-off measure of its weight) of 1. A neutron has no charge but a mass number of 1, as in Table 1.1. The mass and charge of any nucleus can be explained by the fact that nuclei contain one or more protons and zero or more neutrons.

An atom is characterized by two numbers: the atomic number of the nucleus and the mass number of the nucleus. The **atomic number,** Z, is the number of protons contained in the nucleus. Because the atom is electrically neutral, it must have as many electrons (charge -1) outside the nucleus as it does protons

TABLE 1.1. Charge and mass number of the electron, the proton, and the neutron

Particle	Electric Charge	Mass Number
Proton	$+1$	1
Neutron	0	1
Electron	-1	0

in the nucleus. Thus Z also measures the number of electrons of the atom. The mass number is equal to the number of protons in the nucleus plus the number of neutrons in the nucleus.

atomic number, Z, of the nucleus = the number of protons in the nucleus

mass number of the nucleus = the number of protons in the nucleus plus the number of neutrons in the nucleus

From these few facts, many of the atomic properties of the elements can be deduced.

A chemical element consists of atoms with the same atomic number. The atomic number is unique in this regard: it establishes the "identity" of the atom. That element with Z = 1 is called hydrogen. Hydrogen is the lightest element; it obviously has one proton in the nucleus and one electron outside the nucleus. The element with Z = 2 is called helium. It has two protons in the nucleus and, consequently, it has two electrons to remain electrically neutral. Elements with all values of Z from 1 to over 100 are known (Table 1.2). Carbon is that element with Z = 6 (6 protons, 6 electons), oxygen is that element with Z = 8 (8 protons, 8 electrons), and uranium is that element with Z = 92 (92 protons, 92 electrons). Some of the elements that are important in discussions later in the text are in Table 1.2, together with their chemical symbols. The **chemical symbols** are internationally recognized shorthand notations for the elements.

Although the atomic number is unique because it determines the element,

TABLE 1.2.   Some chemical elements and their chemical symbols

Name	Symbol	Atomic Number	Name	Symbol	Atomic Number
Hydrogen	H	1	Arsenic	As	33
Helium	He	2	Strontium	St	38
Carbon	C	6	Cadmium	Cd	48
Nitrogen	N	7	Iodine	I	53
Oxygen	O	8	Cesium	Cs	55
Sodium	Na	11	Mercury	Hg	80
Phosphorus	P	15	Lead	Pb	82
Sulfur	S	16	Radium	Ra	88
Chlorine	Cl	17	Uranium	U	92
Calcium	Ca	20	Plutonium	Pu	94

Note: Over 100 elements are known. The elements given in this table are important in discussion in later chapters.

the mass number is not such a selective measure of the atom. A given element may have more than one mass number, depending upon the number of neutrons in the nucleus. Since neutrons have no charge, their presence does not affect the charge of the nucleus. Atoms of the same element (that is, with the same Z) that have different mass numbers are called **isotopes.** We shall investigate their properties in more detail when we consider nuclear reactions and nuclear power.

Atoms bond together according to well-known rules to form **molecules.** The formation of molecules, or chemical compounds as they are commonly called, does not affect the nuclei of the atoms. Only electrons are involved in chemical bonding; usually, those electrons farthest from the nucleus have the most participation in the bonding. The bonding rules, which are developed in any introductory technical chemistry course, are beyond our scope, but we will of course be interested in many molecules and will often use their chemical formulas, such as H_2O for water, CO for carbon monoxide, and SO_2 for sulfur dioxide. By writing a chemical formula such as H_2O we are indicating that an individual molecule consists of one atom of oxygen (O) and two atoms of hydrogen (H) bonded together in some fashion to form a stable entity. Note that the number of atoms combining to form a molecule is indicated by a numerical subscript. That is, C_2H_5OH indicates that two carbon (C) atoms, one oxygen (O) atom, and six hydrogen (H) atoms are combined in some fashion to form a molecule (ethyl alcohol by name). However, the expression $2 C_2H_5OH$ merely indicates two ethyl alcohol molecules are present; 2 as a prefix does *not* indicate chemical combining—it is merely a multiplier. More knowledge of chemistry than we shall develop is required to predict details and geometry of the bonding, such as whether the molecule SO_2 has the geometry

$$S\text{–}O\text{–}O \quad \text{or} \quad O\text{–}S\text{–}O \quad \text{or} \quad O \overset{\overset{\textstyle S}{\diagup \diagdown}}{} O$$

(The last form happens, in fact, to be the correct structure). However, the knowledge of what atoms are present in the molecule will serve us well.

Chemical compounds are classified into two very general groups, organic and inorganic, depending upon the atoms present in the molecule. Both types are important in the environment. **Organic** compounds contain carbon atoms in combination with one or more of the following: hydrogen, oxygen, nitrogen, phosphorus, or sulfur. The organic group is by far the most abundant. In fact, there are more compounds containing carbon than there are compounds of all the other elements combined. Examples of the organic group are carbon dioxide (CO_2), methane (CH_4), sugars, amino acids, proteins, carbohydrates, wood, plastics, paints, and rubber. Sulfur dioxide (SO_2), water (H_2O), sulfuric acid (H_2SO_4), and table salt (sodium chloride, NaCl) are examples of **inorganic** substances.

Organic Compounds ✓

Organic compounds contain carbon. Several classifications of organic compounds can be made on the basis of chemical composition.

Hydrocarbons. Compounds composed of carbon (C) and hydrogen (H) only are sufficiently important that we will investigate them in more detail. The carbon atom forms four chemical bonds, and hydrogen prefers one bond. An infinite variety of hydrocarbons is possible, beginning with this information and an "erector set" approach to constructing molecules.

The Alkanes—the simplest hydrocarbons. The **alkanes,** or aliphatic hydrocarbons, are chain structures with all carbons connected together with single bonds. They have a general composition that can always be written as C_nH_{2n+2}, where n can be any positive integer (1, 2, 3, etc.). Thus CH_4, C_4H_{10}, and $C_{20}H_{42}$ are all alkanes. The smallest and simplest aliphatic is **methane,** or marsh gas, CH_4. The structure of this molecule is a tetrahedron (a pyramid with four triangular faces) in which the carbon is at the center and the hydrogens are arranged at the apexes of the tetrahedron as in Figure 1.4a. This arrangement satisfies the four-bond requirement for carbon and the single-bond requirement of each of the four hydrogens.

Higher members of the alkane series are formed by increasing the number of carbons. Ethane, the next member, has the formula CH_3CH_3 (Figure 1.4b). In this molecule there is one single carbon-carbon bond and six carbon-hydrogen bonds. Each carbon has four bonds, and each of the hydrogens has one bond. Propane, $CH_3CH_2CH_3$, is the next highest member (Figure 1.4c). In this and succeeding alkane molecules, the tetrahedral arrangement of bonds around each carbon is maintained. Beginning with the four carbon molecules, both

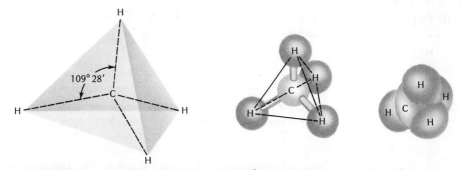

Figure 1.4a. *Methane (CH_4). Two models expressing the same structure. The ball and stick model in the center indicates the position of the nuclei. The space-filling model at the right indicates in an approximate fashion the space occupied by the electron clouds.* [*From R. H. Petrucci, General Chemistry, Macmillan Publishing Co., Inc., New York, 1972.*]

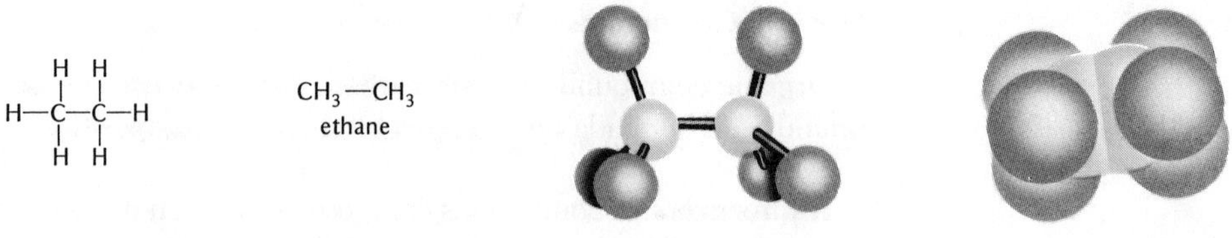

$CH_3—CH_3$
ethane

Figure 1.4b. *Ethane (C_2H_6). [From R. H. Petrucci, General Chemistry, Macmillan Publishing Co., Inc., New York, 1972.]*

$CH_3—CH_2—CH_3$
propane

Figure 1.4c. *Propane (C_3H_8). [From R. H. Petrucci, General Chemistry, Macmillan Publishing Co., Inc., New York, 1972.]*

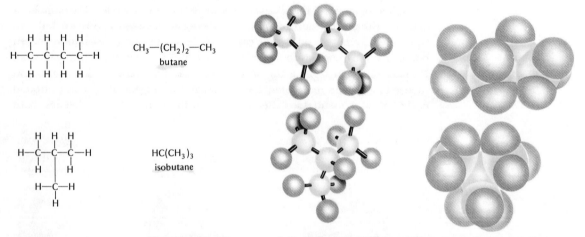

$CH_3—(CH_2)_2—CH_3$
butane

$HC(CH_3)_3$
isobutane

Figure 1.4d. *Structures of the butane isomers. [From R. H. Petrucci, General Chemistry, Macmillan Publishing Co., Inc., New York, 1972.]*

linear and branched chains are possible. The linear, or "straight," chains are termed **normal alkanes** and are prefixed by the letter n. Thus in Figure 1.4d n-butane indicates the "straight chain" form whereas isobutane is branched. A wide variety of branched forms is possible as the chain increases in length.

The alkane hydrocarbons are the primary components of natural gas and most other fuels, as in Table 1.3.

TABLE 1.3.   Some hydrocarbon fuels

Name	Composition	Use
Natural gas	mixture of gases, CH_4, C_2H_6, and C_4H_{10}	industry, electrical power generation, home cooking and heating
Propane	C_3H_8	bottled gas
Butane	C_4H_{10}	bottled gas
Gasoline	mixture of C_8H_{18} and other hydrocarbons	internal combustion engines
Kerosene	mixture of alkanes with about 12 carbon atoms	jet fuel
Paraffin wax	mixture of alkanes with about 16 or more carbon atoms	candles

The Alkenes—Unsaturated chain hydrocarbons. The above alkanes are characterized by a chain structure with *one bond* between each successive pair of carbon atoms. Such compounds are often termed **saturated** to distinguish them from another possible type, the **unsaturated** chain hydrocarbons, which are characterized by *two or three bonds* between carbons. The presence of unsaturation changes the geometry of the chain and usually creates a more reactive molecule. The simplest unsaturated hydrocarbon is ethylene, $H_2C = CH_2$. The two lines between carbons indicate the presence of a double bond. Each carbon still has four bonds each (two to the other carbon and two to hydrogens). Unsaturated bonds can occur at almost any position in the molecule, and chain hydrocarbons with one or more double bonds between carbons are referred to as **alkenes** or **olefins.**

The Aromatics—Unsaturated ring hydrocarbons. Another important class of hydrocarbons consists of molecules in which the carbons form a planar ring structure with many unsaturated (double) carbon-carbon bonds. A flat ring of six carbon atoms with the chemical formula C_6H_6 characterizes the principal aromatic molecule, **benzene** (Figure 1.4e). Each carbon still has four bonds: two to an adjacent carbon, one to the other carbon neighbor, and one to a hydrogen. Experimental data show that the carbons in benzene form a perfect hexagon with all carbon-carbon distances identical. This indicates that the double bonds (which are shorter than single bonds) cannot be localized between specific carbons, as shown at the left of Figure 1.4e. Rather the double bonds are shared equally around the ring in some complicated fashion. The structure at the right of Figure 1.4e indicates that all carbon-carbon bonds in benzene are equivalent; it is universally recognized by chemists as a symbol for benzene.

Other Organic Compounds. Many other elements, including especially oxygen, sulfur, nitrogen, and phosphorus, can combine with carbon to form other organic molecules. Each of the elements has its own preferred number of

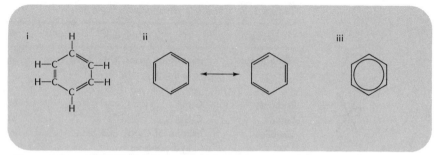

Figure 1.4e. *Benzene (C_6H_6) structural formulas and symbols. The molecule is a perfect hexagon indicating that the double bonds are not fixed between carbon pairs as in i, but are shared equally as noted symbolically by ii and iii. By internationally recognized chemical convention, the structures ii and iii are also used to indicate the structure of benzene. The reader is asked to assume a carbon is present at the line apexes and to add sufficient hydrogens to complete four bonds for each carbon. [From R. H. Petrucci, General Chemistry, Macmillan Publishing Co., Inc., New York, 1972.]*

bonds making generalizations difficult, hence we will merely mention the formulas of some of the more important species. We will use the symbol R to represent a hydrocarbon group.

Alcohols have the general formula $R - OH$. If, for example, $R = CH_3$, then the compound is CH_3OH methyl alcohol (known also as methanol or wood alcohol). If $R = CH_3CH_2$, then the alcohol is CH_3CH_2OH or C_2H_5OH, ethyl alcohol or ethanol or grain alcohol. Ethyl alcohol is the crucial ingredient in alcoholic beverages. Methyl alcohol, on the other hand, is a poison. There is no restriction on R; it can be an aromatic species also. If $R = C_6H_5$, then the alcohol is C_6H_5OH or ⬡ $- OH$ (phenol).

Aldehydes have the general formula $R - \overset{\displaystyle O}{\overset{\|}{C}} - H$ where the carbon has two bonds to oxygen, one to hydrogen, and one to a carbon in the species R. If, for

example, $R = CH_3CH_2CH_2$, then the specific molecule is $CH_3CH_2CH_2\overset{\displaystyle O}{\overset{\|}{C}}H$.

If $R = C_6H_5$, then the specific molecule is ⬡ $- \overset{\displaystyle O}{\overset{\|}{C}}H$. Each carbon still has four bonds, but the aldehydes have chemical properties quite different from those of the alkanes and alkenes.

Organic acids, which are very weak acids, have the general formula $R - \overset{\displaystyle O}{\overset{\|}{C}} - OH$. A **ketone** has the chemical formula $R - \overset{\displaystyle O}{\overset{\|}{C}} - R'$ where R and R' may or may not be identical.

Biologically Important Molecules. The building materials of cells and tissues are carbohydrates, lipids (fatty substances), amino acids, and proteins. These

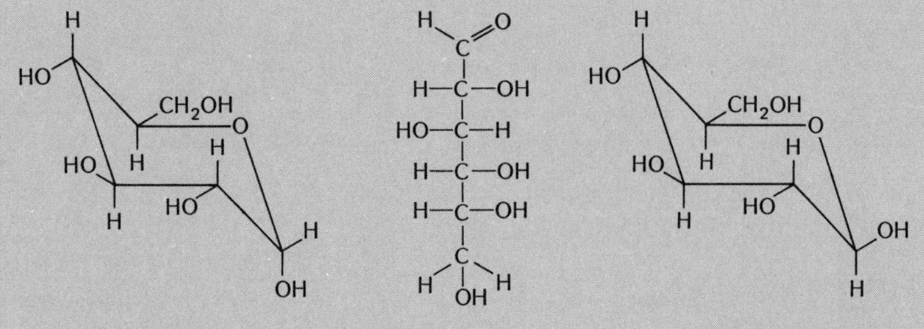

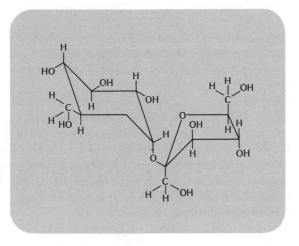

α-D-Glucose
(chair form)

Open-chain form

β-D-Glucose
(chair form)

Figure 1.5a. *Glucose has the chemical formula $C_6H_{12}O_6$ and can exist in three possible forms depending on how the binding occurs within the molecule. In the chair forms, by chemical convention, the carbon atoms that are present at the points where the lines intersect are not indicated.*

molecules often contain nitrogen, phosphorus, and sulfur in addition to carbon, hydrogen, and oxygen.

Carbohydrates, the most abundant naturally occurring class of compounds, includes sugars, starches, and cellulose. Sugars are formed by plants from carbon dioxide and water during **photosynthesis,** a complex process involving sunlight and chlorophyll, an important compound found in green plants. Sugars are composed of carbon, oxygen, and hydrogen. A common, simple sugar, glucose, has the chemical formula $C_6H_{12}O_6$ and the structure given in Figure 1.5a. Table sugar, sucrose, is formed from glucose and another simple sugar, fructose (Figure 1.5b). Cellulose, the principal component of wood and

Figure 1.5b. *Sucrose, table sugar, is formed from glucose and fructose. Glucose is the basis of the chair form at the left; fructose is the basis of the structure at the right.*

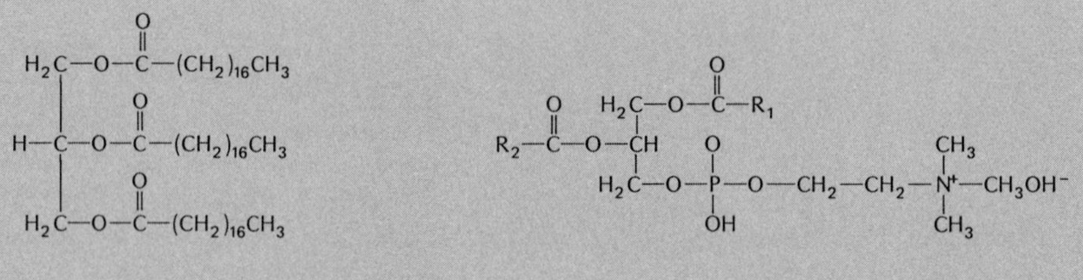

$$
\begin{array}{l}
\text{H}_2\text{C}-\text{O}-\overset{\overset{\textstyle O}{\|}}{\text{C}}-(\text{CH}_2)_{16}\text{CH}_3 \\[4pt]
\text{H}-\text{C}-\text{O}-\overset{\overset{\textstyle O}{\|}}{\text{C}}-(\text{CH}_2)_{16}\text{CH}_3 \\[4pt]
\text{H}_2\text{C}-\text{O}-\overset{\overset{\textstyle O}{\|}}{\text{C}}-(\text{CH}_2)_{16}\text{CH}_3
\end{array}
$$

Tristearin

$$
\begin{array}{l}
\phantom{R_2-\text{C}-\text{O}-}\text{H}_2\text{C}-\text{O}-\overset{\overset{\textstyle O}{\|}}{\text{C}}-\text{R}_1 \\[4pt]
\text{R}_2-\overset{\overset{\textstyle O}{\|}}{\text{C}}-\text{O}-\text{CH} \quad\quad O \\[4pt]
\phantom{R_2-\text{C}-\text{O}-}\text{H}_2\text{C}-\text{O}-\overset{}{\underset{\text{OH}}{\text{P}}}-\text{O}-\text{CH}_2-\text{CH}_2-\overset{\overset{\textstyle CH_3}{|}}{\underset{\underset{\textstyle CH_3}{|}}{\text{N}^+}}-\text{CH}_3\text{OH}^-
\end{array}
$$

Lecithin, a phosphatidyl choline

Figure 1.6. *Examples of lipids. In the phospholipid lecithin, R_1 and R_2 are long-chain alkane hydrocarbon groups.*

paper, is a very large molecule composed of as many as 1500 glucose molecules connected in a chain.

Fats and other molecules that are essential to plant and animal tissues are members of a class of compounds called **lipids.** No precise structural definition is possible for lipids because of the great variety of molecules that constitute this class. Some examples are given in Figure 1.6. One important group in this class is the **phospholipids,** which includes compounds such as phosphatidyl choline. These phospholipids are responsible for the structural integrity of cell membranes.

Amino acids are organic acids that contain the amine group (NH_2^-). The basic structure of an amino acid is

$$
\text{NH}_2 - \text{CH} - \text{COOH} \\
\phantom{\text{NH}_2 - }\underset{\textstyle R}{|}
$$

where a different amino acid exists for each different chemical species that may be present as R. Over 20 amino acids are common; R varies from hydrogen to long-chain aliphatic hydrocarbon groups to aromatic hydrocarbon groups.

Proteins are probably the most complex organic molecules produced in nature. They are present in all living systems and are essential components of skin, hair, muscle, and tissues. Some proteins, called **enzymes,** regulate the rate of biological processes. Other proteins, known as **antibodies,** counteract the effect of invading species and substances. Proteins are formed from combinations of various amino acids, often having 100 to 300 amino acid entities joined together in a definite sequence. It is estimated that the human body contains over 100,000 different proteins.

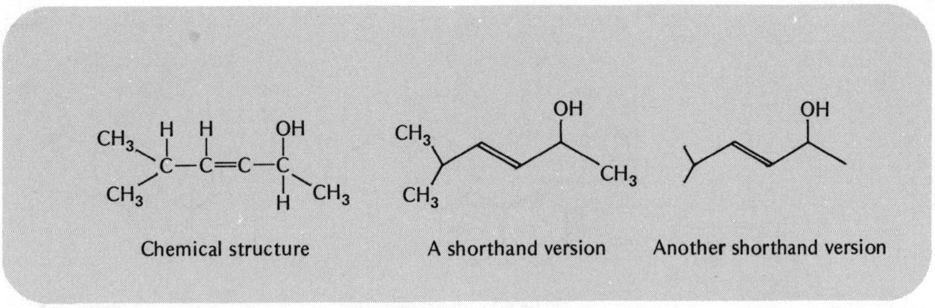

Figure 1.7. *The conventional shorthand form of drawing chemical structures. The reader is asked to assume a carbon is present at each line intersection and, perhaps, at each line terminus. Sufficient hydrogens must be assumed to complete four bonds for each carbon.*

Living organisms make a large number of molecules called **steroids.** These compounds perform many important biological functions. Hormones, which regulate growth, sexual reproduction, and food metabolism, are important steroids. Bile (liver) acids, cholesterol, and cortisone are other examples of steroids.

All steroids have a common chemical skeleton that contains 17 carbon atoms. To simplify the drawing of these and other large structures, a conventional shorthand method has evolved among chemists. Lines are used to represent bonds between carbons, but the carbons themselves are omitted (Figure 1.7). On occasion the hydrogens that bond to the carbons are also omitted. This representation asks the reader to place a carbon at each line terminus and at each line intersection and to fill out the hydrogens consistent with satisfying four bonds for each carbon. We used this scheme to represent benzene (Figure 1.4e) and sugars (Figure 1.5).

The carbon skeleton of the steroids is in Figure 1.8 in shorthand form, together with the structures of three important examples. Cholesterol, the most abundant steroid, is found in the tissues of nearly all vertebrates. It is abundant in nerve tissues, in the spinal cord, and in the brain. Despite its abundance and presumed importance, its specific functions are largely unknown. Cholesterol is present in the blood and is linked to a number of heart and arterial diseases. Cholesterol and its derivatives can deposit on arterial walls, constricting blood flow and contributing to high blood pressure and to hardening of the arteries (arteriosclerosis). Gallstones are deposits of cholesterol; in fact, the name cholesterol derives from the Greek *chole* (bile) and *stereos* (solid).

Testosterone (Figure 1.8) is a male sex hormone which, though produced in minute amounts, controls the development of sex organs and of other male sexual characteristics. Female sex hormones, also steroids, perform analogous functions in the female and are, in addition, the regulators of the reproductive cycle. The birth control pill, a contraceptive discussed in Chapter 2, consists of

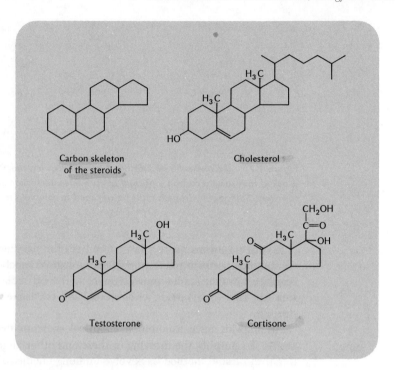

Carbon skeleton
of the steroids

Cholesterol

Testosterone

Cortisone

Figure 1.8.   The carbon skeleton of the steroids and some important examples of this class of compounds.

commercially manufactured chemicals with structures very similar to the hormones produced by the human female.

Cortisone is produced by the adrenal gland. Commercially produced cortisone is used in the treatment of arthritis.

Chemical Equations

It will be necessary to discuss chemical reactions in many portions of the text. Reactions are always described by means of chemical equations; their meaning is very simple. Consider, for example, the chemical equation for the formation of water from the naturally occurring forms of hydrogen and oxygen:

$$2 \, H_2(\text{gas}) + O_2(\text{gas}) \longrightarrow 2 \, H_2O(\text{liquid}) \tag{1.3}$$

As written, this equation indicates that two molecules of hydrogen in the gas phase can combine with one molecule of oxygen (also in the gas phase) to form two molecules of water (liquid phase). Chemical reactions may involve molecules and atoms in any possible physical state. For example, Equation 1.4 indicates that four hydrogen atoms (gas) react with one carbon atom (solid phase) to form one molecule of methane (gas). Equation 1.5 is written to describe the reaction of two hydrogen atoms (H) with ethylene (C_2H_4) to give ethane (C_2H_6).

$$4\ H(gas)\ +\ C(solid) \rightarrow CH_4(gas) \qquad\qquad (1.4)$$

$$2\ H(gas)\ +\ H_2C = CH_2\ (gas) \rightarrow CH_3CH_3(gas) \qquad\qquad (1.5)$$

All chemical reactions are reversible to a certain extent; that is, they may proceed from right to left as well as left to right. Thus Equation 1.3 also indicates that two molecules of liquid water can become two gaseous hydrogen molecules and one gaseous oxygen molecule. Usually one direction or the other of a chemical reaction is strongly favored by nature. That is, if all species listed on both sides of the arrow were mixed in the container, nature would usually prefer ultimately to have most of the atoms combined as the molecules (or atoms) on one side of the arrow. The reaction would proceed in the preferred direction. However, the mere writing of the equation gives *no* indication of which side nature prefers. Equations are by convention written so that the reaction to be discussed proceeds from left to right.

Another important fact about chemical equations is that as written they give *no* indication whatsoever of the *rate* of the reaction. The reaction in equation 1.3, for example, is strongly favored by nature from left to right, meaning that if hydrogen gas and oxygen gas are mixed, nature *prefers* that they react almost completely to give water. However, the rate of the reaction (if the substances are pure) is almost too slow to measure. (It is possible in this particular case, however, to make the reaction proceed at an almost explosive rate by adding some metal dust.) In summary, we should consider chemical equations as a simplified shorthand form used to describe a chemical reaction that can occur. The equation is written so that the reaction of interest to us will occur from left to right, but nothing can be inferred about the rate or about which side of the arrow is preferred by nature.

Reactions in Water

Many of the properties of organic and inorganic compounds can be illustrated by considering their reactions in water. Bonding in organic molecules is usually much less subject to being broken in water than is the bonding in inorganic molecules. When a chemical compound is exposed to or mixed with water, three possible events may occur:

1. The compound may not dissolve or react with water. Rather, if both are liquids, the chemical may form a layer on the water, such as that occurring with oil and water. Usually if this occurs, a *small* amount of the chemical may dissolve in water as in 2, but the solubility is very small.

2. The compound may dissolve in water with little or no change in its chemical structure. If this occurs, often there is a limit to the "solubility" of the species; that is, only so much can dissolve in a given quantity of water. Salt and water, sugar and water, nail polish remover and water are examples.
3. The compound may dissolve in water with an attendant chemical reaction which changes the chemical structure and the chemical nature of the compound. Acids in water are an example (see the following paragraphs).

Most organic molecules, with the exception of the organic acids, behave as described in 1 or 2; that is, they have little reaction with water in general, although they may dissolve to varying degrees. The simpler alcohols, methanol and ethanol, dissolve in water in all proportions but undergo very little chemical change. That is, they exist in a water solution as the chemical species CH_3OH (methanol) and C_2H_5OH (ethanol). Many hydrocarbons, such as oil products and gasoline, behave as in 1, preferring to float on water rather than dissolve and/or react.

Organic acids undergo a slight chemical change, called **dissociation**, when mixed with water. Acetic acid, as in equation 1.6, is an example. The molecule CH_3COOH dissociates

$$\underset{\text{acetic acid}}{CH_3\overset{\overset{\textstyle O}{\|}}{C} - OH} + H_2O \rightarrow \underset{\text{hydronium ion}}{CH_3\overset{\overset{\textstyle O}{\|}}{C} - O^- + H_3O^+} \tag{1.6}$$

losing a hydrogen nucleus (a proton) to a water molecule, forming a **hydronium ion.** (The term **ion** indicates an electrically charged species.) The electron from the hydrogen remains with the CH_3COO^- species, making it a negatively charged ion. The acidic character of acids is due to the presence of the hydronium ion; the more of this species present, the stronger the acid. Organic acids, however, are weak acids because their dissociation is very small. That is, in solution perhaps 99.9999% of the acetic acid molecules are *not* dissociated but are in the molecular form CH_3COOH, while 0.0001% are dissociated into the hydronium and negatively charged ions.

Inorganic compounds (those without carbon) often obey alternative 3 in the list; that is, they react significantly with water. The inorganic acids—hydrochloric (HCl), nitric (HNO_3), and sulfuric (H_2SO_4)—dissociate almost 100% in water, and the resulting large H_3O^+ concentration accounts for the fact that they are very strong acids.

$$HCl + H_2O \rightarrow Cl^- + H_3O^+ \tag{1.7}$$
$$H_2SO_4 + H_2O \rightarrow SO_4^{2-} + 2\,H_3O^+ \tag{1.8}$$

For these acids, there would be almost none of the species HCl or H_2SO_4 in solution. The equilibrium would be shifted almost completely to the right, so that they exist in solution as ions.

Metric System

Derived in the eighteenth century, the metric system of weights and measures is very simple and convenient. It is used throughout the world with the notable exceptions of the United States and Canada, which use the British Imperial system (the British, having seen the error of their ways, have recently begun conversion to the metric system). The scientific literature often uses the metric system. The simplicity of metric units derives from the fact that the basic units are carefully chosen. Furthermore, larger or smaller units are defined in terms of multiplication or division by powers of ten. In contrast, the Imperial system has poorly chosen basic units: the "foot" was originally defined as "36 barleycorns measured from the middle of the ear." In addition, the Imperial system involves mathematically awkward conversions: 12 inches in 1 foot, 3 feet in 1 yard, 5280 feet in 1 mile, and so on.

A rudimentary understanding of the metric system is essential in environmental studies. All scientific data are reported in this fashion, and many articles in the press and popular magazines use it as well. Furthermore, it seems inevitable that the United States will ultimately convert to the metric system.

In the metric scheme, time is in seconds, temperature is in degrees Celsius (°C), mass is in **grams,** length is in **meters,** and volume is in **liters.**

Some comparative Celsius and Fahrenheit (°F) temperature points are in Table 1.4. Note that a 37°C human body temperature is normal and that water does not freeze if the temperature is 1°C! (The Celsius scale is almost identical numerically to the older centigrade scale.)

Table 1.5 summarizes the basic units in the metric and United States version of the Imperial system for mass, length, and volume. We will use these units regularly, and it will prove useful to develop an intuitive feeling for their size. A meter is slighty greater than 1 yard and a liter is slightly greater than 1 quart. A gram is quite small. Over 28 grams are required to make 1 ounce; 454 grams

TABLE 1.4. Comparison of Fahrenheit and Celsius scales

Point	°F	°C
Biscuit baking	450.	232.
Water boiling point	212.	100.
Human body	98.6	37.
Room temperature	68.	20.
Melting point of ice	32.	0.
Absolute zero	−459.7	−273.15

TABLE 1.5. Conversion of metric units into common United States weights and measurements.

Metric Unit (abbreviation)		Equivalent in U.S. Usage
1 gram (g)	=	1/454 pound (lb)
1 meter (m)	=	39.4 inches (in.)
1 liter (l)	=	1.06 quarts (qt)

TABLE 1.6. Common metrix prefixes

Prefix	Abbreviation	Multiplicative Factor	Scientific Notation
Giga	G	1,000,000,000	10^9
Mega	M	1,000,000	10^6
Kilo	k	1,000	10^3
Centi	c	1/100 or 0.01	10^{-2}
Milli	m	1/1000 or 0.001	10^{-3}
Micro	μ	1/1,000,000 or 0.000001	10^{-6}

make 1 pound. Three aspirin tablets weigh about 1 gram, whereas a dime weighs about 2 grams.

An advantage of the metric units is that different weight, volume, and length units are defined in multiples of ten of the gram, liter, and meter. These multipliers have names (prefixes) and symbols as in Table 1.6. For example, the prefix "kilo" means 1000, so a kilogram (kg) is 1000 grams and a kilometer (km) is 1000 meters. The prefix "micro," symbolized by the Greek letter μ, means 1/1,000,000, so a microgram (μg) is one millionth of a gram and a microliter (μl) is one millionth of a liter.

The liter is chosen to be the volume of a cube 10 cm (1 cm is 1/100 of a meter) on a side with a volume of 1000 cubic centimeters (cm^3 or cc). This means that 1 milliliter (ml), one thousandth of a liter, is identical to 1 cm^3. Thus 1 ml = 0.001 l = 1 cc = 1 cm^3.

Although the metric system may appear at first glance to be complex, it is in practice much easier to learn and use than the system used in the United States which has over 80 defined units. A little time invested in practice on the metric scheme will result in almost complete familiarity with its principal features.

Some of the commonly defined United States weights and measures and their metric equivalents are in Table 1.7.

TABLE 1.7.   Some common United States unit to metric unit conversions

---

Length
  1 mile (mi.) = 1.61 kilometers (km)
  1 foot (ft.)   = 30.48 centimeters (cm)
  1 inch (in.)   = 2.54 centimeters (cm)
  1 coed (38–24–36 inches) = 1 coed (97–61–91 cm)
  1 person (5 ft. 10 in.)   = 1 person (1.78 meters)

Mass
  1 ounce (oz.) = 28.4 grams (g)
  1 pound (lb.) = 454 grams (g)
  1 ton   = 908 kilograms (kg)

Volume
1 quart (qt.)   = 0.94 liter (l)
  1 pint (pt.)   = 0.47 liter (l)
  1 fluid ounce (fl. oz.) = 29.6 milliliters (ml)
  1 gallon (gal.)   = 3.76 liters (l)

Units of Pollutant Concentration

The **concentration** of a chemical compound measures the amount present in a given volume or a given weight of sample. Pollutant concentrations are measured and reported in a manner similar to that used for all chemicals. A variety of units are used depending on the concentration and physical form of the system.

In the case of a gas mixture, **volume percent (vol %)** is often used. This is simply the volume fraction converted to a percentage, or

$$\text{volume \%} = \frac{\text{volume of pollutants}}{\text{total volume of sample}} \times 100$$

Volume percent is usually used for large concentrations. For example, if a sample of 100 volume units contains 10 volume units of CO, the CO is present at a concentration of 10% by volume.

Often, however, pollutant concentrations are very small, comprising only one millionth of the volume or weight of a particular sample. For simplicity, such small concentrations are often expressed in the unit of **parts per million** or **ppm**. When the unit is used for gases, **ppm by volume** is used. That is 3 ppm of CO in air means 3 volume units of CO are present in 1 million volume units of air. For liquids and solids, **ppm by weight** is used. A concentration of 5 ppm sulfuric acid in water corresponds, for example, to 5 g of sulfuric acid in 999,995 g of H_2O. One ppm is difficult to visualize. It is equivalent to one apple in 10,000 crates of oranges, or 1 oz of vermouth in 7800 gal of gin, or 1 oz of salt in 62,000 lb of sugar. On occasion we will have to describe even smaller concentrations and will use **parts per billion (ppb)** which is one thousandth of 1 ppm. We should not be deceived into believing that because parts per million concentrations are small they are harmless. As we shall see, parts per million and parts per billion concentrations of certain chemicals have a profound effect on the environment.

Another useful unit is the **weight per volume**—for example, micrograms of pollutant per cubic meter of air ($\mu g/m^3$) (see Metric System in this chapter). This unit is especially applicable to particulate and metal air pollutants.

Which unit is used is a matter of preference or convenience, and conversion from one set of units to another is possible. The natural SO_2 concentration can be expressed as 0.00000002 vol % or 0.0002 ppm or 0.575 $\mu g/m^3$. Oxygen's (O_2) natural concentration can be given as 20.94 volume % or 209,400 ppm. Usually we would use micrograms per cubic meter to discuss SO_2 but volume percent for O_2 because the sizes of the numbers in each case are easier to manipulate.

Measuring Energy and Power

Because work is merely energy that is directed toward some useful purpose, work and energy are measured in the same units. In Chapter 4 we have an extended discussion of these units, but we need to define a few of the more common forms at this time. A **British thermal unit (Btu)** is defined as the amount of energy required to heat 1 lb of water by 1°F. The Btu is commonly used to measure the energy delivered by air conditioners and furnaces. Another common measure of energy is the **defined calorie (cal)**. The calorie is the amount of heat required to heat 1 g of water 1°C. One Btu is about 252 cal. Although use of the word "calorie" is familiar from use in nutrition, the food **Calorie, or Cal** (note the capitalization), is 1000 times larger than the defined calorie (cal). For example, the United States daily per capita food intake is 2000 to 4000 Cal, which is 2,000,000 to 4,000,000 cal. (Nutritionists often omit capitalization of the Cal.) This same food intake corresponds to about 8,000 to 16,000 Btu per day. A gallon of gasoline when burned releases 126,000 Btu. About 50 Btu (12,600 cal) are required to heat a cup of water to the boiling point.

Because power is energy divided by time, power may be measured in calories per second (cal/sec), Btu per second (Btu/sec), Btu per hour (Btu/hr), and so on. Because of convenience, several other units are preferred to measure power. The watt is often used. The **watt** corresponds to 0.06 Btu/sec. Lifting a book from the floor to the table requires about 1 watt, whereas running full tilt can require as much as 1000 watts. The unit of horsepower is often used to measure power, especially in automobiles. One **horsepower** equals 746 watts. The horsepower once provided a crude measure of the power output of a work horse. A toaster uses about 1000 watts, so one horse working very hard could just about provide sufficient power to run a toaster!

Summary   The complexity of the environment precludes obtaining exact solutions to environmental problems from the use of scientific investigations. The principal limitation is that scientific models require approximations that prevent an accurate representation of the environment. However, science must provide the foundation for rational decisions on environmental problems. A rational decision must also include legal, social, economic, and political factors.

The laws of thermodynamics are the most general and accurate of the physical laws. The first law states that energy in the universe can be neither created nor destroyed, merely converted from one form to another. The second law states that, although energy and work can be converted completely into heat, the reverse, the conversion of heat into work, is always done with less than 100% efficiency. The useful potential energy in the world is being inexorably converted to the less useful form of heat. The second law also shows that creation of order by man, which is a necessity for life, results in the creation of disorder in his environment. Thus we are all interconnected in the environment whether we like it or not.

Many chemical elements in nature are continuously cycled through biogeochemical cycles that provide a natural recycling of resources and a natural purification scheme. Man's activities are sufficiently large that they may interfere with the natural cycles for some elements. Because many, though not all, pollutants occur in nature and can be accommodated in the earth's cycles, man's approach to environmental pollution should not be to achieve zero pollution but, rather, to attempt to keep pollutants from overloading the natural purification schemes.

Elementary chemical principles, including organic and inorganic definitions, biological molecules, and chemical reactions, have been covered. These topics are important in later chapters as is the material on the metric system and on units of pollutant concentration.

Review and Study Questions

1. What is the usual development of the scientific approach to a problem? What limits this approach to environmental problems?
2. How can science contribute to a solution of environmental problems?
3. Give an example of a photochemical problem that is important in environmental problems.
4. What factors must be considered if a conversion from gasoline-powered cars to electric cars were made? How would you be affected? Would the employment of any of your family be affected?
5. Distinguish between kinetic and potential energy. Give an example of each.
6. What type or types of energy are possessed by a can of gasoline tossed from a cliff?
7. Define work.

8. How does power differ from energy? What units are used for power?
9. Which law of thermodynamics indicates that the energy present in the universe during creation is still present now?
10. What is a system in thermodynamics usage? How is it distinguished from the surroundings?
11. State the second law of thermodynamics.
12. What is entropy? What is its importance for environmental problems?
13. What are the implications of the second law for man and the environment?
14. Define an ecosystem.
15. What is the importance of the biogeochemical cycles?
16. What is the relevance of naturally occurring levels of pollutants to the limits we should attempt to achieve of man's emissions?
17. What is the smallest building block of an element?
18. Contrast the size, weight, and charge of a nucleus and an electron.
19. What can be inferred about the nucleus from the fact that it is composed of protons and neutrons?
20. What is the significance of the atomic number of an element and how does it differ from the mass number?
21. What are isotopes?
22. What are the simplest hydrocarbons called? For what are they used?
23. Contrast the structure of aromatic hydrocarbons with aliphatic hydrocarbons.
24. Given a choice, which would you prefer *not* to ingest, CH_3OH or C_2H_5OH?
25. How are sugars made?
26. To what chemical species is the strength of an acid attributable?
27. To what chemical property is the difference between strong and weak acids attributable?
28. What is your height in centimeters? In meters? What is your weight in kilograms?
29. A car running 50 miles per hour is going how many kilometers per hour?
30. What is a ppm? What is volume %, how is it defined, and when might it be used?

Questions Requiring Outside Reading

31. How did life dependent upon photosynthesis become established on the planet? (G. Evelyn Hutchinson, in *Scientific American. The Biosphere*, p. 45; see Suggested Outside Reading).
32. Very little of the incoming solar radiation is used by plants. What fate befalls the incident energy which is not used in this fashion? (A. H. Oort, in *Scientific American. The Biosphere*, p. 54; see Suggested Outside Reading).

33. Describe that portion of the nitrogen cycle which involves water. (C. C. Delwicke, in *Scientific American. The Biosphere*, p. 137; see Suggested Outside Reading).

34. Describe the carbon cycle. (B. Bolin, in *Scientific American. The Biosphere*, p. 124; see Suggested Outside Reading).

35. What elements are important to life? (Earl Frieden, "The Chemical Elements of Life," *Scientific American* [July 1972], 52).

36. How are heart disease, cholesterol, and nutrition related? (J. Rinse, *American Laboratory* [July 1973], 25; Roger J. Williams, *Nutrition Against Disease*, Bantam Books, New York, 1971).

37. What are glycoproteins and how are they important to cell function? (N. Sharon, "Glycoproteins," *Scientific American* [May 1974], 78).

38. Starches, carbohydrates, and proteins all possess energy which can be measured in Calories. In the body, the Calories in carbohydrates are utilized quite differently from the Calories in proteins. What accounts for the difference? (Roger J. Williams, *You Are Extraordinary*, Pyramid Books, New York, 1971).

Suggested Outside Reading

BENT, HENRY. "Haste Makes Waste—Pollution and Entropy," *Chemistry* (October 1971), 6. An interesting discussion of the implications of thermodynamics, especially entropy, for biogeochemical cycles, photosynthesis, energy production, and social problems.

HILL, JOHN W. *Chemistry for Changing Times*. Burgess Publishing Co., Minneapolis, Minn., 1972. A more detailed treatment of introductory chemistry. Written for the nonscience student.

MILLER, G. T., JR. *Replenish the Earth*. Wadsworth Publishing Co., Inc., Belmont, Calif., 1970. An excellent nontechnical discussion of ecology biogeochemical cycles, and the elements of thermodynamics.

_____. *Energetics, Kinetics, and Life—An Ecological Approach*, Wadsworth Publishing Co., Inc., Belmont, Calif., 1971. A discussion of the relationships of energy, thermodynamics, and life on the planet.

Scientific American. The Biosphere. W. H. Freeman and Co., San Francisco, 1970. An excellent description of biogeochemical cycles and energy flow in the biosphere. This is the September 1970 issue of *Scientific American*.

2 Population, Resources, and the Quality of Life

... we have concluded that, in the long run, no substantial benefits will result from further growth of the Nation's population. ... We have looked for, and have not found, any convincing economic argument for continued population growth.

Commission on Population Growth and the American Future 1972

Population is an appropriate topic with which to begin a study of environmental problems. We have an expanding population on a planet with a fixed amount of nonrenewable resources, and it is reasonable to assume that the impact of population will increase in the future. The effects of overpopulation could mean exhaustion of many natural resources, a drastically altered world for future generations, political chaos, and perhaps an end to life itself.

The poorer nations of the world have the most pressing population problems. Not only does population grow faster in the poorer nations, but they have inefficient agricultural systems, inadequate health facilities, inadequate educational programs, and other economic handicaps. The result is that these nations are virtually locked into an ever-quickening cycle of poverty, disease, and starvation. Any efforts to break this cycle must begin with a consideration of the implications of population and its growth.

We should not assume that problems resulting from population growth are limited to the underdeveloped nations. Although at first glance it may appear that a nation with so much affluence and so much open land as the United States could afford to give a low priority to its own population growth, such a view is dangerously misleading. Virtually all facets of American life are involved with the rise and fall of our birth and death rate: the economy, the environment, education, health, sexual practices, governmental effectiveness, and life-styles. Furthermore, the impact of the United States resource consumption is a dominant force in the world. With 6% of the world's population, we consume an estimated 30% to 50% of the world's resources, and we are by no means self-sufficient. We presently import about 25% of our gas and oil, 28% of our iron, 58% of our mercury, 96% of our aluminum, 100% of our platinum, and large percentages of nearly all other minerals. Many of these imports are from the poorer nations. Consequently the rich and poor nations of the world are so interdependent that population problems in any one nation have a direct effect on many others. The impact on the world's resources on one American is equivalent to that of 100 or more members of the poorer nations. Population growth is an important problem in any country.

That population and population growth are important factors in our principal topic, environmental problems, is not debated, but debate sometimes arises over the extent of their importance compared to technological misapplications or to excessses in per capita consumption.

Population Dynamics

To study population impact, it is useful to examine the history of population growth and the parameters used to describe and predict population dynamics.

The population change on earth reflects a balance of two competing processes: births and deaths. The net difference between births and deaths over the recent history of the world is shown in Figure 2.1a. The major changes in the world's population reflect changes in the death rate since, until the last two

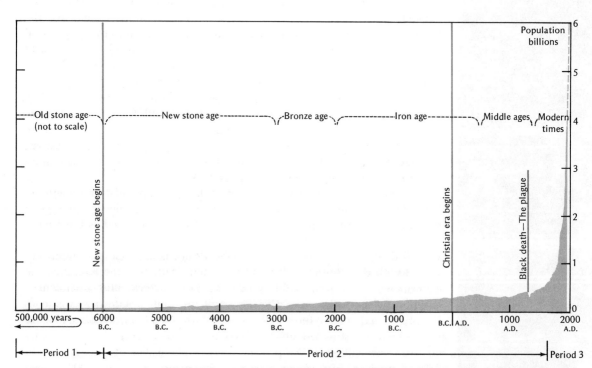

Figure 2.1a. *It has taken all the hundreds of thousands of years of man's existence on earth for his numbers to reach 3 billion. But in only 40 more years population will grow to 6 billion if current growth rates remain unchanged. If the Old Stone Age were in scale, its base line would extend 35 feet to the left! [Reprinted with permission from "How Many People Have Ever Lived on Earth?" Population Bulletin, **18**, (1) p. 7 (February 1962).]*

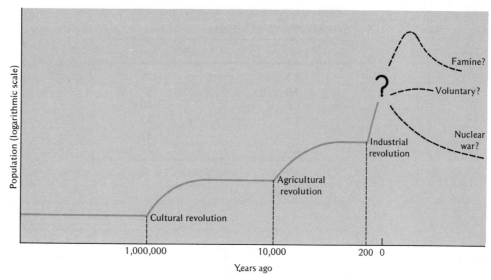

Figure 2.1b. *World population growth plotted to show changes as a result of cultural, agricultural, and industrial revolutions. [Solid line after E. S. Deevey, "The Human Population," copyright 1960 by Scientific American Inc.]*

centuries, the birth rate has been relatively constant. This population growth probably occurred qualitatively, as shown in Figure 2.1b. The humps in the curve in this figure reflect a decrease in the death rate which prompts an abrupt jump in the net population. The initial jump began about 1 million years ago during the **cultural revolution** when the development of man's brain and social order gave him tools and other advantages over the elements and over other animals. About 10,000 years ago the **agricultural revolution** produced another hump. The third jump occurred a few hundred years ago with the coming of the **industrial revolution.**

One obtains a further perspective on population growth by examining doubling times: the time for a population to double in size. If the population has a constant percent increase, its change is identical to the interest earned in a savings account. Table 2.1 shows the doubling time for various compound interest rates and may be used to obtain crude estimates of the doubling time of populations. The numbers in Table 2.1 show the range of doubling times currently exhibited by nations.

Table 2.2 shows the average doubling times required at various periods in history for the estimated world population growth. An examination of this table indicates that not only has the population increased but the **rate of growth** has also increased drastically. For the world to double its population beginning in 1650 it took 200 years; beginning in 1930, however, it took only 45 years; and beginning in 1975 it will take only an estimated 35 years.

TABLE 2.1. Doubling time for various compound interest rates

Annual Increase (%)	Doubling Time (years)
0.5	139
0.8	87
1.0	70
2.0	35
3.0	23
4.0	17

TABLE 2.2. Doubling times for the estimated world population growth

Date	Estimated World Population	Time for Population to Double
8000 B.C.	5 million	1500 years
1650 A.D.	500 million	200 years
1850 A.D.	1000 million (1 billion)	80 years
1930 A.D.	2000 million (2 billion)	45 years
1975 A.D.	4000 million (4 billion) computed doubling time around 1972	35 years

Source: Paul R. and Anne H. Ehrlich, Population, Resources, Environment. W. H. Freeman and Company, San Francisco, 1972, p. 6.

The trend is unmistakable: a relentless march toward shorter and shorter doubling times as the population increases. Such growth is often referred to as **exponential growth.** Figure 2.2 shows an example of pure exponential growth. Because the curve is shaped like the letter J, it is called a J-curve. Figure 2.2 results from a simple calculation that can be demonstrated in the following way. Suppose we double a page of this book over and over again and measure the thickness after each doubling. If the page is 0.004 in. (four thousandths of an inch) thick, then after the first doubling it will be 0.008 in., then 0.016 in., then 0.032 in., then 0.064 in., and so on. After eight doublings it would be about 1 in. thick, after 12 doublings it would be 16 in., and after 24 it would be 1 mile thick. The actual thickness changes very slowly with doubling until we reach the bend in the curve. For the thickness to measure 4000 miles, 36½ doublings are required, and 42 doublings would reach from the earth to the moon (240,000 miles). At this point we are near the bend of the curve and the results are dramatic as the curve "rounds the bend." Only eight more doublings

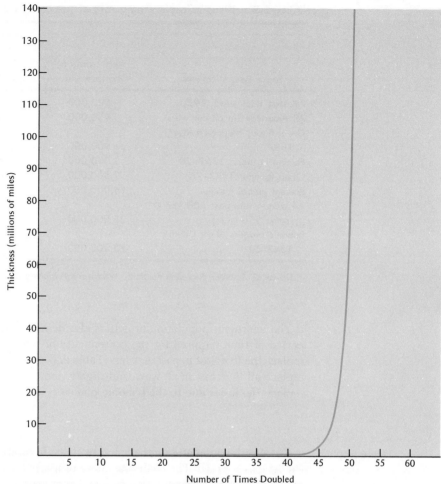

Figure 2.2. *The growth in thickness as a page in this book is doubled successively. The graph represents an exponential, or J-curve.*

(a total of 50) would reach from here to the sun—93 million miles! The impact of exponential growth arises principally because of the ever-growing base.

The world's population growth (Figure 2.1a and Table 2.2) also appears to follow a J-curve. The impact of going around the bend has been termed by Professor Paul Ehrlich of Stanford University a "population bomb," the deadliest explosion in history. Continued growth has produced a world population that is now on the vertical portion of the J-curve. As shown by our model (Figure 2.2), the result of doubling population is catastrophic in this portion of the curve.

TABLE 2.3.   Disaster replenishing rate of population growth

Some Past Disasters	Approximate Number Killed	Present World Population Growth Replaces the Equivalent of This Loss in Approximately
Pakistan tidal wave, 1970	200,000	1 day
All Americans in all our wars	600,000	3 days
Great flood, Hwang-Lo River, 1887	900,000	4½ days
Famine in India, 1769–70	3,000,000	2 weeks
China famine, 1877–78	9,500,000	7 weeks
Present global famine	15,000,000/yr.	2½ months
All wars in the past 500 years (some 250 wars)	35,000,000	6 months
Bubonic plague (Black Death), 1347–51	75,000,000	13 months

Source: G. T. Miller, Replenish the Earth. Wadsworth Publishing Co., Inc., Belmont, Calif., 1972.

The enormous population growth is also demonstrated by considering the period of time required for the present rate of world population growth to replace the lives lost in past disasters (Table 2.3). Only half a year is required to replace all lives lost in all wars, and slightly more than one year is needed to replace the losses due to the bubonic plague.

Demographic Parameters   To describe population growth and to estimate future trends, many statistical devices are used. The **birth rate** (more properly the **crude birth rate**) is the number of babies born per thousand people per year, where as the **death rate** is the number of deaths per thousand people per year. These data are usually compiled for a 12-month period by dividing the number of births (or deaths) in that time period by the estimated population at the midpoint of the period, then multiplying this result by 1000. That is,

$$\text{birth rate (per thousand)} = \frac{\text{number births in year}}{\text{average population that year}} \times 1000$$

The United States birth rate was about 18.2 births per thousand persons during the 12-month period ending April 30, 1971; the death rate was 9.4 per thousand persons.

A third parameter is the **fertility rate,** which measures the average number of births per thousand females aged 15 to 44. This is a closer measure of the actual production capability of the population because it monitors females during their reproductive years. In 1971 the United States fertility rate was 82.3 per thousand females aged 15 to 44.

The **growth rate,** the difference between the birth and death rates, represents the natural rate of change of population (no migration):

$$\text{growth rate (per thousand)} = \text{birth rate} - \text{death rate}$$

In 1971 the United States growth rate was about $18.2 - 9.4 = 8.8$ per thousand, or 0.88%, a typical value for an industrialized nation. Estimates are that the world in 1971 had a birth rate of 34, a death rate of 14 and thus a growth rate of 20 per thousand, or 2%. These growth rates, expressed as percentages, can be used with Table 2.1 to estimate the present population doubling times: 80 years for the United States and 35 years for the world.

For nations or geographic regions, a complete consideration of population growth must include migration; thus the **total growth rate** is used:

$$\text{total growth rate} = \text{birth rate} - \text{death rate} + \text{migration rate}$$

In 1971 the net migration rate was about $+2$ per thousand persons for the United States (see Table 2.4), so its total growth rate was about $18.2 - 9.4 + 2.0 = 10.8$ persons per thousand.

Figure 2.3 shows the United States birth rate since 1910. The decrease in the birth rate from the late 1800s to the Depression reflects a common occurrence during industrialization; all developed nations show a similar trend. This decrease arises from a variety of factors, including increased knowledge of birth control techniques and the availability of contraceptive devices. A further

TABLE 2.4.   Demographic parameters summary

	1971 Figures	
	United States	World
Birth rate: births per thousand persons per year	18.2	34
Death rate: deaths per thousand persons per year	9.4	14
Fertility rate: births per thousand females (ages 15 to 44 years) per year	82.3	—
Growth rate: birth rate — death rate	8.8	20
Total growth rate: birth rate — death rate + migration rate	10.8	20

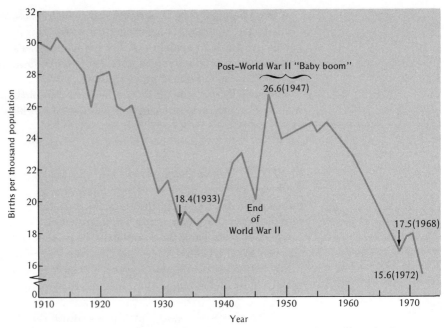

Figure 2.3. *Birth rate in the United States, 1910–1972. After "The Baby Boom Went That-A-Way," Population Profile, March 1967. Used with permission.*

contributor is the fact that, although large families are a virtual necessity in an agrarian society, they may become a burden in a more affluent, industrial society. The death rate also falls during industrialization because of improved sanitary conditions, medical services, education, and social conditions.

Even though the 1972 United States birth rate was the lowest yet recorded, the United States population is still growing and will continue to do so until the total growth rate is zero or negative. If the United States birth rate continued to decline, it might be possible to attain a stable population early in the next century. How such **zero population growth,** or **ZPG,** might be achieved and the effects it might have are discussed later in this chapter. There is no guarantee, of course, that the United States birth rate will remain low in the future. It could increase again as it did following the Depression. In the next section we will consider the principles involved in projecting population changes.

Predicting Population Growth

Birth rates, death rates, and fertility rates are constantly changing, and any long-term (greater than one- or two-year) predictions of population growth based on the assumption that these will not change are subject to large errors.

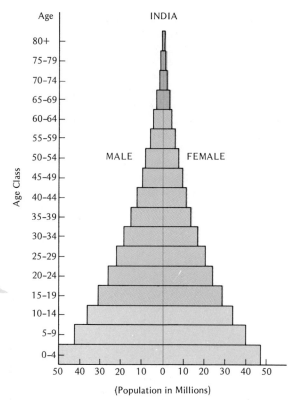

Figure 2.4. *Age composition of population of India in 1970; profile resembles a typical underdeveloped country. [Reprinted with permission from "India: Ready or Not, Here They Come," Population Bulletin,* **26,** *(5), pp. 1–31 (November 1970).]*

In fact these parameters alone are not sufficient for an accurate projection; an additional prerequisite is the age group distribution within the population. This enables a more reliable estimate of the relative impact of the age groups that will move into child production in the upcoming years. Figure 2.4 shows the age structure diagram for India in 1970 which is typical of that of an underdeveloped nation. Since approximately 44% of the population is under age 15, young people predominate. As this group moves into fertility, it will produce enormous stress on the educational, medical, social, and economic system of India, which could barely handle the load of the previous (much smaller) generation.

This age structure is in stark contrast to that of the United Kingdom (Figure 2.5), an industrialized nation whose growth rate has been very low for decades. Only 23% of the United Kingdom population is under age 15, and each age group constitutes about the same percentage of the population. Although the United Kingdom is still growing, the educational, social, and economic demands of each new generation are only slightly greater than those of the

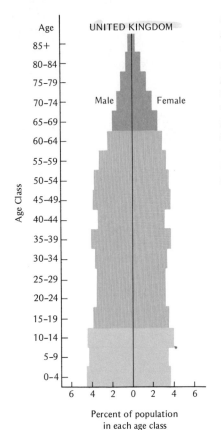

Age

85+
80-84
75-79
70-74
65-69
60-64
55-59
50-54
45-49
40-44
35-39
30-34
25-29
20-24
15-19
10-14
5-9
0-4

UNITED KINGDOM

Age Class

Male   Female

6   4   2   0   2   4   6

Percent of population
in each age class

Figure 2.5. Age composition of population in the
United Kingdom in 1959.
In the United Kingdom the
population is evenly dis-
tributed over the age
spectrum. [Reprinted with
permission from "The
Story of Mauritius from
the Dodo to the Stork."
Population Bulletin, **18**,
(5), p. 38 (November
1962).]

Figure 2.6. Age composition of population of
Japan in 1960. Note nar-
row base profile, caused
by a sharp decrease in
the birth rate. [After
W. S. Thompson and D. T.
Lewis, Population Prob-
lems, 5th ed. McGraw-
Hill Book Company, New
York, 1965.]

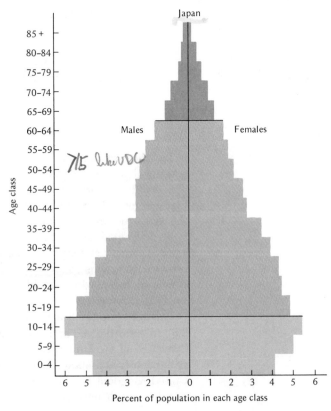

Japan

85 +
80-84
75-79
70-74
65-69
60-64
55-59
50-54
45-49
40-44
35-39
30-34
25-29
20-24
15-19
10-14
5-9
0-4

Age class

Males   Females

6   5   4   3   2   1   0   1   2   3   4   5   6

Percent of population in each age class

previous generation; the slower growth causes much less strain than that which occurs with the faster growth rate in an underdeveloped nation.

Japan has had a decreasing birth rate since 1945 because of industrialization and social changes and thus displayed in 1960 the "pinched off" age distribution (Figure 2.6). Above age 15 it resembles an underdeveloped nation, but the 0 to 14 age group reflects the lower birth rate. Although this is a welcome trend in Japan, there is no guarantee that this "pinch" in the population distribution will continue. If Japan undergoes a period of increasing birth rate, a broader base in the distribution figure will result.

The population distribution in the United States shows the effects of two low birth rate periods: the one that occurred in the Depression and the one we are presently experiencing (Figure 2.3). The "pinch" in the United States curve resulting from the births during the Depression is labeled in Figure 2.7; the bulge in the distribution results from the births of the post–World War II "baby boom." This bulge is the basis for fears that the United States birth rate

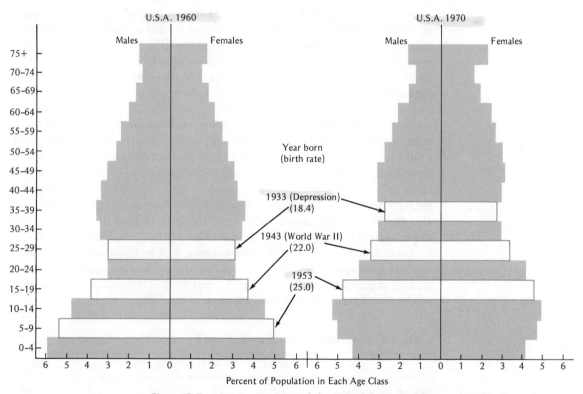

Figure 2.7. Age composition of the United States in 1960 and 1970. [Data from the U.S. Bureau of the Census.]

may increase in the next few years as the children born in the high fertility years of the 1950s, who outnumber those born in previous decades, move into fertility in the 1970s. Even if the fertility rate stays constant, the birth rate will *increase* because of the change in age structure.

Even with all the above data and all the technological advantages of modern computers, predicting future population growth is fraught with difficulties. Historically predictions of future population figures have underestimated population growth. Present estimates project a world population of 5.5 to 7 billion in the year 2000 compared to the present population of about 3.7 billion. It is of course possible that the population growth will slow down and the growth curve will level off, as discussed in the sections that follow.

Consequences of Population Growth

An increasing population puts an increasing demand on the life support system of the planet. It was pointed out by Thomas Malthus in the early nineteenth century that the potential for population growth outstrips the ability to provide for that growth. Until constrained by some voluntary or involuntary force, population expands. Malthus used a mathematical analogy, noting that population growth can be represented by a **geometric series**, doubling at every step (1, 2, 4, 8, 16, 32, . . .). On the other hand, food production is more closely represented by an arithmetic series: 1, 2, 3, 4, 5, . . . To understand his argument, suppose we consider a hypothetical case (Table 2.5 and Figure 2.8) where at some point in time, a relative population and its relative food production base are represented by 1, at Start. The population tends to double itself to 2 (step 1); with effort toward cultivating new land or perhaps using irrigation, food production can also double (step 1). At this point, food production has paralleled population growth. However, the population can again easily double to 4 (step 2). Now, in that period, food production might move to 3, or a 50% increase, but it probably will not double because expanding food production is now much harder. Further expansion of cultivated land area would mean that poorer lands would have to be used (the best went for step 1) and perhaps irrigation water would have to be moved further. As shown in Table 2.5, however, the population can double repeatedly unless constrained (perhaps by food production). Similar arguments apply when contrasting population growth to support areas such as mineral production, fuel production, house construction, and medical services.

TABLE 2.5. Malthus' population and food production predictions

	Start	Step 1	Step 2	Step 3	Step 4
Relative population	1	2	4	8	16
Relative food production	1	2	3	4	5

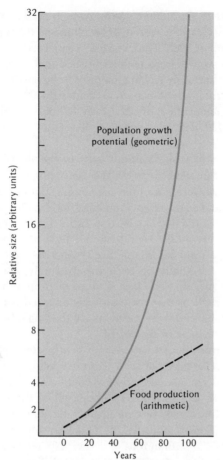

Figure 2.8. A comparison of potential population growth (geometric series) and food production (arithmetic series). The population is assumed to have a 20-year doubling time.

What is the capacity of the earth to support people? Unfortunately there is no simple answer to the question. In the sections that follow we shall consider the consequences of population on various sectors and attempt to establish the constraints these consequences might place on population growth.

Food Production

The UN Food and Agriculture Organization (FAO) has established standard "reference" body weights and standard daily per capita caloric requirements. On the basis of these FAO standards, the President's Science Advisory Committee in 1955 estimated world average calorie needs at a minimum of about 2500 kcal per capita per day. (We are using calories, not Calories—see Chapter 1.) The FAO estimated that food equivalent to 2400 kcal per capita per day was produced in the middle 1960s, enough apparently to just meet the demand. Despite this, however, over two thirds of the world's population had an inadequate diet in 1965. The President's panel estimated that 1½ billion of the people in the world were undernourished (not enough calories per day) or malnourished (seriously lacking in one or more nutrients, principally proteins). Georg Borgstrom, a food expert, estimates that there were 2.4 billion undernourished, malnourished, or deficiently fed persons in 1969 compared to 450 million well-fed people. A variety of reasons contribute to this appalling picture.

To begin with, vast inequities exist in the distribution of food between countries and, in some cases, in the distribution within one country. These result from interacting factors: economics, poor transportation, poor communications, and a lack of governmental action. The poorer nations, of course, bear the brunt of the low nutrition statistics. In the United States, for example, a 2000 kcal daily intake represents a weight-reducing diet for many males. Furthermore, a total calorie picture is somewhat misleading because certain requirements, especially protein, are absolutely essential: an adequate diet *must* include protein. The best sources of protein are those that contain a balance of the 20 or more amino acids required for living cells. Meat, fish, poultry products, and dairy products are the single best sources. But these are luxury foods for most of the world.

In the underdeveloped nations a critical problem is a lack of protein. The principal protein source of such nations is vegetable protein, but *individual* vegetables and grains usually do not contain all of the amino acids necessary for good health and growth. To obtain all of these a variety of sources of vegetable protein should be consumed, but even this is not possible in many of the poorer nations. Ironically the underdeveloped nations export, for economic reasons, 3.5 million tons of desperately needed high protein (fish meal, soybeans, and oil seed cake) and import 2.5 million tons of lower quality protein. A lack of protein causes stunted growth, decreased mental capacity, and lowered disease resistance. Protein deficiencies during the first 3 years of infancy, when the brain grows to 80% of its adult size, are apparently permanent. The effects of protein deprivation during these formative years cannot be reversed by increasing the protein intake in later years. Protein deficiencies, poor health, and reduced mental capacities obviously contribute to the cycle of poverty and suffering in the underdeveloped countries.

Thermodynamics of Food Production

Solar energy provides all our food through an energy input to plants. Through the process of photosynthesis, green plants use some of the incoming solar energy to convert carbon dioxide and water into the large, primarily organic molecules that are required for living organisms. Herbivores eat the plants, and some of the herbivores are consumed by carnivores. The smaller carnivores may, in turn, be eaten by larger carnivores. The organic molecules created by the plants are thus passed "upward" to higher organisms. (Plants, however, are not the only source of biologically important compounds. Most of the higher organisms can make from food a few, though seldom all, of the chemicals necessary for their cells.) All of these higher organisms have the capacity to break down the large molecules into smaller molecules. This releases, for use by the organism, the chemical energy that once bound the atoms into the larger molecules. The animal expends some of this energy in its daily activities and uses some of it for growth or tissue repair. Energy is thus transmitted from plants to man via a "food chain" that consists of several levels, the **trophic levels**: plants, herbivores, small carnivores, large carnivores.

The laws of thermodynamics, discussed in Chapter 1, apply to this food chain. The first law requires that the energy *input* to the chain must equal the total energy *output* (including heat lost to the environment). The second law governs the efficiency of energy use; that is, how much of the energy input is degraded to heat at each trophic level. The food chain is very inefficient; the *useful* energy decreases drastically as one goes up the food chain toward man and the larger carnivores. Typically 80% to 90% of the energy is lost as heat to the environment at each step. In other words, only 10% to 20% of the energy stored in the living tissue is available for transfer to the species at the next level, as in Figure 2.9 which shows actual data from Silver Springs, Florida. This energy flow can also be expressed as an energy pyramid as in Figure 2.10.

Figure 2.11a illustrates a simplified picture of a food chain involving man in which each block represents the useful energy obtained from the larger block on which it rests. One man requires 600 trout in 1 year for survival; the trout require 18,000 frogs that consume 54,000,000 grasshoppers that dined on 2000 tons of grass. Such pyramidal relationships (and the second law) illustrate that at each level, the **biomass** (the living weight) of the producers always exceeds that of the consumers. The production of meat (top of the pyramid) is enormously more expensive in terms of energy than the production of grain (bottom of the pyramid). Meat will not support a very large population. For example, it takes 18,000 lb of alfalfa to raise one steer that might support one man for a year. For the same food chain base, more people would, of course, be supported if they ate further down the chain, as illustrated in Figure 2.11b. By shortening the chain to frogs, 30 people might be supported instead of the one person initially supported by the chain. Although an oversimplification, Figure 2.11 illustrates why most people in the world today must live on a diet of grain rather than meat. The North American diet is 25% livestock, the European diet 17%, and the Asian diet about 3%. Rice is the staple food for over 2 billion people, more than half the world. The protein content of rice is 5% to 13%, but

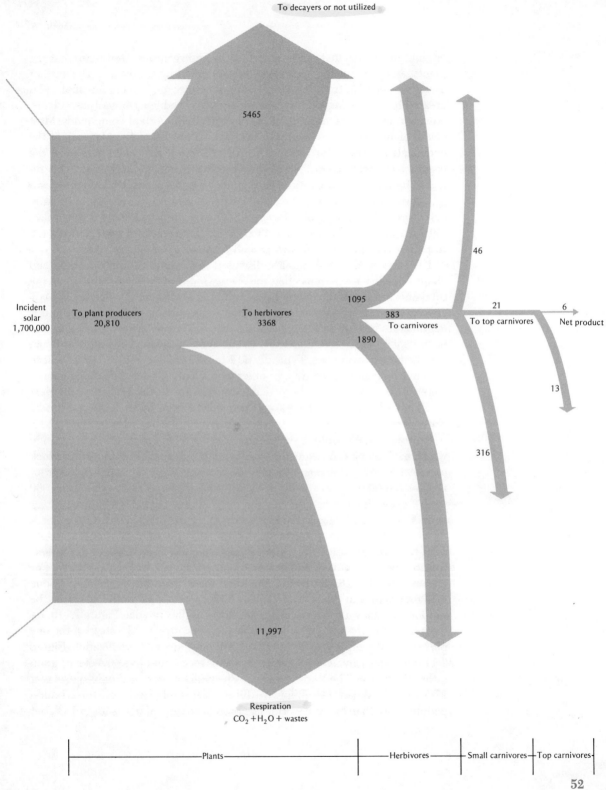

To decayers or not utilized

5465

Incident
solar
1,700,000

To plant producers
20,810

To herbivores
3368

1095

383
To carnivores

1890

46

21
To top carnivores

6
Net product

13

316

11,997

Respiration
$CO_2 + H_2O$ + wastes

Plants —————————————— Herbivores ——— Small carnivores — Top carnivores

Figure 2.9. *The energy flow in a food chain in a large spring in Florida. The widths of the arrows are proportional to the energy in kilocalories per square meter per year. [Data from H. T. Odum, "Trophic Structure and Productivity of Silver Springs, Florida," Ecological Monographs 27, (1957), pp. 55–112.]*

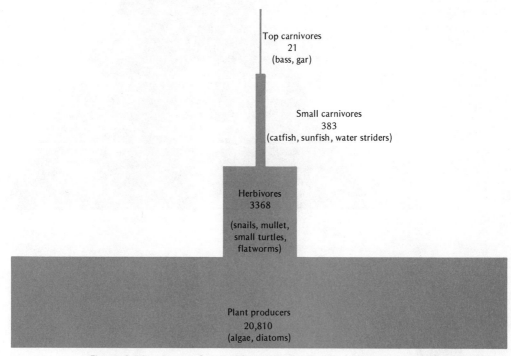

Figure 2.10. *Energy flow in Silver Springs, Florida, in kilocalories per square meter per year. The width of each block is proportional to the energy received from the block on which it rests. [Data from H. T. Odum, "Trophic Structure and Productivity of Silver Springs, Florida," Ecological Monographs 27, (1957), pp. 55–112.]*

the protein present is somewhat deficient in certain amino acids that can be obtained only through the consumption of other sources of protein.

An examination of the nation-by-nation food supplies, especially in Africa, is sufficient evidence that food shortages are probably providing a check on population growth in certain areas.

Figure 2.11a. *Pyramid of numbers in a food chain showing the numbers of each species which might be supported by 2000 tons of grass.* [After G. T. Miller, Jr., Replenish the Earth. Copyright 1972 by Wadsworth Publishing Co., Inc., Belmont, Calif.]

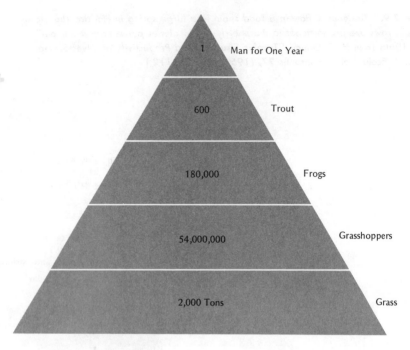

1	Man for One Year
600	Trout
180,000	Frogs
54,000,000	Grasshoppers
2,000 Tons	Grass

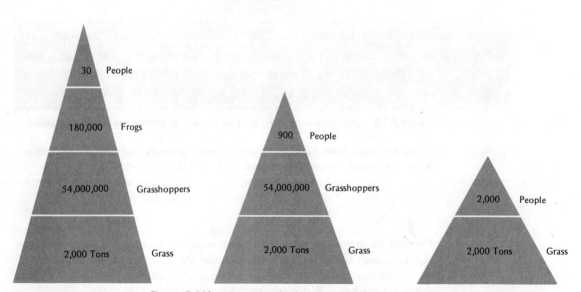

Figure 2.11b. *By eating farther down the food chain a larger population can be supported.* [After G. T. Miller, Jr., Replenish the Earth. Wadsworth Publishing Co., Inc., Belmont, Calif., 1972.]

Prospects for Alleviating the Food Shortage

When one considers increasing the amount of food in the world, a number of obvious suggestions arise. In the sections that follow, we will consider a number of these as well as some, perhaps not so obvious, possibilities.

Increasing Cultivated Acreage. The President's Scientific Advisory Committee report, referred to earlier, estimated in 1967 that the amount of potentially arable (farmable) land on earth was nearly 8 billion acres, more than triple the area actually cultivated and harvested in any given year. The majority of this potentially arable acreage is in Africa, South America, and Australia. However, the possibility of a threefold increase in cultivated acreage should be considered a very optimistic maximum upper limit to the actual gain that is likely. The estimate includes land that would require irrigation; in some areas, for example, water might not be available. Water costs to irrigate arid regions may be $1 per gallon, principally due to transportation; this contrasts with costs of 5¢ to 10¢ per thousand gallons presently paid in agricultural operations. Land expansion on such a scale requires technology, money, and expertise, and these are usually in short supply in the poorer nations. The cost of sample projects in these countries is typically $200 per acre. Furthermore, even if the effort and technology were available, other factors might completely halt a major development project. As one example, the massive erosion that occurs in tropical rain forests after deforestation has halted some modern-day pioneering efforts.

Increasing acreage for food production may conflict with other land use requirements. In the United States, for example, the trend has been toward *less* land for agricultural use. Agricultural land in California is lost to urbanization at the rate of 300 acres/day, and at this rate half of the prime cropland in that state will go to housing and industry by the year 2000.

Finally, unless the rate of population growth is reduced, the world's population will increase faster than new agricultural lands can be developed.

Increasing Crop Yield. Owing to the achievements of Norman Borlaug, increasing crop yield offers perhaps the brightest possibility on our list. Borlaug joined the Rockefeller Foundation in 1944 and went to Mexico to improve food crop production in cooperation with the Mexican government. His achievements in developing high-yielding, semidwarf varieties of wheat—for which he was awarded the Nobel Peace prize—have given rise to the term "Green Revolution." Borlaug utilized to the maximum the Mexican climate, which permits two generations of wheat per year, thus halving the normal period of 10 to 12 years needed to produce a new variety. Also, he experimented with many semidwarf varieties and was ruthless with hybrid material that did not meet expectations. Instead of trying to develop the perfect strain prior to widespread use, he selected his best strains each year and planted them extensively, thus boosting the total wheat production year by year while simultaneously developing new strains.

Borlaug realized that new varieties alone would not suffice and succeeded in encouraging farmers to use and demand, and governments to supply, the fertilizers and irrigation necessary for maximum production. The new grains have several advantages over traditional varieties. They mature early and, unlike traditional varieties, they are relatively insensitive to seasonal variations in the number of daylight hours; these two characteristics permit two or three crops per year. Borlaug's work resulted in a 30% to 100% yield per acre improvement and contributed to a sevenfold increase in the total Mexican annual wheat production since 1945. Borlaug's strains are being tried in India and West Pakistan. Analogous semidwarf strains of rice have been developed by the International Rice Research Institute in the Philippines.

The "Green Revolution" is not without its problems: relatively advanced technology (transportation, fertilizers, machinery, and irrigation) is required, which is seldom available in the underdeveloped nations. Furthermore, as discussed below, the energy required to implement agriculture in this form represents a sizable fraction of that used for all purposes in most of the poorer nations. Furthermore, most of the energy and the fertilizers are derived from petroleum, so future sources will be scarce and prices will probably increase sharply. A potentially serious problem is that the semidwarf strains may not have the broad, time-tested genetic base necessary to withstand diseases, insects, and severe climatic changes. This gives rise to the fear that if the new strains evolve too rapidly, the old strains, with many proved attributes, may be lost forever. If this should happen, and the new strains become disease prone, chaos could result.

Even under the most optimistic conditions, the "Green Revolution" will probably only give breathing room, not a permanent respite from the pressure that population growth exerts on food production.

Still, the accomplishments of the "Green Revolution" are enormous, and it is a measure of Borlaug that he gave his Nobel Prize money (more than $90,000) toward building a new experimental station for crop development.

Food from the Sea. Fewer ideas seem more intuitively obvious to the uninitiated than that which proclaims the ocean is a vast storehouse of food waiting to be tapped. Unfortunately intuition serves us poorly; by many estimates we are already harvesting well over half of the maximum *sustainable* harvest of the ocean! In 1969 J. H. Ryther of the Woods Hole Oceanographic Institute estimated that the maximum sustainable fish harvest was about 100 million metric tons. In 1969 the actual harvest was about 70 million tons. Ryther's estimates arise from a consideration of the fish production of the sea coupled with the idea that about half of this could be harvested annually for an indefinite period without danger of overexploitation and the decline in fish population that would result. Ryther's data in Table 2.6 indicate that the richest

TABLE 2.6.   Maximum sustainable fish harvest as estimated by Ryther

Area	Percent of Ocean	Area (square kilometers	Annual Fish Production (metric tons)
Open ocean	90	326,000,000	1,600,000
Coastal regions (including offshore)	10	36,400,000	240,000,000
Total annual fish production			241,600,000
Amount available for sustained harvesting			100,000,000

portion of the ocean is near the coasts. Some of Ryther's assumptions have been criticized, and he openly notes that many unknowns are involved. However, it is important to realize that even if his results are in error, they would have to be ten times too low before the annual fish harvest would even approximate the annual billion ton grain harvest produced on land. Thus, while the oceans are a source of high protein, their total yield will probably remain below land-produced protein. Another note of concern is that the richest region of the ocean, the estuarine area (that near the coasts), is the region most affected by pollution and the encroachments of man. Thus expansion of man's presence significantly decreases the potential yield of the ocean.

Protein Concentrates. Another potential food source is a high protein supplement, often termed fish flour or fish meal or fish protein concentrate (FPC). This potential food source is derived from rough fish and possesses a protein content such that 10 g (less than ½ oz) costing less than 2¢, would provide the daily animal protein requirement of the average human being.

In 1950 a process was developed by the VioBin company to make fish protein concentrate, one of many such products, palatable to human beings. FPC is a refined version of fish meal, long used for animal feed. It is a gray powder, almost completely odorless and tasteless, that can be made from fish not usually used for food. It is stable, it does not require refrigeration, and its lack of character renders it inoffensive as an additive to a variety of foods.

Recent efforts at large-scale (2 million pounds at 42¢ per pound) manufacture of FPC under the auspices of the Agency for International Development failed for a variety of reasons. The contractor attempted to use hake as a source, but found the supply variable. The United States Food and Drug Administration (FDA) standards had to be met, and their requirement of 1-lb bags instead of bulk packaging was an added expense. In addition, this discouraged interest on the part of food processors who prefer large lots. Other problems also inhibit widespread use of FPC. Its price is higher than that of local plant protein. As we shall see in the next section, the energy required to produce FPC will probably be a barrier to widespread use. Furthermore, its bland character actually limits its use, since it has nothing to recommend it as a flavoring agent, a texturizer, a binder, or a preservative. If poured into Kool Aid, for example, FPC simply lies on the top.

Several other varieties of high protein concentrate have been tried, but much of the research is presently supported by the United States government with only limited industrial interest. The difficulties involved have led the international agencies to favor other sources of protein, such as the cheap and adaptable soybean.

Energy Requirements for Food Production. A wide variety of agricultural practices, efficiencies, and yields exist in the world. Previously we considered the fundamental dependence of food production upon solar energy which permits photosynthesis in plants. In the industrialized nations with large agricultural outputs, a considerable energy "subsidy" is also added to increase food production. This subsidy arises by virtue of the energy required to manufac-

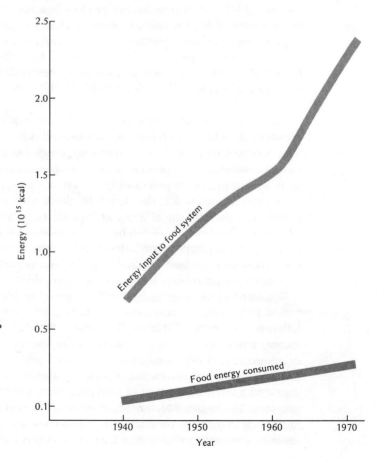

Figure 2.12. *Energy use in the United States food system, 1940 through 1970, compared to the kilocalories of food consumed. The kilocalories of food actually produced are estimated to differ from the consumption by at most 20% to 30%. [After J. S. and C. E. Steinhart, Energy: Sources, Use and Role in Human Affairs. Copyright 1974 by Duxbury Press, North Scituate, Mass. Reprinted with permission.]*

ture and use machinery, fertilizer, crop-drying barns, and pesticides. This energy input has increased annually in the developed nations while the man-hours devoted to agriculture have declined drastically. Man's efforts have increased crop yields, but, as we shall see, the improvements in production are not proportional to the energy expended. Because of the present and impending food shortages in the world, it is worth analyzing the energy requirements of technological agriculture with a view toward applying these techniques to agriculture in the poorer nations.

United States food production will serve as a model for agricultural production in developed nations. Enormous changes in modern agriculture have occurred in this century. There has been a marked decrease in farm workers, the ratio of nonfarmers to farmers having increased from 10 in 1930 to over 50. Agricultural productivity, however, has increased considerably, at a cost of an increasing energy subsidy. The energy expenditure has in fact increased faster than the resulting increase in food production (Figure 2.12). Figure 2.13 illustrates the diminishing returns that have been reached on energy investment in

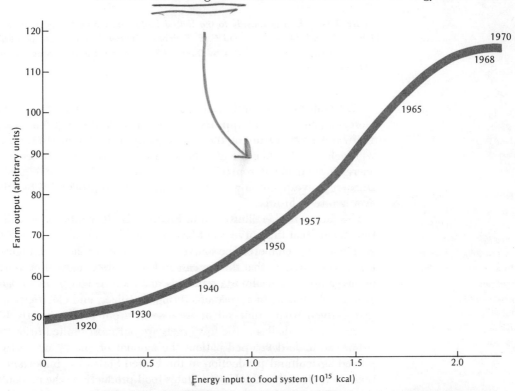

Figure 2.13. *Farm output as a function of energy input to the United States food system, 1920 through 1970. [After J. S. and C. E. Steinhart, Energy: Sources, Use and Role in Human Affairs. Copyright 1974 by Duxbury Press, North Scituate, Mass. Reprinted with permission.]*

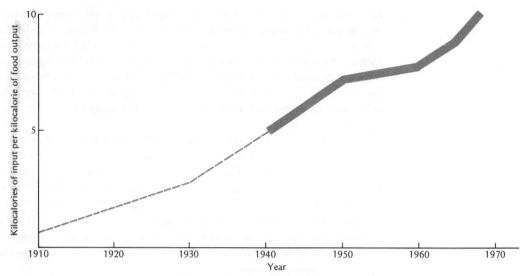

Figure 2.14. *Energy subsidy to the United States food system needed to obtain 1 kcal of food. [After J. S. and C. E. Steinhart, Energy: Sources, Use and Role in Human Affairs. Copyright 1974 by Duxbury Press, North Scituate, Mass. Reprinted with permission.]*

United States agricultural production. The curve showing farm output with increasing energy input appears to be leveling off after rapid increases in the 1940s and 1950s. Doubling the energy input beginning in 1971, for example, will yield nothing like the 30% increase in farm output that occurred when energy input doubled from 0.5 to 1.0 (1935 to 1945). That is, it requires more of an energy investment now to increase agricultural production by 1 kcal than ever before in history.

This fact is further illustrated in Figure 2.14. It is interesting and possibly threatening that this curve is not leveling off. We appear to be increasing our energy input even more. The sources of this input are shown in Figure 2.15. It is important to note that the increase in United States agricultural production in this century is due almost exclusively to a fossil fuel (coal, oil, or natural gas) subsidy! Fuel, machinery manufacturing, fertilizer, and electricity are all either derived from crude oil or use energy from predominantly fossil fuel sources. As we shall see, the fossil fuels are our most limited resources.

For the underdeveloped nations the amount of energy necessary for improved agricultural production in the United States has important implications. It is clear that the United States food production scheme could not be shipped intact to foreign nations. For India to feed her people at the United States level using United States techniques would require an energy input greater than India now uses for all purposes.

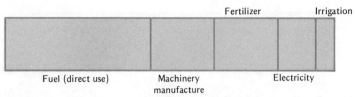

Figure 2.15. *Energy requirements in United States agriculture. The length of the bar is proportional to the energy requirement. [From J. S. and C. E. Steinhart, Energy: Sources, Use and Role in Human Affairs. Copyright 1974 by Duxbury Press, North Scituate, Mass. Reprinted with permission.]*

Feeding the world in the United States fashion would pose a hard question: where will the energy come from? Even exporting part of the industrialized nation food production apparatus may not work. The "Green Revolution," which is an attempt to accomplish this, requires an extensive energy input, and this requirement may ultimately limit its use in the poorer nations.

Energy Requirements of Specific Foods. The importance of energy suggests a look at specific foods to discover which provide the most efficient return on invested energy, that is, the largest number of digestible kcal resulting from a given energy input. In Figure 2.16 a variety of foodstuffs are displayed arranged according to the number of kcal of energy expended to produce 1 kcal of food. The vertical scale thus measures the energy "price" to be paid per kilocalorie of food. At the bottom of the figure are the crops grown exclusively using solar energy with little energy subsidy by man except his muscles. As one proceeds up the figure, the energy requirements increase. High-protein foods—such as beef, fish, and eggs—are highly **energy intensive**—that is, they require greater energy input per kilocalorie output than do, say, the cereal grains. Several crops, such as corn and potatoes, are shown with two energy ratings. These represent the two extremes under which these crops can be grown, with extensive energy input and with moderate energy input. Distant fishing requires more fuel than coastal fishing and is therefore more expensive in terms of energy.

Other important points in the figure need emphasis. Because of the energy required, food sources in the upper half of the table are not and probably will not be available to the poorer nations. In fact, if the world were to consume these at the per capita rate of the United States, 80% of the world's energy consumption would be needed just to provide food. Fish protein concentrate does not appear to be a likely source of protein for the underdeveloped nations to produce for themselves because it is second only to feedlot beef in energy requirement. High-quality protein at the lowest energy price is found in

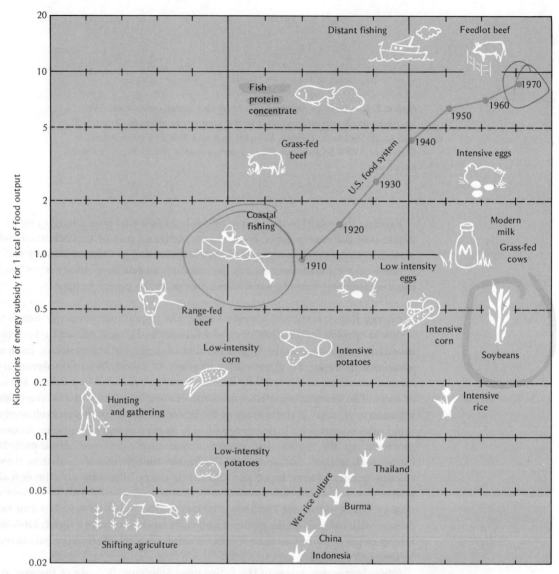

Figure 2.16. *Energy subsidies for various food crops. The energy history of the United States food system is shown for comparison. [From J. S. and C. E. Steinhart, Energy: Sources, Use and Role in Human Affairs. Copyright 1974 by Duxbury Press, North Scituate, Mass. Reprinted with permission.]*

coastal fishing but, as pointed out previously, we threaten that region with pollution and development. The position of the soybean may be crucial. It has a better amino acid balance for nutrition than most cereal grains, and because of its low energy price might be an important contributor to the world's protein supply.

The implications of Figure 2.16 are grave. Primitive cultures often invest 0.1 kcal of energy to produce 1 kcal of food. Industrial nations, however, now require 10 kcal of energy to produce 1 kcal of food, and there is every evidence that the energy cost will increase. As the costs of energy increase, food costs will of course also increase. This increasing cost presents an ever-increasing barrier to the underdeveloped nations in their quest for food, especially protein.

Resource Limitations

Predicting the lifetimes of resources is an enormous task. Accurate predictions rely on establishing the amounts of known and estimated reserves. Such data are often proprietary information of private corporations and hence are seldom available for public use. Furthermore, the economic factors involved, such as retail prices and the oil depletion allowance, make it necessary to examine closely the data that are made public. The predictor's crystal ball must also enable him (or her) to predict the effects of population growth (uncertain at best), changes in per capita consumption, and the development of new technology. Finally, the occurrence of "feedback" effects must be recognized, such as recycling stimulated by increasing prices arising from resource depletion.

Many estimates of resource lieftimes have been made. A computerized model of world growth, by Dennis Meadows and Jay Forrester, yielded the resource lifetimes shown in Table 2.7. This study attempted to include the changes in population, in per capita consumption, and in resource discovery. It did not include changeovers to other technologies that will inevitably occur as the fuels and minerals become scarce and expensive. This study indicates a reserve lifetime of about 20 years for petroleum and natural gas. Many other resource lifetime estimates have been made, but in all of them petroleum reserves are at most about one century, whereas natural gas is limited to a few decades. Coal, our most abundant fossil fuel, has an expected lifetime of over one century. The 20- to 70-year reserves of metals such as gold, copper, and lead will have serious implications for future technology, as will the reserves of such elements as molybdenum, which is needed to change iron into steel. Obviously some new reserves will be found, but the consumption rate will also increase. These estimates, if even remotely accurate, forecast significant changes in life-style within the next few decades because of resource depletion accompanied by the substitution of other materials.

TABLE 2.7. Depletion of resources

Resource	Estimated Global Reserves (Years)	Projected Depletion
Gold (26%)	9	1981
Silver (26%)	13	1985
Tin (24%)	15	1987
Petroleum (33%)	20	1992
Copper (33%)	21	1993
Lead (28%)	21	1993
Natural gas (63%)	22	1994
Tungsten (22%)	28	2000
Aluminum (42%)	31	2003
Molybdenum (40%)	34	2006
Platinum group (31%)	47	2019
Nickel (38%)	53	2025
Iron (28%)	93	2065
Chromium (19%)	95	2067
Coal (44%)	111	2083

* United States consumption as percent of world total.

Source: After D. H. Meadows, et al., The Limits to Growth. *Universe Books, New York, 1972.*

Worldwide problems arising from the disproportionate share of the world's resources used by the United States are also indicated in Table 2.7. The numbers in parentheses are the United States consumption as percentages of the world total. If such inequities are brought into focus throughout the world, the prospects of continued importation into the United States may not be bright.

Voluntary Population Limitations

Even a cursory glance at the data given in the previous sections of this chapter on population growth, food production, and resource limitation leads almost irrevocably to the conclusion that the only viable alternative to future catastrophe is the limitation of population growth. If we accept this, we immediately encounter three questions: When? Who? How?

The question of "When?" is easily answered: we must begin immediately. There is a time penalty in the sense that anything not done sooner will be harder to accomplish later. The question of "Who?" requires the answer: all nations, all peoples. Although the immediate problems of the underdeveloped countries may deserve priority because they are more pressing, we cannot use this as a reason to neglect efforts in the developed countries. The per capita resource impact of the developed countries is so large that limitation must be

Figure 2.17. *Drawn by Steve Brodner in the* Cooper Pioneer. *This cartoon won first prize in the 1974 Population Cartoon Contest of the Population Institute. The contest was designed to increase public awareness of population problems.* [*Reprinted with permission.*]

urged in them as well. For the most part, efforts within a nation must focus on all segments of the population. A nation's population growth derives its character from the properties of the *majority* groups, and statistically one can derive the greatest impact from changes in this group. Limiting the population growth of a minority group in a nation such as the United States would have little impact on the present overall population growth, since any one minority group constitutes a small fraction of the total population.

The remainder of this chapter will be devoted to the toughest of the three questions: "How?" There is a certain agreement on the desirability of lowering birth rates but not on the method nor on how fast nor on how far. By any of its names—birth control, population control, or population limitation—this is an extremely sensitive political, legal, social, and religious issue. There is no easy method nor is there a single best way to achieve a decrease in population growth. Many obstacles stand in the way of a solution to this problem. A variety of tax incentives, economic penalties, educational programs, family planning programs, and social structure changes have been proposed. Some of these are voluntary and some are involuntary. Most are not new; they have been around for some time. If they are not being tried it is because they have not been accepted or are too expensive either in terms of money or medical personnel. An extended discussion of these proposals is beyond the scope of this text. We shall, however, consider some of the technical aspects of birth control agents and techniques.

Birth Control Methods for the Developed Countries

A variety of birth control methods are available today. Some prevent the meeting of the sperm and ovum whereas some prevent the development of the fetus after fertilization. Some of these methods and devices are more effective than others, and each has advantages and disadvantages that may make it more or less suitable for a particular couple at a particular stage of life. Beyond these contraceptives, for a couple whose family is complete, there is sterilization. For the male, especially, this is a simple procedure that prevents the possibility of fathering a child and has no other known significant effect.

The Condom

Usually promoted as a means of avoiding venereal disease, the condom is also one of the most popular means of birth control. Usually made of rubber, the condom is a very thin sheath that fits tightly over the penis during intercourse and retains the semen after ejaculation. Its advantages lie in its simplicity of use and its availability. The failure rate is low, especially if the man has been instructed in its proper use. Care is especially required to ensure against spillage of semen at the time of withdrawal. Defective condoms are seldom encountered and can be guarded against by inspection before use. Unlike other devices, condoms do not require fitting or prescription by a doctor. But many men and women complain that they detract from the full enjoyment of intercourse: sensation is reduced and foreplay is interrupted in order to apply them.

The Diaphragm

This device, essentially a rubber cap with a rubber-clad rim of flexible spring steel, is designed to fit over the cervix, where it acts as a barrier to sperm. It is inserted into the vagina before intercourse, and is left in place for several hours afterward. Before insertion it is coated on the edges and underside with a spermicidal jelly or cream to prevent any sperm from passing around the diaphragm. A well-fitted, properly used diaphragm is a highly effective contraceptive, but it is relatively complicated to use compared to the condom and some other methods. To ensure a proper fit, it must be prescribed by a doctor, who also instructs the woman in its placement and use. When in position it does not in any way interfere with either partner's enjoyment of sexual relations.

The Cervical Cap

Like the diaphragm, the cervical cap bars entrance of sperm into the uterus. It is made of plastic or metal, fits tightly over the cervix, and may be left in place for long periods of time. It needs to be removed only for menstruation. If properly fitted, the cap is extremely effective. Its main disadvantage is the difficulty of placing it correctly.

Spermicidal Agents

A variety of spermicidal jellies and creams are available that can be deposited in the upper vagina with special applicators. Although these agents are less effective than the devices discussed in the preceding paragraphs, they have the advantage of being easier to use than mechanical contrivances, and they require neither fitting nor prescription by a doctor.

Foam tablets, aerosols, and suppositories are similar to the jellies and creams, and operate on the same principles. The foam varieties, perhaps because they are more thoroughly dispersed in the vagina, may be more effective than the others.

Rhythm

Also referred to as "periodic abstention," rhythm is the only method of birth control now sanctioned by the Roman Catholic Church. The basic idea is to abstain from sexual relations during the several days each month when a woman might be capable of conceiving. The difficulty is that this period is often hard to determine, particularly in women with irregular menstrual cycles. To avoid conception the couple must abstain from coitus for at least two days before and one-half day after ovulation. Unfortunately the occurrence of ovulation can be determined only after the event, and not too accurately even then. When ovulation has taken place, the woman's temperature rises about half a degree and drops again when her menstrual period begins. The time of ovulation must be predicted on the basis of carefully kept records of her previous menstrual and temperature cycles. To allow an adequate safety margin, several additional days should be included both before and after the estimated fertile period. For the very careful, this extended period without intercourse approaches complete absention. (Abstention, incidentally, is an extremely effective method when used, which it apparently is not in today's life-styles!) The rhythm method is one of the least effective birth control methods. Approximately one woman in six has a cycle so irregular that the system will not work for her at all. Surveys indicate that, particularly in the United States, many Catholics use some other form of birth control despite the lack of Church sanction.

The Pill

The modern steroid oral contraceptive, generally known as "the pill," is the most effective means of birth control generally available today, other than sterilization and abortion. When taken without fail according to instructions under medical supervision, it is nearly 100% effective.

The pill functions as a contraceptive by stopping the process of ovulation: the release of an ovum (egg) from the female ovary.

The female reproductive cycle is only partially understood. It is believed that the cycle originates with the release of a hormone, the **follicle stimulating hormone,** from the pituitary gland located at the base of the brain. This hormone, a protein (Chapter 1), stimulates the growth of follicular tissue on the ovary. The follicular tissue produces the egg and releases a very important female sex hormone, **estradiol.** Estradiol, a steroid (Chapter 1), performs two important functions. It reduces the secretion of the follicle stimulating hormone, thus preventing the simultaneous production of a large number of eggs. In addition, estradiol initiates the growth of uterine tissue, beginning the preparation of the uterus for the possible reception of a fertilized egg.

When the egg is mature, it is released from the ovary into the fallopian tube, which leads to the uterus. After the egg is released, the follicular tissue undergoes an important change, converting into a structure called the corpus luteum, which secretes a steroid female sex hormone **progesterone.** Progesterone, also known as the pregnancy hormone, prevents the production of any more eggs in the ovary by blocking the release of the follicle stimulating hormone from the pituitary gland. It acts similarly to estradiol in this regard. Progesterone has another important function: it causes the final development of the uterine wall for possible reception of a fertilized egg.

If fertilization is to occur, it must occur in the fallopian tube; the fertilized egg then migrates down the tube to the uterus. If the fertilized egg attaches to the uterus successfully, the corpus luteum continues to function throughout pregnancy, releasing progesterone, which has a role in the development of the fetus. Through hormonal control, the menstrual cycle is suspended until the fetus is completely developed and birth occurs. If the egg is not fertilized, the corpus luteum breaks down, progesterone secretion stops, and menstruation occurs. The built-up uterine wall sloughs off during menstruation, following which the pituitary initiates a new cycle.

Over 20 years ago it was discovered that prevention of egg production could be accomplished by setting up and supporting a hormonal state of pseudo-pregnancy by swallowing steriods similar to estradiol and progesterone. The pill contains these synthetic—that is, commercially made—hormones. The particular steroids used in any version of the pill may vary. Figure 2.18 shows the chemical structures of the two natural hormones and of two synthetic hormones: **mestranol and norethynodrel.** The addition of the **acetylenic group** $(-C \equiv CH)$ is the major change in the molecular structure. This is an important change because it makes the steroids orally active; that is, they have an effect when taken orally. Estradiol and progesterone, the natural hormones, can be made commercially. Interestingly they must be administered by injection to block ovulation. They are *not* orally active, presumably because they are destroyed by chemical action in the digestive tract.

The pill is taken daily for 20 to 21 days of the 28-day cycle, beginning on the fifth day after the onset of the menstrual period. The steroids may be administered sequentially or in a combined form. In the sequential system, the estradiol analog is administered alone during the early part of the cycle, with the progesterone analog added only during the latter part. The pills have the effect of regularizing the menstrual cycle to exactly 28 days, even in women who have never had regular cycles before. Moreover, menstrual flow is noticeably reduced, or even occasionally suppressed altogether. Most women consider these effects advantageous.

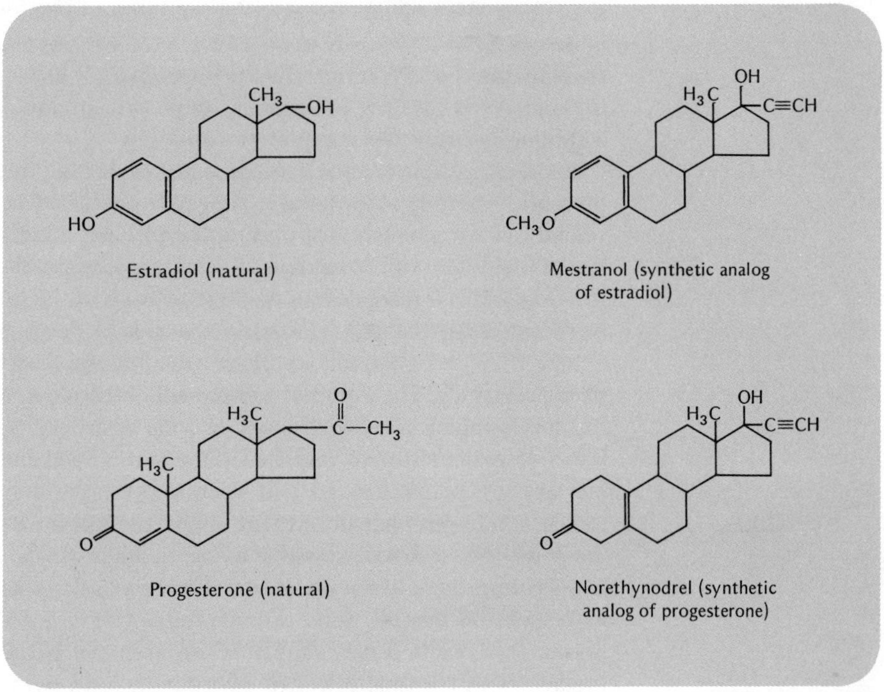

Figure 2.18. *Natural steroid female sex hormones (left) and their synthetic analogs. The analogs are present in the pill.*

As is inevitable with any drug, particularly a hormonal drug, there are undesirable side effects. Most of these, however, last only a few months or can be alleviated by adjusting the relative concentrations of the hormones. Many of the side effects resemble symptoms of pregnancy, which is not surprising because the pill creates a pseudopregnancy state. The most common side effects are tenderness and swelling of the breasts, weight gain and retention of fluid, nausea, depression, irritability, changes in complexion, and bleeding. About one in four or five women taking the pill experiences one or more of these symptoms.

The pill was tested on large numbers of women prior to its release to the public in 1962. In its use by millions since then, the pill has produced no serious medical problems for the overwhelming majority of women, although the results of using it over an entire reproductive span of 30 years or so are still unknown. Over the short term it is less hazardous than undergoing full-term pregnancy, although physicians generally do not prescribe the pill for women who have histories of liver disease, cancer, or venous diseases.

Whether the pill plays a role in inducing cancer is still unclear. Early results from a study sponsored by Planned Parenthood of New York were inconclusive. They showed that certain "precancerous" changes in the cervix were more common in women using oral contraceptives than in a control group using diaphragms. Whether this "precancerous" change would lead to cancer in women who use the pill remains unknown. In any case, the precancerous condition is easily and completely curable.

Another possible hazard, increased blood clotting (thromboembolism), also presents a complicated picture. Research published in England in 1968 revealed that women who were using the pill had a higher chance of dying of thrombophlebitis (inflammation of veins, together with blood clots) or pulmonary embolism (blood clots in the blood vessels of the lungs) than women who were not using the pill. However, the risk of death from thromboembolic disease while using the pill is still considerably less than that of death resulting from pregnancy. The estradiol component of the pill appears to be responsible for development of these thromboembolic disorders. Newer versions contain lower doses of mestranol, and the United States Food and Drug Administration has strongly recommended that doctors prescribe only low mestranol pills.

There are some indications that taking the pill for a long time may extend the period of a woman's sexual activity; in other words, a middle-aged women who is using the pill may be as sexually active as a woman five years younger who is not on the pill. Some women may even find that they have increased libido. This reaction may simply be the result of not having to worry about pregnancy at the time of coitus, or it may have a hormonal basis.

The advantages of an oral contraceptive are obvious, even apart from the advantage offered by its effectiveness. Its use is far removed in time from the act of intercourse, and there are no mechanical devices or chemicals except the pill to deal with. On the other hand the woman must remember to take the pill each day, which requires a fairly high degree of motivation. The chances of pregnancy increase with each forgotten pill.

Much more time will have to pass and much more data will have to be gathered before conclusive statements can be made about the risks of cancer and thromboembolism involved in the use of oral contraceptives. The risks must be weighed against the benefits, and the long-term risks are still unknown. From what we know now, it appears that for most women the risks can be minimized through close supervision by an alert physician. Obviously continued monitoring of the long-term effects of the pill (or any future hormonal contraceptives) is essential.

The Intrauterine Device (IUD)

The intrauterine device, or IUD as it is generally known, is a plastic or metal object placed inside the uterus and left there for as long as contraception is desired. It comes in a variety of shapes, each having its own advantages and disadvantages relative to the others. Exactly how these devices work is uncertain. One possibility is that they prevent or disrupt the implantation of the embryo after conception, or they may interfere with fertilization by stimulating the ovum to travel very rapidly through the fallopian tubes.

The advantages of an IUD are several; the primary one is that once in place it can be forgotten. There are no pills to remember, no contraceptive materials to deal with. This is a great advantage for an individual whose lack of motivation, educational background, or financial resources would make other forms of birth control unreliable or beyond her means. The device costs only a few cents. It must be inserted and subsequently checked by a physician or a paramedical person.

About 10% of women spontaneously expel the device, sometimes without knowing that they have done so. Some women have to remove the IUD because of such side effects as bleeding and pain. Such problems usually disappear immediately after the device is removed. Less frequently, the IUDs may be associated with pelvic inflammation, though there is some question whether the device is primarily responsible or only aggravates a preexisting condition. There is no evidence that IUDs lead to the development of cancer.

Comparison of Contraceptive Devices

The various methods of birth control are compared in Table 2.8. The effectiveness of techniques is calculated on the basis of 100 woman years—that is, the number of women per hundred who will become pregnant in a 1-year period while using a given method. Among 100 women using no contracep-

TABLE 2.8.   Failure rates of contraceptive methods

Method	Pregnancy Rates for 100 Woman-Years of Use	
	High	Low
No contraceptive	80	80
Foam tablets	43	12
Jelly or cream	38	4
Douche	41	21
Diaphragm and jelly	35	4
Condom	28	7
Coitus interruptus	38	10
Rhythm	38	0
Steroid contraception (the "pill")	2.7	0
Abortion	0	0
Intrauterine contraception (average)	about 2	

Source: After Berelson et al., Family Planning and Population Programs. Univer-sity of Chicago Press, 1966.

tion, 80 can expect to be pregnant by the end of one year. The failure rates are based on actual results, no distinction being made whether the method failed or whether the individuals were careless in using it. The lower rates are generally achieved by highly motivated individuals under close medical supervision.

In addition to these conventional methods of birth control, certain "folk methods" exist in our culture which have in the past been used mainly by ingenious teenagers. These include douches with soft drinks and condoms derived from plastic wrapping materials. Despite the ingenuity they reflect, they cannot be recommended. Their effectiveness is unknown, although that of the plastic condom may be quite high.

Sterilization

Another solution, sterilization, exists for those who, for any number of reasons, wish to rid themselves of concern about contraceptives.

The female operation, called a salpingectomy or a tubal ligation, involves severing the fallopian tubes so that the ovum cannot pass through. This operation, which usually requires hospitalization for a few days, is best done immediately after childbirth. The hormonal system is left intact. Ova are still brought to maturity and released, and the menstrual cycle still occurs. This operation has become common in middle-aged women, and perhaps because of its familiarity, adverse psychological reactions are rare.

The analogous male operation, a vasectomy, is simple and is sometimes performed in the doctor's office using a local anesthetic. The procedure, which usually takes less than an hour, consists of cutting or tying off the vas deferens thus preventing the passage of sperm into the penis with the ejaculate. The vasectomy has not been common until recent years when the number performed has jumped drastically. A vasectomy has no commonly known side effects; sperm are still produced but prevented from leaving the body. Sexual performance is not decreased; a vasectomy is not castration.

In both cases, sterilization does not hamper sex life. In fact, sexual appetites often increase because of the removal of the worry over unwanted children. In the few cases in which psychological problems develop, they usually are found to have grown out of disturbances that existed before the operation.

Many individuals considering sterilization are hesitant to take so final a step, and questions concerning reversibility almost inevitably arise. For men, reports on reversibility are varied. Some studies indicate 40% to 70% reversal. On the other hand, some surgeons advise that the operation be considered irreversible. New techniques may improve this. The woman's operation can be reversed in 25% to 66% of the cases. An important fact is that in actual practice only a small percentage of sterilized persons ever desire reversal.

For men who still have lingering doubts about taking the final step of sterilization, it is now possible to preserve a sample of sperm in a frozen sperm bank for at least 2 years and perhaps longer. In the future this period can probably be extended. Thus, if a second wife years later wishes children, they can be provided by artifical insemination. Another possibility is that live sperm can be removed from the father's testes and used for insemination.

Sterilization is legal in the United States, but in many places it is still difficult to find a doctor who is willing to perform these operations. This arises in part from old traditions and from a fear of lawsuits. A release is often required before surgery. The legal barriers appear to be dropping and voluntary sterilization is far more obtainable than in the recent past.

Abortion

Despite the fact that induced abortion is illegal in much of the world, it is apparently the most common form of birth control in all parts of the world, even the developed nations. Reliable estimates indicate that nearly 8% of all potential mothers in any year undergo abortions.

Abortion is the arrest of pregnancy. The medically approved method of inducing abortion in the early stages is through a simple operation known as dilation and curettage (scraping) of the uterus, which removes and destroys the fetus. This operation is preferably done no later than the twelfth week of pregnancy. After the sixteenth week the procedure is considerably more complicated. A newer, evidently safer, method for the early months involves the use of a vacuum device in place of curettage. It was invented in China and further developed in the Soviet Union. This device is being generally used in many eastern European countries, in China, in the Soviet Union, and in England. It has been introduced in the United States and in Japan.

When performed under appropriate medical circumstances by a qualified physician, abortion is safer than a full-term pregnancy, but if it is delayed beyond the twelfth week the risks of complications or death rise considerably. Approximately 20 to 100 deaths per 100,000 women per year occur as a result of pregnancy, whereas the legal abortion case fatality rate in New York City hospitals in 1971 was about 7 deaths per 100,000 abortions for gestation periods of less than 12 weeks and about 21 per 100,000 abortions for gestation periods over 12 weeks.

The risks increase drastically when the abortion is illegal, the amount of increase depending upon the circumstances. These may range from self-inducement with a knitting needle or, almost equally dangerous, unsterile help from untrained or semitrained people to reasonably safe treatment by a physician in a hotel room or a clandestine clinic. In the United States, as in many other parts of the world, bungled illegal abortions have been the greatest single cause of maternal deaths, accounting for between 400 and 1000 deaths per year.

TABLE 2.9. Legal therapeutic abortions (1967 and 1968 data unless otherwise noted)

Country	Birth Rate per 1000 Population	Abortion Rate per 1000 Population	Ratio Abortions per 1000 Live Births
Romania (1965)	14.4	56.70	3,940
USSR	18.2	26.09	1,435
Hungary	15.1	19.40	1,346
Japan	19.3	7.90	387
Poland	16.3	3.76	232
Sweden	14.3	1.22	79

Source: H. Rudel, F. Kincl, and M. Henzl, Textbook of Birth Control, Macmillan Publishing Co., Inc., New York, 1970.

Where legal, abortion can be an extremely effective constraint on the birth rate. Table 2.9 lists data on abortion rates in countries where abortion is legal. During the first year (1971) that abortion was legalized, 200,000 legal abortions were performed in New York hospitals. The maternal death rate in New York, which includes deaths from abortion, dropped by more than one-half.

In the eastern European nations, abortions outnumber births. The Romanian government became so concerned that in 1966 it instituted severe legal restrictions under which abortion was permitted only in women over the age of 45 who already had four children! As a result the birth rate doubled within one year (1967), but by 1968 it again dropped by nearly 30% indicating the Romanian population was taking recourse to other means, perhaps including illegal abortions.

The illegal abortion rate is difficult to ascertain reliably. Estimates in 1967 ranged from 100 to 600 per thousand live births (United States), to 700 to 1000 (Italy), to 1500 to 2000 (Belgium and Austria).

The legal barriers to abortion are becoming less stringent. The greatest obstacles to freely available safe abortion in most countries arise from religious concerns. The view of the Roman Catholic Church is that abortion is murder because the embryo, from the moment of conception, is a complete individual with a soul. Another view with considerable support is that abortion may promote promiscuity.

Birth Control for the Underdeveloped Nations

In most poorer nations, often only 10% to 20% of the adult population is familiar with birth control, and knowledge is often restricted to withdrawal. Although surveys indicate that interest is high, especially among couples with three or more children, apparently only a fraction of those who know about birth control practice it.

Family planning programs in the underdeveloped countries are usually carried out in clinics or in mobile units. These often actively recruit clients, using whatever mass communication technique seems feasible from radio and television to puppet shows and traveling troupes. India, in 1952, was the first underdeveloped country to establish a family planning program, and a 1965 reorganization produced a much stronger stance. Vasectomy and IUD programs are very common, but other contraceptive methods are also available. In the 1960s a number of other countries, including Pakistan, Indonesia, the People's Republic of China, and Taiwan, initiated programs. Wide variations occur from nation to nation, but all face formidable problems. A shortage of educational, communication, and transportation networks is combined with limited and already overburdened medical personnel and facilities. In underdeveloped countries childbirth seldom occurs in a hospital, thus removing medical contact at an opportune moment. Superstitution and fears of discrimination are additional sources of opposition. Inasmuch as the impetus for birth control is supplied by Western medical practice, the prevalence of many non-Western medical traditions is also a factor.

Development of New Chemical Birth Control Agents

The development, rapid acceptance, and widespread use of the pill and IUDs during the late 1950s and early 1960s gave rise to optimism that new chemical birth control agents might be expected shortly. This has not occurred. Although such techniques as immunological approaches, male contraceptive pills, "once-a-month" pills, "once-a-year" implants for females, additives to water, and other exotica are often discussed, a realistic view gives the definite impression that new developments should not be expected for widespread use until after the mid-1980s.

There are several reasons why the development of new birth control techniques has been slow. To begin with, the human reproductive process is extremely complex, and our knowledge of even the rudiments of the process is far from complete. Furthermore, no animal has a reproductive system sufficiently similar to that of humans to provide a completely reliable testing of new agents. For example, dogs have a semiannual reproductive cycle and are quite sensitive to steroid hormones, whereas the most commonly employed primate—the rhesus monkey—differs from man in the timing and mode of embryo implantation and in the production and metabolism of sex hormones.

The technological and financial base required to test and develop a new agent exist only in the developed nations. Unfortunately these are the nations with the smallest population growth rate and therefore the least pressure to develop new methods. Consequently governmental and private resources in these nations are usually applied to more visible needs.

If the new agent comes from a developed nation, then the development and testing will be subjected to regulation by the appropriate agency in that nation. In recent years the trend has been toward tightening the licensing tolerances and procedures for drugs.

Closely related to the lack of knowledge about human reproduction and stricter drug regulations is the difficulty of testing drugs on human subjects. The pill, for example, was tested in Puerto Rico. Although such testing is closely regulated, it often receives a less than enthusiastic response in the underdeveloped nations. The basis for protest may arise from opposition to birth control, from concern about the health of those tested, from concern about genocide, or from a fear of "Yankee Imperialism"; the fact is that such opposition exists in both the richer and the poorer countries and must be considered in any new development program.

FDA Requirements

The United States Food and Drug Administration (FDA) regulates drug (and chemical contraceptive) testing and licensing in the United States. No drug can be lawfully administered to humans in the United States without an Investigative New Drug (IND) exemption from the FDA; chemical contraceptives have requirements that exceed those of medicinal drugs. In the past decade the FDA has tightened its restrictions, and several well-publicized examples lend weight to this posture. For example, despite heavy drug industry pressure, such procedures prevented the licensing of the tranquilizer thalidomide, which later proved to be severely teratogenic (caused birth defects). On the other hand, no drug is completely safe, and nearly all drugs have side effects. Tighter controls may hinder new advances if development is halted too soon, say, at the initial signs of side effects. No appeal is possible from many FDA decisions in the early testing stages. Carl Djerassi, a biochemist at Stanford University and the Syntex Corporation, who was a developer of the pill, feels that the FDA reviewer's incentive to recognize benefits is now overshadowed by his incentive to avoid the risk of approving a drug and later having to defend his decision to the media and perhaps to Congress. Djerassi contends that "if the present climate and requirements had prevailed in 1955, oral contraceptive steroids would still be a laboratory curiosity in 1970."

An Example: A Chemical Abortion Agent

Suppose we consider, first from an ideal viewpoint and later from the present "state of the art," the development of a new birth control agent suitable for use in all nations. Ideally such an agent should separate coitus from contraception; the success of the pill and the IUD stems in part from such an advantage which maximizes the use pattern. In addition the agent should be cheap and sufficiently simple that self-application is possible. Any agent requiring regular injections, for example, would have limited use in the poorer nations as would any method, such as the pill, that required daily attention. To be useful in all segments of the world, the agent should be useful to those who may be totally ignorant of all reproductive physiology except the knowledge of pregnancy. Ideally the agent should also have no side effects, but experience dictates that this property will be virtually unattainable; at best we should hope to minimize these.

An agent that might satisfy the above criteria could be a pill, taken once a month, or alternatively, taken during the first months of pregnancy, which would induce menstruation and hence, an abortion, whether pregnancy had occurred or not. Such a chemical is termed an **abortifacient**, and its action might take one of several paths. One method might be the destruction of the corpus luteum, whose hormonal secretions control embryo development in the uterus. A chemical causing this is a luteolytic agent. For such an abortifacient, Djerassi compiled the procedure required to obtain FDA approval. Djerassi's estimate begins after suitable chemical compounds have been identified (this identification is by no means an easy task). Testing on rats, dogs, and monkeys must be done for periods as long as 10 years. Studies of toxicity are needed, and the chemicals must be carefully tested to establish whether they might create birth defects. An economically feasible chemical synthesis must be developed, and such aspects as tableting procedures and safe storage periods must be established. All testing is reviewed at each stage by the FDA. In 1970 Djerassi estimated that only one of every 25 compounds might survive the testing and receive permission for marketing. The entire procedure would cost an estimated *$20 million* and require nearly *18 years!* The possibility exists, of course, that all of the compounds could be rejected by the FDA on the basis of test results.

Several possible candidates for just such an agent have been suggested, and we shall now consider their present status. The most highly publicized candidates are from a class of organic acid (Chapter 1) derivatives called prostaglandins. Discovered in the 1930s, prostaglandins are naturally occurring substances thought to be critically involved in the regulation of a variety of human physiology and body functions including reproduction physiology, muscle expansion and contraction, lipid metabolism, and central nervous system activity. The chemical structures of these compounds were developed in the 1950s. Research in this area has been intense in the last 10 years, despite the high cost of isolating the compounds: the human body synthesizes only a few tenths of a milligram of prostaglandins per day.

Two prostaglandins, designated $PGF_{2\alpha}$ and PGE_2 (Figure 2.19) are abortifacient agents in humans and are considered potential birth control agents. $PGF_{2\alpha}$ is claimed to be luteolytic on rhesus monkeys, but conflicting evidence exists with respect to the action of $PGF_{2\alpha}$ and PGE_2 in humans. They may hasten uterine contractions, which would be a decided practical advantage because in the event abortion did not occur this mode of action would be less likely to create fetal damage than the action of a luteolytic agent. This absence

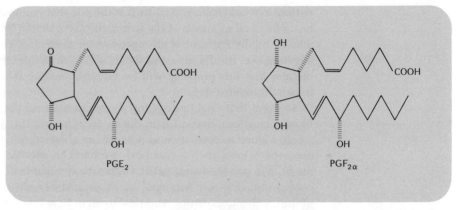

Figure 2.19. *Two of the many known prostaglandins. The dashed lines indicate that the attached group is oriented below the plane of the ring, therefore pointing into the page.*

of teratogenic activity would simplify testing procedures. $PGF_{2\alpha}$ appears to have a higher incidence of "nonserious" side effects such as nausea, diarrhea, and vomiting. It raises the blood pressure, whereas PGE_2 lowers the blood pressure.

By 1971 these had been tested on fewer than 1000 women in separated cases, and there had been no clinical coordination of the studies. Success rates of 40% to 90% had been encountered. At present both prostaglandins must be administered by either intravenous or intrauterine injection. The material cost alone in 1971 was $100 per woman with the intravenous method (100 milligrams required) and $1.00 per woman with the intrauterine shot, and both methods require medical or paramedical aid. No orally active versions have been developed, and prospects appear dim because of the very rapid metabolic destruction of the natural prostaglandins.

Thus the prostaglandins, though offering hope, do not at present meet our "ideal" specifications, but improvements will no doubt occur in the next few years. In particular, new syntheses and new natural sources of prostaglandins may provide cheaper sources of these compounds and stimulate even wider study. It was recently discovered that a rich source of prostaglandins exists in a sea plant, the sea whip or sea fan, which grows in coral reefs off the coast of Florida.

Other Possible New Techniques

Female Contraceptives

Current interest, in addition to the prostaglandins, involves a postcoitus, or "morning after," pill. These contraceptives would inhibit either embryonic development or implantation in the uterus of the fertilized egg. The latter appears the more effective at present; the compound **diethylstilbestrol** (DES) has been used in this approach. In both rhesus monkeys and humans, DES will inhibit implantation, but the compound is not effective after implantation has occurred. Recently DES has been conditionally approved by the FDA for limited use under medical supervision. It is not intended for use on a regular basis but can be used in emergency or rape cases. Because it is a carcinogen, a cancer inducer, DES must be considered with care. The FDA in the early 1970s banned its use in animal feeds, in which it had been used as a fattening agent, when it was discovered that DES was sometimes present in beef after processing. A federal court in 1975 reversed this decision, permitting DES to be used once again. In medical use, the benefits arising from its use in specific emergency cases offset the very small possibility of inducing cancer in the patient.

Male Contraceptives

Progress in this area has been slow because even less is known about male than about female biology. Future advances are not expected to be rapid because in the entire United States existing facilities can evaluate only two male contraceptive drugs at a time.

Spermatogenesis, the development of sperm, occurs in the gonads, and in man this requires 10 to 11 weeks. An additional 1 to 2 weeks is required for sperm maturation and transport through the epididymis, a convoluted tubular organ close to the surface of the testes. The mature sperm are then discharged, together with other secretions, through the penis.

Inihibition of spermatogenesis has been achieved by a variety of compounds, some of which act on sperm development directly and some of which operate at the level of control mechanisms in the brain and pituitary gland. Since sperm production is a continuous process, sustained treatment would probably be required because the absence of sperm in the ejaculate would occur from 2 to 3 weeks to 2 to 3 months after medication.

Suggestions for Improving Birth Control Development

A number of proposals have been made by Djerassi to facilitate the development of new birth control agents:

1. Conditional approval by FDA should be allowed for clinical testing after major toxicity and teratology studies have been completed. During this time, the agent could be marketed but some of the profits would be used for structured follow-up studies of sizable populations of patients. The FDA could monitor such results and concentrate on those patients who do poorly while taking the drug.
2. Appeal procedures to FDA decisions should be established wherein a permanent body of independent experts (appointed by the National

Academy of Sciences) exists to whom appeal is possible and whose decisions would be binding on the FDA.

3. Government-industry interaction should be improved. Rising costs in the $20 million range mean fewer pharmaceutical companies will risk research capital on contraceptives when the chances of good development are less than those of other drugs. Some or all of the phases could be government supported, with companies obligated to repay the government on a royalty basis.

4. Establish primate research centers to obtain definitive information on the reproductive systems of the higher apes to improve data extrapolation to humans. This suggestion arises because of the limitations on human testing discussed previously. One concern which has been voiced about this proposal is the impact it would have on the already declining population of the primates.

Our survey of present and future birth control methods indicates that much needs to be done. The 18-year time lag required to develop a new chemical technique means more than a billion humans would be added to our planet before the agent was available. As Bernard Berelson, president of the Population Council, has stated, "What will be scientifically available, politically acceptable, administratively feasible, economically justifiable, and morally tolerated depends upon people's perceptions of consequences." That is, the more evident the crisis to the public, the more responsive will be society's action.

Zero Population Growth in the United States

Increasing environmental awareness and the decreasing birth rate in the United States have focused concern on zero population growth (ZPG): attainment of a stable population. Such a condition would obtain when the total growth rate is zero; a population decrease would of course occur if the total growth rate were negative. Actually in the United States the birth rate has always greatly exceeded the death rate even during the Depression years. If ZPG were to be attained at all, it would presumably be a slow procedure. One widely publicized method of achieving ZPG is the encouragement of smaller, that is, two-child families with the urging of adoption of additional children for parents who desire larger families.

As discussed earlier in the chapter, a projected growth model is a complicated function of population, age structure, and future fertility behavior. Figure 2.20 shows the calculated population growth in the United States as a function of family size. Maintaining a one-child per family difference causes enormous population changes a century hence; a two-child family would yield an ultimate population of 350 million, whereas a three-child family would yield nearly a billion! Demographers often use the **total fertility rate** to measure the approach to a stabilized population. This hypothetical parameter can

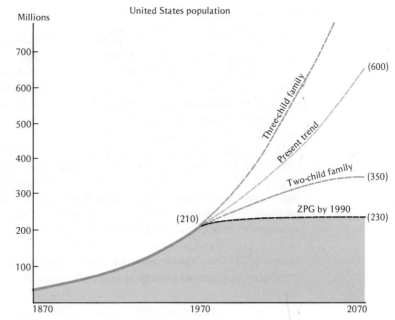

Figure 2.20. *Population projections in the United States. [From Population and the American Future, The Report of the Commission on Population Growth and the American Future, March 1972.]*

be considered for our purposes as the total number of children produced by a woman during her reproductive years. ZPG could be attained if the total fertility rate achieved and remained at the value of 2.11. This level, termed the **replacement level of fertility,** would replace the parents and compensate for premature births. After the total fertility rate dropped to this point, the population would initially continue to grow as the pre-15 year old age groups reached productivity. If the 2.11 figure were maintained, however, the age distribution would change, and the death rate would slowly rise owing to the larger fraction of older persons in the population. Ultimately the birth and death rates would be equal, and the population would stabilize, assuming no immigration occurred.

In the United States the total fertility rate was 3.7 in 1957 at the height of the baby boom; it dropped to 2.5 in 1968; and in 1972, for the first year in history, it dropped below the replacement level to a value of 2.03. This of course does not mean that the United States has achieved ZPG because those under 15 have yet to make their contribution. The ultimate population that would be attained at ZPG depends on when the total fertility rate fell permanently below 2.11. If the United States total fertility rate stayed below 2.11, the population would ultimately stabilize at about 290 million persons around the year 2030, assuming no immigration. One question that must, in fact, be answered if ZPG is to be achieved is whether immigration is to be continued at the present rate.

The Impact of ZPG

It has become almost a cliché in the United States to equate growth, especially local population growth, with progress and the quality of life. This position is often strongly promoted by local business and fraternal organizations. A careful analysis, however, yields few facts to support such a conclusion on a nationwide basis. A principal factor is the economic impact of decreased or stabilized population growth; economists disagree on the necessity of population growth for economic stability.

In July 1969 President Nixon appointed the Commission on Population Growth and the American Future to examine the meaning and consequences of population growth. The Commission was formed to assess the impact that growth will have upon the American future and the desirability of forming a national policy on population. After enlisting many of the nation's leading scientists in more than a hundred research projects and after collecting testimony in public and private hearings, the Commission issued its final report in March 1972. The Commission took an emphatic stand in favor of slowing and stabilizing population growth in the United States.

> In the brief history of this nation, we have always assumed that progress and "the good life" are connected with population growth. In fact, population growth has frequently been regarded as a measure of our progress. If that were ever the case, it is not now. There is hardly any social problem confronting this nation whose solution would be easier if our population were larger.
>
> We have looked for, and have not found, any convincing economic argument for continued national population growth. The health of our economy does not depend on it. The vitality of business does not depend on it. The welfare of the average person certainly does not depend on it. In fact, the average person will be markedly better off in terms of traditional economic values if population growth follows the two-child projection rather than the three-child one.

Because of the ingrained view that fewer people might mean fewer jobs, a poorer economy, and stagnation, it is instructive to look in closer detail at some of the Commission's findings relating to ZPG discussed relative to Figure 2.20.

The rate of population growth will have a significant effect on per capita income. The Commission's research indicated that in the year 2000, per capita income may be as much as 15% higher under the two-child than under the three-child population growth rate. The main reason for the higher per capita income under the two-child projection is the shift in the age composition resulting from slower population growth. A secondary reason is that with lower birth rates the percentage of women in the labor force is expected to rise somewhat faster than it would otherwise. Taken together, these trends mean relatively more workers and earners and relatively fewer mouths to feed. As income increases, people show an increased preference for services, such as education and health services, as compared with manufactured goods. So, the population of the year 2000 will boost its consumption of services faster than its consumption of manufactured goods.

Will a slower rate of population growth hurt specific industries, particularly those that manufacture products for young people? Does it threaten jobs? Although it is certainly true that there would be a faster increase in the sales of certain products—for example, baby foods and milk—under conditions of higher population growth, it is also true that other products and services—for example, convenience foods and airline travel—would be favored by the faster rise in per capita income associated with slower population growth rates. More important, it does not appeas, for several reasons, that a lower population growth rate will cause serious problems for any industry or its employees.

First, regardless of the rate of population growth, total income, and hence demand, are expected to rise. Second, slower population growth will cause total as well as per capita income to be higher over the next 10 to 15 years than would a more rapid population growth rate. Third, it is important to note that, under the two-child family projection, there is no year in which there would be *fewer* births than there were in 1971. In other words, a gradual approach to population stabilization would not reduce demand from current levels for any industry studied. The Commission studied the effect of the two-child and three-child population projections on demand for housing starts, mobile homes, domestic cars, imported cars, men's suits, frozen foods, power boats, credit, furniture and household equipment, food and beverages, beer, clothing and shoes, steel, dishwashers, railroad and airline travel.

Beyond the next 10 to 15 years, the adjustments businesses must inevitably make to changes in consumer tastes and technological developments should far exceed the problems of adjusting to a lower population growth rate. For example, the loom tender in the diaper factory was hurt more by the competition from synthetic disposables than by the recent decline in births. Large fluctuations in birth rates will require larger adjustments by businesses than will smaller ones; still there can be fluctuations around a three-child as well as a two-child growth rate.

A slower population growth causes a gradual increase in the percentage of old people and a decline in the percentage of youth—hence, a rising average age of the population. The same process also causes the labor force to age. Concerns have been expressed that an older labor force will lack the energy, flexibility, and imagination of a younger one. These concerns are further reasons to support programs desirable on other grounds, such as the provision of continuing education to our labor force. Indeed, in the light of the rapid changes occurring in all aspects of life, the idea that education should be completed at ages 18, 22, or even 30, is clearly out of date. In declining communities, small businesses will not do as well economically as they would if there were more people around—some adjustments will be required. But other changes, which are unpredictable today, will require far more important adjustments by individuals, as well as by entire industries, than will those required by a stabilizing population.

A reduction in the rate of population growth would bring important economic benefits and would provide the opportunity to improve the quality of life. For example, the Commission felt that the general improvement in income would assist in reducing poverty but would not eliminate it. Strong recommendations were made to utilize the increased income to improve social conditions, health services, and human resource development.

Advantages of Small Families

A trend toward ZPG means a trend toward smaller families. Evidence exists that such a trend may be beneficial for the family members. Much research relevant to the economic, psychological, and cultural advantages of smaller families has been conducted. A child requires a tremendous emotional and financial investment on the part of his parents. A smaller family affords a greater opportunity for interaction between parent and child, although of course the opportunity is not always seized.

Several studies of elementary and high school children indicate that the youngster in a small family gets along more happily with his siblings as well as with his parents, is less likely to suffer emotional upsets, and is much less likely to end up in a mental hospital than a child from a larger family. Studies conducted by the National Institutes of Mental Health indicate that nursery school children from small families are more vigorous and independent and require less attention than those from larger families. Child creativity is apparently greater in smaller families: teenage boys judged most creative in the Westinghouse talent search usually come from two-child families, only rarely from larger ones. A Scottish study indicated that the number of children in a family affects intelligence at each economic level, with children from smaller families consistently registering higher scores on intelligence tests.

Finally, the Institute of Life Insurance recently estimated that the cost of rearing a child to age 18 for a family with an annual income of $6600 amounted to over $23,000, not including college. Thus each child requires a financial investment approximately equal to one home.

Recommendations of
the Presidential
Commission

From its findings, the Presidential Commission on Population and the American Future formulated a number of recommendations in addition to population stabilization: an increase in population planning research and services, an increase in sex education, an increase in child care activities, reform of adoption laws, elimination of sex discrimination, no increase in immigration, and the creation of a National Institute of Population Science within the National Institutes of Health.

Summary   The growth of the world's population is a major problem. Unless greatly reduced, population growth will result in drastic effects: depleted resources, inadequate food supplies, altered living conditions, and political chaos. The present problems are the most pressing in the poorer nations where the exploding population combines with inefficient agriculture to produce a cycle of poverty and despair. However, population growth in the prosperous nations cannot be ignored. The resource consumption per capita in the richer nations is often 100 times greater than that of the poorer nations so that the impact on the world of the members of prosperous nations is considerably greater. Furthermore, because most of the resources used by the richer nations are imported (often from the poorer nations), population growth in any nation has an effect on many other nations.

Historically a characteristic of population growth is a relentless trend toward shorter and shorter doubling times. Population growth has followed an exponential or J-curve, and even shorter doubling times are predicted for the future. Demographic parameters combined with population distribution charts can provide a very crude estimate of future population growth. The poorer nations will experience the biggest impact in the near future. Typically 40% of the population in these nations is under 15 years of age; when this group moves into fertility, an enormous strain will be placed on the already inadequate facilities of these nations. In the prosperous nations the distribution of population among age groups is more uniform. The movement of the adolescent group into fertility will have less impact on the system because this group constitutes perhaps only 20% of the population.

Population has the potential to grow faster than food production. Although food production is an enormous problem, a more critical aspect is protein deficiency. The energy required to produce a high quality protein and the resulting cost of the high quality protein almost preclude its use by the poorer nations in a competitive economy.

Several methods have been proposed to improve food production. Cultivated acreage could be increased threefold at most, but soil erosion, water shortages, and growing season limitations would keep production increases well below a factor of three. The development of dwarf grains, the Green Revolution, provides some hope, but considerable technology is required which may limit its use in the poorer nations. Food from the sea has definite limitations; we already harvest over half of the estimated harvest we could take without depleting the world's fisheries. Food production in the industrialized nations requires an enormous energy subsidy; these techniques cannot be transferred to the poorer nations without a drastic increase in their energy base.

The quantity of the resources on the planet is fixed, and population growth combined with increasing per capita consumption have depleted some of our fuels and minerals to the point where their lifetimes are much less than a century at present rates of consumption. Oil and natural gas are the fuels in shortest supply; most of the estimates place their lifetimes at a few decades, unless reductions in consumption rates occur. Coal is sufficiently abundant to last for over a century; but many metals, such as gold, lead, and copper, will be exhausted in two decades, unless their use patterns are altered.

Considerations of the interrelated problems of population growth, food production, and resource depletion lead to the conclusion that the problems would be more easily soluble if the population growth rate were reduced or brought to zero. In the affluent nations sufficient contraceptive methods are available to accomplish this, but in the poorer nations the prospects are dim. Efforts to develop useful, cheap birth control agents for use in all nations encounter increasingly stringent regulations for the testing and licensing of new drugs. Estimates are that development of a chemical abortion pill which met all of the FDA requirements would require 18 years and $20 million.

ZPG in the United States would be achieved by early in the next century if the total fertility rate stayed at or near the 1972 level and if immigration were reduced. ZPG would result in increased per capita income, less stress on job markets, and an opportunity to solve some of the more pressing social problems. Because population stabilization would be gradual and because in any one year under ZPG more babies would be born than were born in 1971, the effects on business and labor would be less than those that occur from usual competitive changes. The Presidential Commission on Population Growth and the American Future strongly recommended in 1972 that the population growth in the United States be slowed and that ZPG be encouraged.

Review and Study Questions

1. Why should the developed nations be concerned about the growth of their populations?
2. When did each of the three estimated major increases in world population occur and what were the reasons for the increases?
3. What changes have occurred in the population doubling time in the last three centuries?
4. What is a J-curve?
5. What is the growth rate? How does it differ from the total growth rate?
6. For what reasons does the death rate of a nation decrease during industrialization?

7. Is it realistic to expect that the solution to the population problems of the poorer nations can be achieved by causing each of them to become industrialized?

8. What characteristic of the United States population structure distribution leads to the fear that the birth rate may increase beginning in the late 1970s?

9. What mathematical representations did Malthus use for population growth and for food production?

10. Why is it inadequate to consider the world's nutritional state on the basis of total food calories produced per year?

11. Describe in approximately ascending order the elements of the earth's food chain.

12. Does an exclusive diet of rice provide sufficient protein for nutritional requirements?

13. What techniques were used by Borlaug to develop the Green Revolution?

14. What problems have been forecast for the Green Revolution?

15. By how much could the world's fish catch be increased without seriously decreasing the fish supply by overexploitation?

16. Which types of food require the largest energy subsidy per food calorie produced? Which require the least?

17. On what energy sources are the United States agricultural gains since 1925 dependent?

18. List the principal birth control techniques available in the United States. Which ordinarily require medical supervision or attention?

19. What is the role of estradiol in female reproduction?

20. Why does the pill use chemical compounds whose structures are slightly different from the hormones estradiol and progesterone?

21. Is sterilization legal in the United States?

22. What is the sterilization operation for males? What are the known effects?

23. When, if ever, has abortion provided a check on population growth?

24. What barriers exist to the development of new chemical birth control agents?

25. Identify: Norman Borlaug, Carl Djerassi, J. H. Ryther.

26. About how much time is estimated to be required to develop an abortifacient that meets FDA requirements?

27. What were the recommendations in the 1972 final report of the Presidential Commission on Population Growth and the American Future?

28. Contrast the expected per capita income in the United States of a two-child versus a three-child family population projection.

29. Suppose you take a job that pays 1¢ the first day, doubles to 2¢ the second day, doubles this to 4¢ the third day, and so on. This is an example of growth with a short doubling time. How much do you earn for the month (30 days) if your pay is doubled each day?

Questions Requiring Outside Reading

30. Did celibacy and infanticide ever provide limitations on population growth? (W. L. Langer, "Checks on Population Growth: 1750–1850," *Scientific American* [February 1972], 33.)

31. What specific economic proposals have been made that would encourage small families? (B. Berelson; see Suggested Outside Reading.)

32. Animal crowding experiments invariably result in a breakdown of the usual behavior of the animals. What are the implications for humans living in urban environments? (O. R. Galle, W. R. Gove, and J. M. McPherson, "Population Density and Pathology: What are the Relations for Man?," *Science 176* [April 7, 1972], 23.)

33. One danger of man's activities is his tendency to overexploit his resources until they are depleted. What has happened to the qopulations of the great whales and why? (See Chapter 9.)

34. How did the bubonic plague affect population growth in Europe? (W. L. Langer, "Black Death," *Scientific American* [February 1964].)

35. What major changes have occurred in the United States government's policy on contraceptives? (F. S. Jaffe, "Public Policy on Fertility Control," *Scientific American* [July 1973], 17.)

36. What predictions are made for the total labor force under the two- and three-child projections of Figure 2.19? (*Population and the American Future;* see Suggested Outside Reading.)

37. How much fresh water per day is provided by nature and how much of this will man require by 1980? (G. Borgstrom, *Too Many*, Macmillan Publishing Co., Inc. 1969.)

38. What is the optimum world population? (H. R. Hulett, *Bioscience*, March 1970; H. H. Iltis et al., *Bulletin of the Atomic Scientists*, January 1970.)

39. What is the feeling of Blacks in the United States as regards population growth and ZPG? (W. A. Darity, C. B. Turner, and H. J. Thiebaux, "Race Consciousness and Fears of Black Genocide as Barriers to Family Planning;" C. V. Willie, "A Position Paper," *Perspectives from the Black Community, PRB Selection No. 37*, Population Reference Bureau, June 1971.)

40. What changes would occur in the relative sizes of minority and majority groups in the United States under ZPG? (E. B. Attah, "Racial Aspects of Zero Population Growth," *Science* 180 [1973], 1143.)

41. How does aspirin work? Are prostaglandins involved? (John E. Pike, "Prostaglandins," *Scientific American* [1971], 84; *Chemical and Engineering News*, August 7, 1972, p. 8.)

42. Abortion as a means of birth control has at present the most impact on what portion of a scale that rates nations on a socioeconomic-cultural scale? (C. Djerassi; see Suggested Outside Reading.)

43. Disagreement exists on whether population growth or technological changes bear the major responsibility for environmental pollution. What data are used to support each view? (See articles by Barry Commoner and by P. R. Ehrlich and J. P. Holdren, *Bulletin of the Atomic Scientists*, May and June 1972.)

Suggested Outside Reading

BERELSON, BERNARD. "Beyond Family Planning," *Science* **163** (February 7, 1969), 533. Reviews proposals that have been made to solve the population problem. Includes estimates of the utility of the proposals plus political, economic, and legal limitations.

CHEDD, GRAHAM. "Hidden Peril of the Green Revolution," *New Scientist* **22** (1970), 171. Expresses concern about possible loss of the broad genetic base, disease-resistant crop strains if the Green Revolution is adopted on a wide scale.

CLOUD, PRESTON. "Realities of Mineral Distribution," *Texas Quarterly*, **11** (1968), 103.

——— (ed.). *Resources and Man*, W. H. Freeman and Co., San Francisco, 1969. These present data on the resource limits of the earth, both on land and in the sea.

DJERASSI, CARL. "Birth Control After 1984" *Science* **69** (September 4, 1970), 941; "Fertility Control through Abortion," *Bulletin of the Atomic Scientists*, January 1972, p. 9. These present Djerassi's views on birth control agents, their present chemical limitations, and their developmental costs. The role of the FDA is discussed and some suggestions for improvement in the regulation and testing of drugs are made.

EHRLICH, ANNE H., AND PAUL R. *Population, Resources, Environment*, W. H. Freeman and Co., 1972. A treatment of world population, resource, and environmental problems. The Ehrlichs have strong views on the impact of population growth and of technology on the world. These views are stated forcefully, and many alternatives to present policies are suggested.

HOLT, S. J. "The Food Resources of the Ocean," *Scientific American* (September 1969), 178. Estimates of the potential of the ocean as a food source and of what might be done to improve its production.

JOHNSON, ERIC. "Is Population Growth Good for Your City?." Zero Population Growth, 4080 Fabian Way, Palo Alto, California 94303. An analysis of costs to the individual which accrue as the population of a city expands.

Population Bulletin. Population Reference Bureau, Inc., 1755 Massachusetts Avenue, N.W., Washington, D.C. 20036. The Bureau publishes articles and data on world population growth. The *Bulletin* is a major source of authoritative data.

Population and the American Future. The final report of the Commission on Population Growth and the American Future. Available from the U.S. Superintendent of Documents. An excellent description of the energy subsidy required for U.S. food production and of the implications of this for the world.

Scientific American. "Human Population," September 1974. The entire issue is devoted to population and its resultant problems: history of growth, food production, reproductive physiology, genetics, and transfer of technology to underdeveloped nations.

WATTENBURG, BEN. "The Nonsense Explosion," *The New Republic,* April 4 and 11, 1970. An article which argues that concern about U.S. population growth is overstated, that concentration on this issue allows us to relax our concern on other, more pressing problems.

WILKES, H. G. AND S. WILKES. "The Green Revolution," *Environment,* **14,** October 1972, p. 32. Discusses implications of the Green Revolution, especially the possibility of the loss of older, time tested genetic pools in grains.

"World Population Conference, 1974." *Bulletin of the Atomic Scientist 30,* June 1974, pp. 10–35. A special report on that conference emphasizing the problems of third world nations and the views of various nations on population policies. Includes an article by Carl Djerassi on fertility control in China.

3 Air Pollution

Nobody can be in good health if he does not have all the time, fresh air, sunshine, and good water. . . . The Great Spirit . . . made . . . sunlight to work and play. . . . The white man does not obey the Great Spirit.

Chief Flying Hawk of the Sioux Tribe

Air Pollutants   A useful definition of an air pollutant is a compound added directly or indirectly by man to the atmosphere in such quantities as to affect humans, animals, vegetation, or materials adversely. Air pollution requires a very flexible definition that permits continuous change. When the first air pollution laws were established in England in the fourteenth century, air pollutants were limited to compounds that could be seen or smelled—a far cry from the extensive list of harmful substances known today. As technology has developed and knowledge of the health aspects of various chemicals has increased, the list of air pollutants has lengthened. Even water vapor might be considered an air pollutant under certain conditions.

As we saw in Chapter 1, many of the more important air pollutants, such as sulfur oxides, carbon monoxide, and nitrogen oxides, are found in nature. As the earth developed, the concentrations of these pollutants were altered by various chemical reactions; they became components in the biogeochemical cycles. These serve as an air purification scheme by allowing the compounds to move from the air to the water or soil. On a global basis, nature's output of these compounds dwarfs that resulting from man's activities. However, man's production usually occurs in a localized area, such as a city.

In this localized region, man's output may be dominant and may temporarily overload the natural purification scheme of the cycles. The result is an increased concentraton of noxious chemicals in the air. The concentrations at which the adverse effects appear will be greater than the natural levels, the concentrations that the pollutants would have in the absence of man's activities. The actual concentration need not be large for a substance to be a pollutant; in fact the numerical value tells us little until we know how much of an increase this represents over the concentration that would occur naturally in the area. For example, sulfur dioxide has detectable health effects at 0.08 parts per million (ppm), which is about 400 times its natural level. Carbon monoxide, however, has a natural level of 0.1 ppm and is not usually a pollutant until its level reaches about 15 ppm (concentration units are discussed in Chapter 1). The composition of pure air, or the average natural levels of its components, is given in Table 3.1.

91

TABLE 3.1. Composition of clean, dry air near sea level

Component	Content vol. %	ppm	Component	Content vol. %	ppm
Nitrogen	78.09	780,900	Nitrous oxide	0.000025	0.25
Oxygen	20.94	209,400	Hydrogen	0.00005	0.5
Argon	0.98	9,300	Methane	0.00015	1.5
Carbon dioxide	0.0318	318	Nitrogen dioxide	0.0000001	0.001
Neon	0.0018	18	Ozone	0.000002	0.02
Helium	0.00052	5.2	Sulfur dioxide	0.00000002	0.0002
Krypton	0.0001	1	Carbon monoxide	0.00001	0.1
Xenon	0.000008	0.08	Ammonia	0.000001	0.01

Source: Cleaning Our Environment, American Chemical Society, Washington, D.C., 1969.

Note: The concentrations of some of these gases may differ with time and place, and the data for some are open to question.

The Cost of Air Pollution

Before we discuss details of air pollution, we should consider estimates of the economic cost of air pollution. Some effects are easily assigned a dollar value as in Table 3.2. Many other effects of air pollution are known, but assigning a precise value to their economic impact is difficult. Illness from air pollution results in hospital and medical bills, in time lost from work, and in low productivity. Home maintenance and painting are required more frequently as air pollution increases. Hospitals, research laboratories, and industries are often forced into the added expense of building air purification facilities to obtain clean air for their use. At what price does one tabulate the psychological effects of odors, haze, and eye irritants on the population of an entire city? At what price does one tabulate the years by which a human life is shortened, or the years of suffering of an emphysema patient?

Air pollution also wastes valuable resources. Some are lost directly, as pollutants, and some are lost indirectly—as, for example, when more electricity is generated because street lights turn on earlier when skies are darkened by pollution. Air pollution threatens the complex natural cycles on which all life

TABLE 3.2. 1968 estimate of annual United States air pollution costs per year

Painting of steel buildings	$ 100,000,000
Laundering and cleaning of fabrics	800,000,000
Car washing (pollution caused)	240,000,000
Agricultural crops and livestock	500,000,000
Total	$1,640,000,000

Source: Department of Health, Education and Welfare

depends, and undesirable ecological changes may well be the most dangerous long-range result of air pollution.

Effects such as these easily total several billion dollars per year. Early in the 1970s the Environmental Protection Agency (EPA) estimated that the total cost to the United States of air pollution was nearly $25 billion per year, over $100 for each person in the nation. In this total, the costs of human illness and death were estimated at $9 billion. These economic figures combined with other aspects of air pollution serve as a definite impetus for our study.

Ironically the costs of *reducing* air pollution also total several billion dollars per year. The preceding partial list of the costs that result from the effects of air pollution stands in contrast to the often-heard claim that the United States cannot afford the economic cost of air pollution control.

Sources of Air Pollution

Table 3.3 gives the United States man-made sources of air pollution in total tonnage. The pollutants listed are those we will be most concerned with in this chapter: carbon monoxide (CO), sulfur oxides (SO_x), total suspended particulates (TSP), nitrogen oxides (NO_x), and hydrocarbons (HC). Over 95% of the emissions under Transportation are due to the automobile. Almost all of the emissions under Fuel combustion, stationary sources, are due to electrical power plants. The automobile is the major polluter, contributing over half of all air pollutants by weight. Stationary source fuel combustion, principally power plants, ranks second (21%), followed by industry (17%). One rather simple generalizaton can be made from these data. Three fourths of our air

TABLE 3.3.   Estimated emissions of air pollutants by weight in United States in 1971 (millions of tons)

	CO	TSP	SO_x	HC	NO_x	Total	Percent
Transportation	77.5	1.0	1.0	14.7	11.2	105.4	50.6
Fuel combustion, stationary sources	1.0	6.5	26.3	0.3	10.2	44.3	21.3
Industrial processes	11.4	13.6	5.1	5.6	0.2	35.9	17.2
Solid waste disposal	3.8	0.7	0.1	1.0	0.2	5.8	2.8
Miscellaneous	6.5	5.2	0.1	5.0	0.2	17.0	8.1
Total	100.2	27.0	32.6	26.6	22.0	208.4	100.0
% change from 1970 to 1971	−0.5	+5.9	−2.4	−2.6	0.0		

Source: Environmental Quality, *Fourth Report of President's Council on Environmental Quality, September 1973.*

pollution is created by a need for power, either to move vehicles or to create electricity. The statement "all power pollutes" is often applied in discussion of air pollution.

Table 3.3 presents pollutants and sources on the basis of the total tonnage of pollutants. This does not necessarily imply that the auto, the major polluter, has the worst impact; such a decision would require an analysis of the effects of the individual pollutants. One of the many possible estimates of the relative effects of the various pollutants is shown in Figure 3.1. This figure displays the source and tonnage data of Table 3.3 on a percentage basis. The right-hand portion of the figure shows the relative effects of the pollutants on the basis of toxicity as calculated by Professor Lyndon Babcock of the University of Illinois. Thus, although half of the total man-made air pollution in the United

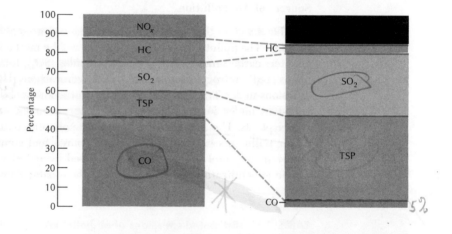

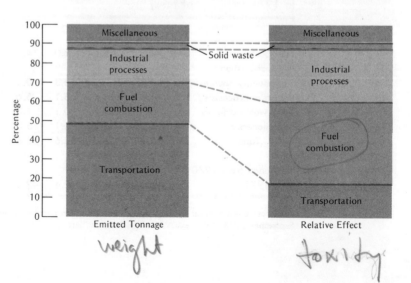

Figure 3.1. Air pollution emissions on a percentage basis by weight and by relative toxicity (Babcock model). [From Environmental Quality, Fourth Report of President's Council on Environmental Quality, September 1973.]

States is in the form of carbon monoxide, it accounts for only about 5% of the overall toxic effects of air pollution. The Babcock method more than triples the importance of particulate emissions and doubles the significance of the sulfur oxide emissions. Consequently, the effluents from power plants have, on a relative basis, greater impact than the emissions from transportation.

Toxicity is not, however, the only aspect of air pollution with which one could derive a relative effects chart similar to Figure 3.1. In the next section we discuss in detail the individual pollutants and their effects.

Carbon Monoxide

CO, the most abundant of the air pollutants, is produced in combustion processes when carbon (C) from fuels reacts with a limited amount of oxygen (O_2) from the air. Chemically this may be written as $2\,C + O_2 \rightarrow 2\,CO$. As given in Table 3.3, man created an estimated 100 million tons of CO in 1971 in the United States, three fourths of which came from transportation; the vast majority of this tonnage came from automobiles. Figure 3.2 indicates the major sources of CO. Actually the total tonnage is uncertain because many CO reactions in the atmosphere have only recently become known. The yearly average CO *global* concentration of 0.1 ppm has apparently not changed for decades, which is surprising because man's activities release sufficient CO to increase this by 0.03 ppm per year. Since man's actions could double the CO atmosphere level in 4 years, the conclusion to be drawn from the unchanging CO concentration is that CO is being removed from the atmosphere as fast as it is created.

In recent years much has been learned about CO reactions in the environment. On a global basis nature dwarfs man's output, producing at least ten times as much. This vast natural CO production is coupled with several rapid CO removal mechanisms (or "sinks") to keep the CO concentration constant with time. A study done by the Argonne National Laboratory indicated that the natural production has several sources whose individual contributions show seasonal variations. The hydrocarbon methane (CH_4) appears to be a dominant source. Methane, produced by the decay of vegetation, can be converted to CO in the atmosphere by reaction with hydroxyl radicals, by reaction with oxygen atoms, or by reaction with oxygen molecules after the absorption of solar radiation, a process known as photolysis.

The processes that remove CO from the air are also varied. CO can be converted to CO_2 (carbon dioxide) by hydroxyl radicals in the air and by microorganisms in the soil. Other possible removal mechanisms are fungi (which convert CO to CO_2), decomposition of CO by the sun, and perhaps a variety of processes involving the ocean.

Although overshadowed by nature on a global basis, man is by far the

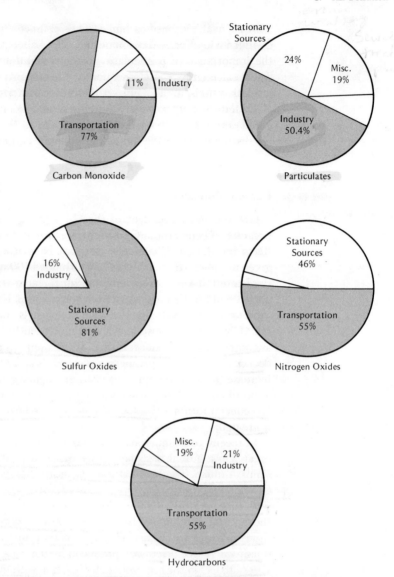

Figure 3.2. *Air emissions in the United States by major sources (1971). [From Environmental Quality, Fourth Report of President's Council on Environmental Quality, September 1973.]*

dominant producer of CO in urban areas. In heavy traffic, sustained levels of 50 ppm or more (500 times the natural level) are common, and concentrations of several hundred parts per million are sometimes encountered.

CO is a colorless, odorless, tasteless gas that fortunately is not known to have adverse effects on vegetation, visibility, or material objects. Its dominant environmental impact appears to be toxicity to man and animals, which arises from its well-known competition with O_2 for hemoglobin in red blood cells, as in equation 3.1. Hemoglobin (Hb) carries O_2 in the form of oxyhemoglobin

(HbO_2) to all body tissues. Since CO binds to Hb 200 times more strongly than O_2, even a low concentration of CO in the air has a greatly magnified effect on the body because it converts oxyhemoglobin into carboxyhemoglobin ($HbCO$). The resulting low O_2 level in the blood affects the heart and brain. Exposure for 8 hr at 30 ppm of CO can cause nausea and headache through a 5% loss in HbO_2. Smokers who smoke a package of cigarettes per day also inhale CO and may have HbCO blood levels of 5% or more.

$$CO + \underset{\text{oxyhemoglobin}}{HbO_2} \longrightarrow \underset{\text{carboxyhemoglobin}}{HbCO} + O_2 \tag{3.1}$$

Oxygen depletion for extended periods can of course result in permanent damage or even death. Death occurs after 10 hr exposure to air containing 600 ppm of CO. The temporary effects of exposure to small concentrations of CO (10 to 200 ppm for a few hours) can be reversed by having the victim breathe oxygen from a respirator. Chemically this exposure to increased oxygen reverses equation 3.1, so that the HbCO is changed to HbO_2.

Sulfur Oxides

Sulfur oxides (SO_x) may be the most serious pollutants. Although sulfur can exist in many chemical forms in the environment, the dominant form in the air is sulfur dioxide, SO_2. H_2S, hydrogen sulfide, is produced naturally by the decay of a variety of organic substances such as vegetation but is also produced by many industrial processes. H_2S has been traditionally identified in chemistry classes as that foul-smelling substance responsible for the odor of rotten eggs; however, the urbanization of the United States has made this identification a remote one for most students. Sulfates (SO_4^{2-}) from salt spray and SO_3 (sulfur trioxide) are other sulfur compounds in the atmosphere from natural sources. H_2S can be converted to SO_2 in the atmosphere,

$$H_2S \xrightarrow[\text{(air)}]{O_2} SO_2 + H_2O \tag{3.2}$$

and this accounts for about 80% of the SO_2 present in the air. SO_2 can be readily converted to SO_3 and to sulfuric acid (H_2SO_4), a strong acid, by natural processes as in equation 3.3.

$$SO_2 \xrightarrow[\text{sunlight}]{O_2 \text{ (air)}} SO_3 \xrightarrow[H_2O]{air} H_2SO_4 \tag{3.3}$$

Because of the possible conversion of sulfur in any form to either SO_2, SO_3, or SO_4^{2-}, it is common to refer to *all* sulfur emissions as SO_x.

The global man-made sulfur pollutants are estimated to be 100 million metric tons per year, about half as much as nature produces. Indications are that by the year 2000 man will exceed nature's output into the air. Man's emissions, of which 95% are SO_2, arise principally from the combustion of coal and other fossil fuels to generate electricity (Table 3.3 and Figure 3.2). Note that SO_x is unusual in that automobiles are *not* a major source. Sulfur in gasoline creates problems in internal combustion engines, and refining processes were designed long ago to reduce the concentration of sulfur in the gasoline fraction. Gasoline typically contains about 0.05% sulfur, but even this small percentage might create an air pollution problem if it were converted into sulfuric acid emissions. Evidence exists that certain auto emission control devices might increase the conversion of sulfur in fuel to sulfuric acid aerosols. Such an occurrence might result in sulfuric acid levels in excess of allowed standards in the vicinity of highways.

The environmental impact of sulfur emissions is as severe as that of any pollutant. As noted above, sulfur in any of its forms usually becomes SO_2 or H_2SO_4, and we shall concentrate on the effects of these. The natural level of SO_2 is about 0.0002 ppm, and in urban areas of the United States concentrations several hundred times this are often observed. SO_2 is a colorless, nonflammable gas that has an acrid taste and a pungent irritating odor above 3 ppm. H_2SO_4 is a strong acid that can attack paints, metals, vegetation, and fabrics such as nylon, cotton, and rayon. Aerosols (discussed later with particulates) containing sulfuric acid provide a means of wide dispersal in air; gravitation or rainfall returns the H_2SO_4 to earth. Such processes can raise the acidity level of an area with a drastic impact on the aquatic life. Tests have shown that 0.03 ppm SO_2 plus high particulate levels can cause unprotected steel panels to lose 10% of their weight in 1 year (presumably due in part to the H_2SO_4 formed from SO_2).

Marble and limestone, both forms of calcium carbonate ($CaCO_3$), are attacked by SO_2 and H_2SO_4. They react as

$$CaCO_3 + H_2SO_4 \rightarrow CaSO_4 + CO_2 + H_2O \qquad (3.4)$$

limestone   calcium
or marble   sulfate

Many historical monuments, buildings, and works of art have deteriorated more rapidly in recent years because industrialization has led to increased SO_x emissions. Italy (Figure 3.3) and Greece face particular problems in the weakening and deterioration of historically important ruins.

The health hazards of air polluted with sulfur oxides have been demonstrated in a number of air pollution disasters: Donora, Pennsylvania (1948); London (1952), and many other years in the 1950s; New York; Osaka; and Rotterdam. The London disaster of 1952, which took 4000 lives, is the single

Figure 3.3a. The impact of 50 years of air pollution is evident in photos of this Renaissance Italian fresco of the Madonna. The right photo, taken in 1970, shows the erosion that has occurred since the left photo was taken in 1920. [Courtesy of the Italian Art and Landscape Foundation, Inc.]

Figure 3.3b. Sulfuric acid damage to a marble statue from the Cathedral in Milan. [Courtesy of the Italian Art and Landscape Foundation, Inc.]

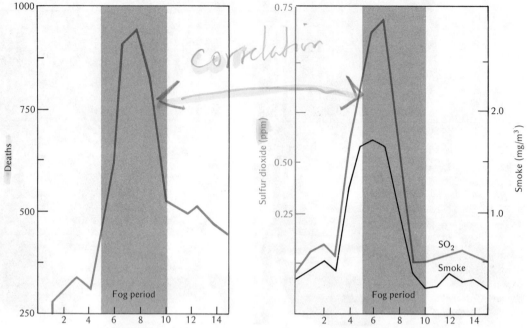

Figure 3.4. *Sulfur dioxide and smoke concentrations and daily deaths in London Administrative County before, during, and after the great smog of December 5 to 9, 1952. [After E. T. Wilkins, Journal of the Royal Sanitation Institute **74**:1 (1954). Crown copyright.]*

greatest air pollution toll on record. Figure 3.4 demonstrates the correlation of deaths with SO_2 level during this disaster. Evidence of adverse health effects becomes apparent when the concentration of SO_2 rises above about 0.04 ppm (115 $\mu g/m^3$), which is 200 times greater than the natural level (Figure 3.5). In 1968 the *daily* mean SO_2 concentration was 0.11 ppm in Chicago, 0.08 in Philadelphia, and 0.04 in Washington, D.C. All are at or above the SO_2 level that might cause more older persons to be admitted to hospitals for respiratory diseases and above which deaths from bronchitis, emphysema, and lung cancer increase.

SO_2 is a major cause of damage to trees, shrubs, plants, and crops (see Figure 3.11). It can injure plant tissues, discolor leaves, stunt growth, and reduce crop yields. Many plants are more sensitive than humans or animals to this pollutant. Plant damage has been noted as far as 50 miles downwind from a smelter. Chronic plant injury and leaf drop occur at annual mean SO_2 concentrations of 0.03 ppm. Exposure for 8 hr to 0.3 ppm of SO_2 is sufficient to injure some species of trees and shrubs; Chicago had 20 such days in 1968.

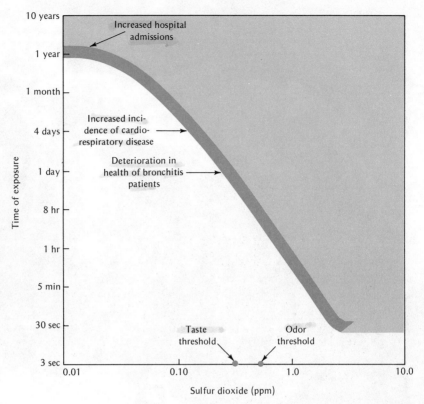

Figure 3.5. *Health effects of SO₂ for various exposure times and concentrations. The solid line indicates the beginning of the shaded region in which deaths have been reported in excess of normal expectation. [Adapted from "Air Quality Criteria for Sulfur Oxides," National Center for Air Pollution Control, 1967, U.S. Department of Health, Education, and Welfare, Public Health Service.]*

Finally, sulfur compounds, including many that are more chemically complex than those listed above, are responsible for most of the odors created in industrial operations.

Particulates

This classification includes liquid drops or small pieces of solids suspended in the atmosphere. A wide range of sizes is possible. The smaller particulates are called aerosols. The largest aerosols are just visible with a good microscope. Examples are the finer components of tobacco smoke, insect dusts, and large viruses. Table 3.3 shows that particulates are produced by combustion and industrial processes in addition to erosion and weathering. Another recently discovered source is sewage treatment plants, which produce aerosols of water containing coliform and other bacteria.

Aerosols are so small that they settle very slowly, often only a few hundred feet per year, because they undergo the random motions characteristic of gas

Figure 3.6. Particulate and aerosol emissions such as those from the stacks above (burning automobile batteries in Houston) create health hazards, darken cities, and soil buildings. The City Hall in Cincinnati (right) was photographed while undergoing cleaning. [From: Above—EPA DOCUMERICA-Marc St. Gil. Right—U.S. Environmental Protection Agency.]

molecules themselves. They can grow by condensation and can be transported for hundreds of miles, often internationally. There is some evidence that particulates can alter climatic conditions and modify the weather through moisture condensation mechanisms.

Aerosols and particles can act on the environment through chemical action (for example, H_2SO_4 aerosols), abrasive action, soiling action, and various effects arising from light scattering. Many of the effects discussed for the other pollutants are enhanced by the presence of particulates.

Figure 3.6. *Continued*

A significant aspect of airborne particulates is that they can scatter and absorb sunlight, thus reducing visibility. Cities receive on the average 15% to 20% less solar radiation than rural areas, and the reduction can become as high as one third in summer and two thirds in winter. Specific effects depend on particle size, but typically particulates at 150 $\mu g/m^3$ can reduce visibility to 5 miles and reduce sunlight by 20% to 70%. Chicago's annual daily mean particulate level in 1968 was 165, and Denver's was 157. Most United States cities exceeded 130. The lowered sunlight levels mean increased resource consumption (and more air pollution) because of increased demand for electrical lighting.

One type of particulate matter of increasing concern is asbestos, a general term used for a number of fibrous minerals. The minerals are a complicated mixture of the oxides of silicon, magnesium, iron, aluminum, calcium, and sodium, together with a small amount of water. Inhalation of large amounts of asbestos dust produces symptoms of chronic systemic poisoning, respiratory ailments, and lung cancer. It may be the size of the particles, rather than their composition, that is important. Studies of insulation workers from 1943 to

1971 have shown an increased incidence of lung and gastrointestinal cancers. Asbestos comes from industrial operations, construction, and from the automobile. Almost all the asbestos used in brake linings in automobiles is eventually converted into airborne particles.

Nitrogen Oxides

Nitrogen (N) forms eight compounds with oxygen, but only three (Table 3.4) are known to be important constituents of the atmosphere. Nitrous oxide (N_2O) has been used as an anesthetic ("laughing gas") and occasionally as the propellant spray in aerosol spray cans. The other two nitrogen oxides arise from combustion. Nitric oxide (NO) is the form initially produced in combustion in air when N_2, elemental nitrogen, combines with O_2:

$$N_2 \underset{\text{(from air)}}{\overset{O_2}{\xrightarrow{\hspace{0.6cm}}}} \underset{\text{burn}}{NO} \qquad (3.5a)$$

$$NO \overset{O_2 \text{ or}}{\underset{\text{ozone}}{\xrightarrow{\hspace{0.6cm}}}} NO_2 \qquad (3.5b)$$

NO_2 is then formed rapidly by the oxidation of NO by either O_2 or ozone (O_3) as in equation 3.5b. Nitric acid (HNO_3), a strong acid, can be formed by the interaction of NO_2 with H_2O. Because the nitrogen oxides change rapidly in the air from one form to another, nitrogen oxide emissions are commonly referred to as NO_x.

Man-made sources cannot account for the major fraction of NO_2 that enters the global atmosphere; the natural sources are unknown but biological processes in the soil are one possibility. Man is the dominant source in urban areas, however, with combustion from transportation (51%) and industrial operations (46%) being the major sources (Figure 3.2).

Until recently it has been difficult to detect and measure nitrogen oxides in air accurately, consequently less is known about them than about the other pollutants. Nevertheless, it is clear that they can cause serious injury to vegetation. At 25 to 250 ppm, NO_2 causes defoliation, whereas a level of 0.3 to 0.5 ppm for 10 to 20 days inhibits the growth of tomato and bean seedlings (the natural level is 0.001 ppm).

TABLE 3.4.   Major nitrogen oxides in the air

N_2O	nitrous oxide	most abundant (0.25 ppm globally); relatively inert, not man-made
NO	nitric oxide	man-made in combustion
NO_2	nitrogen dioxide	man-made; triggers smog reaction

In man, exposure to NO_2 at 3 ppm for 1 hr causes bronchio-constriction, and short exposure to high levels (150 to 200 ppm) produces death from lung damage.

In addition to the reactions already mentioned, nitogen oxides, in the presence of sunlight, can react with hydrocarbons to form photochemical oxidants. This process, smog formation, is discussed later in the chapter.

Hydrocarbons

As discussed in Chapter 1, an endless variety of hydrocarbons (HC) can be formed from carbon (C) and hydrogen (H) atoms. Nearly all of the simpler saturated and unsaturated alkanes as well as the simpler aromatic hydrocarbons are common in the atmosphere. Many of the aromatics composed of several rings, such as benzo(a)pyrene and 2-naphthylamine, are carcinogens (cancer inducers) and are obviously very important if present in the air (Figure 3.7). Aromatic compounds with several rings are often called "condensed aromatics." It is important to realize that only a few of all the higher weight aromatics are carcinogens. The process by which cancer is caused is unknown, but even a slight structural difference in a molecule can mean a change from noncarcinogenic to carcinogenic properties, as is the case for 1- and 2-naphthylamine (Figure 3.7). The location of the NH_2 group is crucial: 2-naphthylamine *is* carcinogenic, but 1-naphthylamine is *not*.

Establishing whether a compound is carcinogenic is a very difficult process. Testing is done with animals, and in order to separate the effects on the one tested chemical from all other effects, high concentrations are used. Extrapolations of these data to effects of lower concentrations and to effects in humans are fraught with difficulty. Usually a substance known to be carcinogenic at high dosages is assumed to be carcinogenic at lower dosages; at present, no example is known where carcinogeneity is determined by concentration. Since

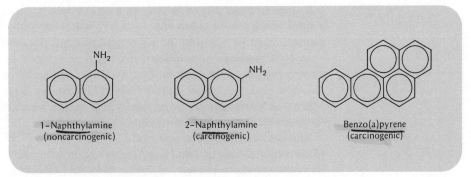

Figure 3.7. *Condensed aromatic hydrocarbons.*

many carcinogens cannot be completely avoided, "safe levels" of exposure must be established wherein no observable increase in the cancer rate of a large population sample would be observed.

Testing at high dosage in humans is seldom done. Thus, research involves the isolation of the effects of one substance, at often low air pollutant concentrations, from the myriad of other natural or environmental effects that may occur. A further difficulty arises from the fact that a 10- to 20-year "latent period" often occurs between exposure and development of cancer. This places obvious difficulties in the path of a researcher who attempts to establish the source and nature of the carcinogen. Much of our early knowledge of the causes of cancer came from studies in which the disease was linked to a person's occupation, which in turn was linked to some chemical he came in contact with daily. For example, it was first noticed in 1775 that persons employed as chimney sweeps in England had a higher rate of skin cancer than the general population. It was not until 1933, however, that benzo(a)pyrene was isolated from coal dust and shown to be carcinogenic. Since then this compound has undergone considerable study.

In 1970 a National Research Council committee reported to the Environmental Protection Agency (EPA) on the health aspects of air pollution, especially the incidence of lung cancer. The study noted that cigarette smoking alone could not account for the increasing incidence of lung cancer or for the fact that cancer rates in urban areas are twice as high as in rural areas. The committee cited data on benzo(a)pyrene prevalence in the urban atmosphere and estimated that the lung cancer death rate rises 5% with each additional microgram of benzo(a)pyrene per cubic meter of air. Annual nationwide emissions of benzo(a)pyrene amount to 500 tons from heat and power generation, 600 tons from refuse burning, 200 tons from coke production, and 20 tons from motor vehicles. Benzo(a)pyrene is also found in cigarette smoke and is thought to be the principal carcinogenic agent responsible for the higher lung cancer rate in smokers as opposed to nonsmokers. The Public Health Service estimates that nonsmokers living in an average-sized American city inhale daily the same quantity of benzo(a)pyrene as would be obtained from smoking a third of a pack of cigarettes. In a city where air pollution is heavy, this estimate must be raised to the equivalent of a full pack a day.

In 1974 evidence accumulated that vinyl chloride, a key chemical in the plastics industry, was a carcinogen. An Italian researcher reported that vinyl chloride inhalation induced a rare form of liver cancer in rats. The deaths of at least 20 industrial workers from such a liver cancer were attributed in 1974 to their several years exposure to vinyl chloride.

Vinyl chloride has the structure $H_2C=CHCl$, in which chlorine (Cl) attaches to carbon with a single bond (see Chapter 1). It is the starting product for the formation of many plastics, including polyvinyl chloride, or PVC, the clear plastic used for food wrapping. It is also used in aerosol products such as shellac-based hair sprays, paint sprays, and insecticides. An estimated 6500 industrial workers have been exposed to high concentrations of vinyl chloride in the air, but the risk to consumers from vinyl chloride in food containers and aerosol sprays has not been established. The EPA began establishing standards in 1974 for vinyl chloride air concentrations in industrial plants.

Man produces only about 15% of the total hydrocarbons on a global basis. Natural sources include trees and plants that emit turpentine, pine oil, and thousands of other hydrocarbon fragrances into the air. Bacterial decomposition of organic matter often emits large amounts of methane (or marsh gas), CH_4. The automobile produces over half of man's emissions (Table 3.3 and Figure 3.2).

At levels currently measured in urban areas, no adverse human effects are known to be caused by the hydrocarbons themselves, with the exception of the carcinogens. Hydrocarbons do undergo reactions to produce photochemical oxidants that are important in smog formation.

Photochemical Oxidants

Although these compounds do not have similar chemical structures, they are often considered as a group because of similar chemical properties, specifically their action as chemical oxidizing agents. Examples are ozone (O_3) and peroxyacylnitrates (PAN). Most are man-made or are formed in the atmosphere from other man-made emissions. Ozone, however, is formed naturally by lightning and by electrical discharges in electrical transmission lines. PANs are important (and probably the most noticed) components of photochemical smog, as discussed later in this chapter.

Oxidants have several adverse effects. They can directly affect the lungs causing respiratory irritation and, possibly, changes in lung function. Oxidants also cause subjective eye irritation. They are extremely toxic to many kinds of plants (see Figure 3.11) and can weaken such materials as rubber and fabrics.

Meteorological Aspects

The atmosphere above the earth is layered. The region nearest the earth, extending 7 to 10 miles up, is the **troposphere.** Above it lies the **stratosphere** (10 to 30 miles), the **mesophere** (30 to 50 miles), and the **ionosphere** (50 to 350 miles). The pressure and density of air are greatest at sea level. With increasing altitude the pressure decreases, and the air thins rapidly.

The troposphere contains 75% of the total amount of air in the atmosphere and is the warmest region. In the troposphere, the temperature decreases as altitude increases, the rate of temperature decrease being about 3°F per 1000 ft. Because of density differences, warm air tends to rise, whereas cool air tends to sink. This natural process causes vertical currents that are essential to cloud formation and precipitation. These natural processes also afford a mixing between airborne pollutants and the surface of the earth, thus facilitating the incorporation of the pollutants in the biogeochemical cycles, nature's purification scheme.

The stratosphere, however, in contrast to the troposphere, becomes warmer with altitude. Since the warm air is above the cold, there are no vertical air currents, so the stratosphere tends to be a very stagnant layer with pollutants released there tending to remain for years. Serious consideration would have to be given to possible pollutant concentration buildup that may occur over the years if a steady source of pollution were introduced into the stratosphere.

Thermal Inversions and Air Pollution

The air immediately above the surface of the earth, the troposphere, usually exhibits on a global basis the general behavior described. For smaller local regions, however, meteorological conditions can cause short-term variations from this behavior. One of these variations, the **thermal inversion,** is crucial to the impact of air pollution. An inversion in the troposphere occurs when the cool air lies below warmer air. When this happens, a stagnant condition is created. When the inversion is at the ground level, for example, the cold, dense air remains at the surface of the earth and the warm air is higher. This is the *reverse* of the usual troposphere conditions. This reversal, or inversion, re-

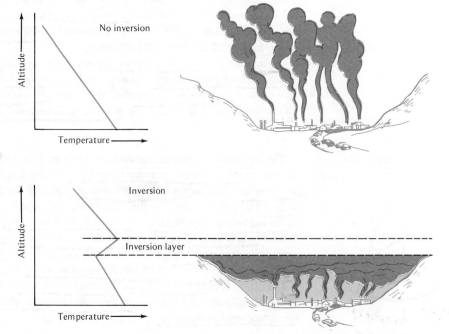

Figure 3.8. *A temperature inversion, in which a layer of warm air overlies a layer of cooler air, traps air pollution close to the ground. [From Population, Resources, Environment: Issues in Human Ecology, 2nd ed., by Paul R. Ehrlich and Anne H. Ehrlich. W. H. Freeman and Company. Copyright ©·1972, reprinted with permission.]*

moves the tendency for vertical air currents. Inversions are common at night when the earth's surface and the adjacent air cool rapidly. The inversion layer, which may be several yards or several miles thick, will last until some natural process, possibly morning sunlight, alters the temperature pattern of the lower atmosphere.

Of significance to air pollution is the fact that the inversion prevents vertical mixing and thus traps pollutants (Figure 3.8). The pollution concentration thus increases until the inversion is dissipated. Urban areas obviously suffer most because their pollutant output is largest.

Thermal inversions occur naturally and were observed by Indians in America long before the coming of the white man. The Indians noted that under certain conditions smoke from campfires did not rise indefinitely but often spread at a particular altitude. Such incidents were sometimes used for weather prediction.

Energy Balance on Earth

The earth receives energy from the sun. This continuous energy input means that the earth's temperature would slowly rise unless a heat loss from earth occurred. The earth's temperature is determined by the net balance of two processes: continuous energy input from the sun (heating), and heat energy radiated into space by the earth (cooling). The yearly average temperature of the earth has been relatively constant (to within a few degrees) for thousands of years indicating that the heating and cooling are well balanced. Any alteration of the heating rate or the cooling rate would be expected to result in a change in the average temperature of the planet, which might have a drastic effect on the climate and ecosystems. For example, an ice age might be triggered if the *yearly average* temperature dropped by a few degrees and remained there for several years. Melting of the ice caps might occur if the earth's temperature were increased only a few degrees. The melting, which would take centuries, would raise the ocean levels 200 ft, inundating most of the world's major cities.

To discuss possible effects that might alter temperature, we must first delineate the differences between the energy that comes from the sun and the energy that is lost from the earth. The earth is heated by ultraviolet (UV) and visible radiation (energy) from the sun, but is cooled through a loss of infrared (IR) energy. Both are part of the **electromagnetic spectrum,** but they exhibit quite different properties. The various regions of the electromagnetic spectrum are measured by the wavelength of the radiation (Figure 3.9). The wavelength is used to characterize all wave motions. A rock dropped into a pool creates concentric waves that spread outward, and the wavelength of these waves is the measured distance between adjacent wave troughs. Electromagnetic radiation includes many important forms of energy, including microwaves, radiowaves, light, and x rays.

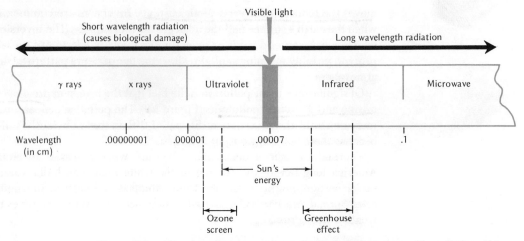

Figure 3.9. *Wavelength spectrum in centimeters of electromagnetic radiation. The locations of various parts of the spectrum having environmental impact are shown, including the wavelength range of the ozone screen and the carbon dioxide (CO_2) greenhouse effect.*

A large portion of the sun's radiation consists of UV and visible light (Figure 3.9). UV light, which is not visible to the human eye, is the radiation responsible for sunburn. The visible region is so named because it is the only portion that can be seen by the human eye. As the sun's radiation approaches the earth, a small but very important amount of the very shortest wavelength (highest energy) portion of the UV is absorbed by ozone in the upper atmosphere. This "ozone shield" exists as a band 10 to 25 miles above the surface of the earth in the stratosphere. It performs a valuable function in that it protects living organisms from the harmful effects that would result if the highest energy portion of the sun's radiation reached the surface of the earth. (See the section "Chlorofluorocarbons, Aerosol Sprays, and Ozone," later in the chapter for additional discussion on the importance of the ozone shield.) The rest of the sun's energy enters the atmosphere and provides energy (heating) to the earth in the form of UV and visible light.

The earth radiates or loses energy in the IR part of the spectrum (Figure 3.9). This heat radiation is exhibited by all bodies. Although IR energy is not visible to the human eye, most of us have encountered it—IR energy is utilized in heat lamps. Unlike sunlight, IR radiation *cannot* pass freely through the earth's mantle of air. Water vapor and CO_2 both absorb IR radiation, and thus act as "a blanket" to trap and retain heat energy radiated from the surface of the earth. Without this blanket, the planet would be subject to very low temperatures at night, as is the moon, for example, which has almost no atmosphere. Typical average temperatures for the earth without an atmosphere are predicted to be $-40°C$ $(-40°F)$ as compared to the present $15°C$ $(60°F)$.

Possible Man-Induced Climatic Changes

Man has the ability to alter the atmospheric levels of pollutants and thus in principle to alter the net heat balance of the earth. H_2O and CO_2 allow UV light to pass freely, but absorb IR radiation. If the amount of H_2O and CO_2 in the atmosphere were increased, this would *increase* the blanket (less heat loss) which would result in a temperature rise. This is termed the "greenhouse effect."

Greenhouse Effect. The greenhouse effect takes its name from the warmth of greenhouses, a warmth occurring in part because glass transmits most of the sunlight but absorbs most of the IR radiation. A blanket is thus created by the glass, and the temperature under it is higher. Similar effects might occur on a wider scale in the atmosphere if either the H_2O or CO_2 levels or both were increased. Man's impact on the H_2O levels in the troposphere is small. Furthermore, the H_2O cycle (Chapter 6) is very efficient at keeping the H_2O concentration constant in this region. CO_2, on the other hand, is rather unreactive in the atmosphere, although it does dissolve readily in the oceans (eventually forming limestone). Extensive combustion of fossil fuels has increased the atmospheric CO_2 levels from about 290 ppm in the mid-1880s to about 322 ppm in 1970. If the present rate of increase is maintained, the CO_2 level in the atmosphere early in the next century will be about double that which existed in the mid-1800s. A recent calculation indicated that, if no other changes occurred, doubling the CO_2 concentration would increase the earth's temperature by about $1°C$ $(2°F)$.

Particulates. Increasing the amount of particulates in the air, however, could result in a temperature *drop*. Particulates reflect UV and visible light, so that less heat reaches the earth. The IR losses are unaffected, however, so the net effect is a cooling of the earth. Recent calculations indicate that a fourfold increase in atmospheric particulates might reduce the temperature by $3.5°C$ $(7°F)$. If maintained for several years, this drop would probably be sufficient to trigger a new ice age. Man's activities since 1900 have almost doubled the particle content of the global atmosphere.

Chlorofluorocarbons, Aerosol Sprays, and Ozone. As mentioned earlier the ozone layer, which surrounds the earth in the lower stratosphere, performs an important function: it shields the planet from strong ultraviolet radiation which, if it reached the earth, might possibly cause an increase in skin cancer, a change in vegetation, and a change in global weather. It is generally agreed that life developed on the planet only after the protective ozone shield was formed.

Solar radiation interacts with naturally occurring oxygen (O_2) to form ozone (O_3) in the stratosphere; however, sunlight also interacts with the ozone itself to destroy it. The result is believed to be a stable natural concentration of ozone in the stratosphere.

Since the early 1960s concern has been expressed that chemicals released by man might act to deplete the ozone layer and thus expose the earth to increased UV radiation. The particular threat posed by the chlorofluorocarbons, substituted hydrocarbons in which chlorine and fluorine replace hydrogen, has only recently been recognized. The two most common chlorofluorocarbons are **Freons:** Freon-11 ($CFCl_3$) and Freon-12 (CF_2Cl_2). [In both of these, fluorine (F) and chlorine (Cl) replace hydrogen in the methane molecule.] Freon-11 is widely used as the propellant in aerosol spray cans, and Freon-12 is the refrigerant gas in air conditioners and refrigerators.

Concern that these might deplete the ozone layer comes from the extreme inertness of these compounds to ordinary chemical reactions. They are man-made, not integrated in the biogeochemical cycles, and apparently have no natural sinks; they are not absorbed or degraded in the ocean or in the soil. Thus virtually the entire production (1 million tons of each annually) ultimately enters the air. Because no chemical reactions destroy these molecules in the lower atmosphere, it is contended that they will eventually drift into the stratosphere. Here they can be broken down: UV light can degrade the chlorofluorocarbons, releasing highly reactive chlorine atoms, which can trigger a rapid conversion of ozone to oxygen.

Because Freons have been manufactured since 1930 and because the time required for these molecules to diffuse to the stratosphere is as long as 50 years, the total effects of their release may not be known for half a century. The chlorofluorocarbons are apparently now ubiquitously distributed in the lower global atmosphere. Although the data are scanty, there was in 1974 little evidence that they had diffused into the stratosphere. A large pool of Freons released into the lower atmosphere, however, guarantees a steady flow of chlorine atoms that might cause ozone depletion should the molecules reach the stratosphere.

Initial research on ozone depletion, conducted at the University of California at Irvine by Dr. F. S. Rowland and at the University of Michigan by Dr. C. Cicerone and coworkers, predicts the maximum effects will occur in 50 to 80 years. These researchers caution that continuing to use these compounds may greatly increase future ozone depletion. Typical estimates are that, if the manufacture of these compounds were stopped immediately, the ozone would be depleted by about 5% in 1990. However, if production were to continue to increase at the present 20% per year, a 30% depletion of ozone by 1990 might result. The specific effects of such a change can only be crudely estimated. A National Academy of Sciences panel suggested that a 5% depletion of ozone might produce an additional 8000 cases of skin cancer in the United States. Estimates by medical researchers have placed the figure at upwards of 60,000. Less clear is the impact on vegetation. Some plant growth rates might decrease, but some others might increase; plant mutation rates would probably increase. Weather modifications might occur if ozone depletion were not uniform around the globe.

The possible depletion of the ozone level and the potential effects that might result are obviously highly speculative, and much more needs to be known. Although some research is being done, the complexity of chemical reactions in the upper atmosphere probably means that it will be a long time before we will know with any assurance the effects of our interference with the atmosphere.

The SST and Ozone. The supersonic transport (SST) may have a strong impact on the global environment. Because the plane operates in the stratosphere, its emissions may remain in the atmosphere for many years before slowly being drawn into the biogeochemical cycles. When the SST was being promoted in the United States in the 1960s, it was estimated that by 1985 a fleet of 500 SSTs would be flying 7 hr/day at an altitude of about 12 miles. Each would consume enormous quantities of fuel: 66 tons/hr aloft. Furthermore, the emissions of per plane hour were expected to be approximately 83 tons of water, 207 tons of CO_2, 3 tons of CO, and 3 tons of NO. The H_2O emissions were the focus of concern because they would nearly double the water vapor concentration in the region of heavy flying. It was feared this might lead to increased haziness and cloud cover. In addition it was suggested that the NO_x emissions might deplete the ozone shield through chemical reactions. Professor Harold Johnston of the University of California at Berkeley calculated that significant amounts of ozone could be removed by reaction with NO to give NO_2 and O_2. Johnston estimated that about half of the ozone might be removed by this method.

Because of the large number of variables and unknowns involved in a global environmental problem, such calculations are only approximate. Inasmuch as the potential effects are catastrophic, such considerations should give us pause before we embark on an extended development of the SST. As often happens with technological developments, the SST was (and still is) being pressed forward before its full global impact was known. The United States version was halted (perhaps temporarily) in 1971 when funding in Congress was denied. Involved in that decision were many considerations: economics, impact of the noise and sonic boom, excessive fuel consumption, and private versus government financing of new technology. The French-British Concorde and the Russian SST have been operating since the early 1970s.

Future Climates. Global climate impact of such factors as CO_2 increases, particulate increases, and SST effects can be estimated only approximately. One key factor, for example, is whether increasing the average temperature of the earth will increase or decrease the cloud cover. The answer is not known because it is impossible to construct an accurate model of the earth's global system. Some researchers feel that the earth's meteorological balance is so massive that it will never be affected by man. For example, an initial greenhouse effect might lead to increased evaporation of water. More clouds might form which would decrease the sunlight sufficiently to offset the original greenhouse impact. The net result would be no appreciable change in the earth's average temperature. A 1% increase in cloud cover, for example, would offset a 50% increase in CO_2 concentration.

Figure 3.10 shows the recent variations in the temperature of the earth. It is not possible to say what has caused the variations. It has been suggested that the increase from 1920 to 1945 was due to CO_2 increases (greenhouse effect), whereas the decrease since 1945 is due to increased particulates. There is, however, no proof that this suggestion is correct. The changes may represent some natural cyclical variation.

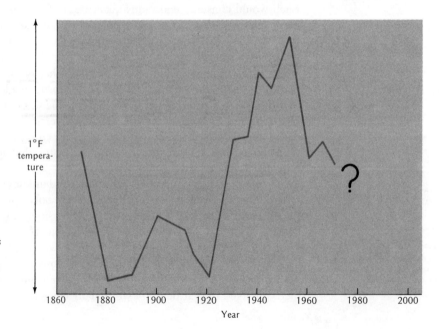

Figure 3.10. *Variations in temperature in the Northern Hemisphere for the past century are too erratic to project a definite trend.*

Oxygen Depletion

In the past, concern has been expressed about the possibility that man's activities might lead to a depletion of oxygen. Combustion consumes oxygen, converting it to CO_2. In addition the decrease in vegetation due to man's agricultural and industrial activities raises questions of whether this reduction in the amount of oxygen provided by plants threatens man's existence. A precise chemical balance exists between the oxygen created in photosynthesis, the oxygen used in plant decay, and the oxygen used in fuel consumption.

Plants use precisely as much oxygen to respire and decay as they created in photosynthesis. Photosynthesis has occurred for millions of years. Our present massive oxygen reservoir arises from burial of organic matter (plants) in fossil fuels and sediments which effectively removes them from the oxygen-consuming decay scheme. A thorough analysis is beyond our treatment, but the results indicate the amount of oxygen in the atmosphere will not be seriously affected by any of man's activities. Burning of all fossil fuels, for example, would reduce the amount of oxygen by only about 20%. Other, more important, problems will surface long before man's combustion of fossil fuels or man's other activities seriously reduce the oxygen levels.

Smog

Smog is probably the most famous aspect of air pollution. The word was coined in the early 1900s to describe the "smoke and fog" of London pea soupers but is now used generally to describe air pollution in urban areas. Two different scientific classifications of urban pollution have evolved: London or "classical" smog and Los Angeles or "photochemical" smog. The two forms are quite different in terms of both formation and effects.

London smog is due principally to the presence of sulfur oxides and particulate matter accompanied by thermal inversions that permit a buildup of high pollutant levels. Poor visibility and health problems due to SO_2 characterize the presence of this form, which is usually worse during winter months. London has been subjected to many smog formations over the years, but the worst occurred on December 5 to 9, 1952, when the visibility was measured in yards and the SO_2 and smoke levels were five times greater than normal. This episode resulted in an estimated 4000 excess deaths in London. The data in Figure 3.4, discussed earlier in this chapter, were recorded during this period. Donora, Pennsylvania, experienced a "classical" smog disaster October 26–31, 1948. Death rates and hospitalization rates for persons over 55 years of age increased sharply.

Photochemical smog, first noted in Los Angeles in the mid-1940s, is a complex phenomenon whose instigation requires sunlight; hence the use of the term "photochemical." Characteristic symptoms are a brownish coloration of the atmosphere, reduced visibility, plant damage, eye irritation, unpleasant odors, and respiratory distress. Photochemical smog is usually at its worst during the summer months at midday. The initial scientific characterization of Los Angeles smog was done by Professor A. J. Haagen-Smit of the California Institute of Technology. Because of its complexity, smog formation is not completely understood despite intensive research. A great deal is known, however, about the ingredients necessary for smog formation and about the mechanism that triggers its formation. In addition the principal end products have been identified and their effects studied. Between the triggering process and the end products, however, are many chemical reactions whose roles are not completely defined. These reactions may occur consecutively or competitively. Researchers often deal with 150 to 200 chemical reactions, which involve many of the more than 1000 compounds that have been identified in the air over Los Angeles.

The prime ingredients for these reactions are hydrocarbons and nitrogen oxides, produced principally by the automobile and some stationary sources. Solar radiation triggers an initial reaction with NO_2, releasing products that can combine with organic molecules. A series of complex reactions then follows.

A greatly simplified model, which accounts for most of the properties of smog formation, is as follows. **Primary pollutants,** such as NO_2 and varied hydrocarbons, are present in the air and are necessary starting ingredients for smog formation. The triggering of smog formation occurs when NO_2 absorbs UV light from the sun and is "photolyzed" into nitric oxide and an oxygen atom ($O\cdot$).

$$NO_2 \xrightarrow{\text{sunlight}} NO + O\cdot \qquad \text{smog trigger reaction} \qquad (3.6)$$

The oxygen atom is a very reactive species: it can react with nearly any other molecule (especially O_2 or hydrocarbons) it encounters. It initiates a series of reactions whose products are termed **secondary pollutants** because they occur only after the photochemical (trigger) initiation.

In particular, oxygen atoms can react with oxygen molecules to form ozone (O_3), an oxidant discussed earlier in this chapter. If oxygen atoms encounter hydrocarbons, then other very reactive species termed **free radicals** are created; these are indicated by the use of an asterisk (°) in the secondary pollutant formation equations. The free radicals can react with a wide variety of other species, but their reactions with hydrocarbons and NO_2, shown in equations 3.7 through 3.11, are the most significant, because they create many of the important end products of smog. Aldehydes, ketones, and, especially, peroxyacyl nitrates (PANs) are known eye irritants that are responsible for most, though not all, of the impact of smog. PAN and ozone are chemical oxidants that can react with a number of organic substances including vegetation, rubber, and fabrics. Although the natural oxidant level is about 0.01 to 0.03 ppm, peak levels in Los Angeles of 0.10 to 0.15 ppm were recorded over 30% of the time during the mid-1960s. Many other United States cities, including Denver, St. Louis, Philadelphia, and Washington, D.C., exceed 0.10 ppm during peak periods. The highest oxidant levels tend to occur during the summer and early fall months when the solar radiation is most intense.

$$O\cdot + O_2 \rightarrow O_3 \qquad\qquad \text{ozone formation} \qquad (3.7)$$
$$O\cdot + HC \rightarrow HCO° \qquad\qquad \text{free radical formation} \qquad (3.8)$$
$$HCO° + O_2 \rightarrow HCO_3° \qquad\qquad \text{free radical formation} \qquad (3.9)$$
$$HCO_3° + HC \rightarrow \text{aldehydes, ketones} \qquad \text{end product formation} \quad (3.10)$$
$$HCO_3° + NO_2 \rightarrow \text{peroxyacyl nitrates (PAN)} \qquad \text{end product formation} \quad (3.11)$$

Figure 3.11a. *Air pollution and smog have lasting effects on all living organisms. The lungs collect particulate matter and also serve as an entry point into the bloodstream for other pollutants. A healthy alfalfa plant (left) offers a contrast to a plant (right) exposed for 4 hrs. to air containing 20 parts per hundred million of ozone. [Cartoon by Swan in Industrial Research. Copyright 1972, reprinted with permission. Plant damage photo, U.S. Department of Agriculture.]*

Figure 3.11b.   SO₂ destroys the chlorophyll-bearing cells of leaves. The affected leaves develop white areas between the veins. Leaves at the right are healthy. [Courtesy U.S. Department of Agriculture.]

The principal impact on humans appears to be eye irritation, which results when oxidant concentrations exceed 0.1 ppm. Asthma attacks sometimes increase when levels exceed 0.13 ppm; nasal and throat irritation occurs at 0.3 to 0.6 ppm. At present toxicological and epidemiological data relating to photochemical smog are scarce. There is no evidence of irreversible change in the human respiratory system from exposure to ozone concentrations commonly found in urban air. Experiments on animals, however, suggest that such changes can occur after 1 year of exposure at higher levels, about 1 ppm. There have been no known episodes of photochemical smog that have produced increased mortality, unlike the examples of the London and Donora SO₂ disasters. It does appear that smog worsens the impact of respiratory diseases such as bronchitis, asthma, and sinusitis. Respiratory system irritation appears to be worse for those who undergo vigorous exercise. High school athletes in competitive events show reduced performance when oxidant levels have been higher earlier in the day.

Figure 3.11c. At relatively low levels, pollution causes chlorotic dwarf disease in the eastern white pine. The trees in this picture are 15 years old, but the ones in front have the dwarf disease, caused, it is believed, by either sulfur dioxide, ozone, or an interacting mixture of both. [Courtesy U.S. Department of Agriculture.]

Smog damage to plants and vegetation (see Figure 3.11) including many cash vegetable crops, is well known. The most notable injury is "silverleaf": a shiny, oily effect on the leaf's underside. Most of the vegetation effects of smog are due to ozone and are sufficiently widespread that no one point source is evident.

Los Angeles has a number of characteristics that lead to increased smog problems: many automobiles, industrial development, sunny skies, inland mountains that impede the movement of pollutants, and a strong tendency toward thermal inversions. The inversion layer is often located just below the mountain tops and consequently forms the "lid" of a box in which smog can form.

Photochemical smog is now known in most urban areas of the United States.

The Automobile

As noted previously, the automobile is the major source of air pollutants by tonnage, and is the dominant producer of CO and hydrocarbons (HC). The internal combustion gasoline engine has fundamental characteristics that lead to these pollutants; in particular, it requires **incomplete combustion** of the fuel. This characteristic serves to distinguish internal combustion engines from external combustion engines, steam engines being an example, which allow **complete combustion.** Combustion, the burning of fuel, is described chemically by an equation relating the combining of the fuel (a hydrocarbon) with oxygen:

$$2\ CH_3\!-\!\underset{\underset{CH_3}{|}}{\overset{\overset{CH_3}{|}}{C}}\!-\!CH_2\!-\!\overset{\overset{CH_3}{|}}{CH}\!-\!CH_3 + 25\ O_2 \rightarrow 16\ CO_2 + 18\ H_2O \qquad (3.12)$$

complete combustion of isooctane (gasoline)

The equation indicates that 2 isooctane molecules react with 25 O_2 molecules to yield 16 CO_2 molecules and 18 H_2O molecules. To get complete combustion, there must be 25 (or more) O_2 molecules for each 2 isooctane molecules. If this condition obtains then equation 3.12 indicates that the products are CO_2 and H_2O. Suppose, however, one has *insufficient* O_2 (less than 25 molecules in equation 3.12). Under these conditions, *incomplete* combustion occurs so that isooctane may combust as in equation 3.13.

$$2\ CH_3\!-\!C(CH_3)_2\!-\!CH_2\!-\!C(CH_3)_2 + 7\ O_2 \qquad \rightarrow$$

(limited O_2)

$$3\ CH_3\!-\!CH_2\!-\!CH_3 + 6\ CO + 6\ H_2O + CO_2 \qquad (3.13)$$

propane

incomplete combustion of isooctane

The end products of incomplete combustion are HC, CO, H_2O, and CO_2. Thus the internal combustion gasoline engine and any other engine that requires incomplete combustion will *always* produce the pollutants CO and assorted HC.

A further aspect of *all* engines is that, regardless of whether they operate on complete or incomplete combustion, if they get O_2 from the air they will inevitably cause the reaction of nitrogen from the air to form NO_x upon combustion.

$$\text{fuel} + \text{excess } O_2 \text{ (air includes } N_2) \rightarrow CO_2 + H_2O + NO_x \quad (3.14)$$
complete combustion

$$\text{fuel} + \text{limited } O_2 \text{ (air includes } N_2) \rightarrow CO + HC + H_2O + NO_x \quad (3.15)$$
incomplete combustion

Thus equations 3.14 and 3.15 are the best general representation of complete and incomplete combustion in the presence of air.

As is well known, the presently used internal combustion engine burns the gasoline in a cylinder. The heat from the fuel is transformed to kinetic energy when the heated gas expands and moves the piston that serves as the bottom of the cylinder. The temperature during combustion is relatively high, varying from 3000°C (5400°F) in the bulk of the gas to 200°C (390°F) near the cylinder surface. Exhaust gases are typically at 1000°C (1800°F) but cool rapidly upon expansion into the tailpipe.

Many variables of engine design and operation can affect both the type and quantity of pollutants. The surface to volume ratio of the combustion chamber, the spark timing, and the speed of the crankshaft rotation are some of the important factors. The air to fuel ratio in the carburetor determines the completeness of combustion. The amount of air and the amount of fuel available as well as the instant of ignition with respect to piston position are all adjustable through carburetor and other engine adjustments. This means that the quantity of emissions may be adjusted as well.

An air to fuel ratio of 14.5 (or more) to 1 by weight is necessary for complete combustion. That is, if in the combustion chamber the air/fuel ratio exceeds 14.5, the combustion will be complete and will tend to occur as in equation 3.14: most of the HC will be burned but NO_x will be produced. If the air/fuel ratio is less than 14.5, then by equation 3.15, HC, CO, and NO_x will be produced. Figure 3.12 indicates the wide variation in engine emissions that occur as the air to fuel ratio is varied. The CO concentration drops rapidly as the ratio approaches 14.5. The HC values can be much greater than those listed depending on the condition of the car, but their overall behavior is similar to that shown. Although the HC emissions drop as 14.5 is approached, they are not affected as drastically as CO because the nonuniform combustion conditions in the cylinders preclude the burning of all the HC, especially in the

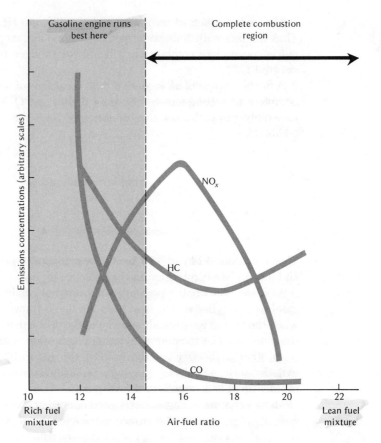

Figure 3.12. *Effect of air/fuel ratio on exhaust emissions. The vertical line indicates the value (14.5) necessary for complete combustion.*

region between the piston and the cylinder walls where the flame does not reach.

Another important point in Figure 3.12 is that *minimization* of CO and HC emissions may *maximize* NO_x. Similar "trade off" behavior between one pollutant and the others occurs almost invariably in the internal combustion engine as the temperature, engine tuning, and engine speed are varied. Thus all emissions from internal combustion engines cannot be simultaneously reduced through engine adjustment, although emissions of one or two of the pollutants may be reduced.

The gasoline engine usually operates best within the left-hand region of Figure 3.12, that is, with an air/fuel ratio of less than 14.5. The actual value varies with the speed of the car, but typical values are given in Table 3.5. Obviously the relative values of emissions depend upon driving mode. Although wide individual variations occur, an auto in urban use often operates 50% of the time under either acceleration or deceleration conditions, both of which maximize HC and NO_x emissions. The actual emission values, however,

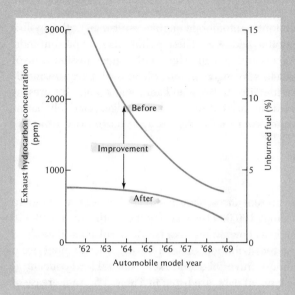 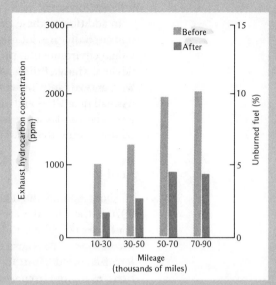

Figure 3.13. *Hydrocarbon emission tests before and after tune-up by engineering students at University of Michigan. Although cars mechanically too unfit to tune were eliminated, the 50 that were tuned and tested were a typical cross-section of vehicles. [From Popular Science, August 1970, p. 101. © 1970 Popular Science Publishing Company, Inc. Reprinted with permission.]*

are strongly dependent upon engine tuning, as is evident in Figure 3.13. These data are from a University of Michigan study wherein automobiles selected at random off the streets had emissions measured before and after an "on the spot" tuneup. The tuneups were the usual ones used with 1970 model automobiles and, as is evident, the emissions are reduced by factors of 2 to 4 by

TABLE 3.5. Automobile emissions under different driving conditions

Operation	CO (vol. %)	HC (ppm)	NO$_x$ (ppm)	Air:Fuel Ratio
Idle	6	750	30	11–12:1
Acceleration	5	600*	3000	12–13:1
Cruise	3	350	1200	13–14:1
Deceleration	4	4000	600	11–12:1

Note: The data are typical for late 1950s model cars with no emission controls. Wide individual variations may occur.

* Depends very strongly upon rate.

tune-up! Another aspect of Figure 3.13 is that emissions increase as the engine mileage increases.

In addition to these pollutants, automobile engines emit many other pollutants initially present as gasoline additives. These pollutants may present individual environmental problems and may interfere with some emission control devices. Many of these, such as scavengers (remove lead oxide), deposit modifiers, antioxidants (preserve fuel life), antirust and antiicing agents, are present in small quantities and are, at the present state of knowledge, relatively harmless. The most severe hazard from fuel additives appears to arise from the additive tetraethyl lead (TEL).

Lead

About 80% of the TEL present in gasoline is emitted as lead aerosols. In 1970 in the United States about 180,000 tons were put into the air (about 116 lead per 200 gal gasoline). This constitutes about 98% of all man-made lead aerosols and has caused a redistribution of lead in the environment from localized ore deposits to almost universal occurrence. Natural lead concentrations, estimated from geochemical data, are listed in Table 3.6. An increase of the lead level by factors of 10 to 10,000 has occurred in many portions of the environment. There is little disagreement that man's activities (principally the automobile) have caused this dramatic increase during this century, with a dramatic rise since 1940. Chronological ice layers in northern Greenland show that lead levels have increased from 0.0005 μg lead/g of ice in 800 B.C. to 0.08 in 1940 to 0.2 in 1965. Similar increases are found from the analysis of elm tree rings in urban areas. The levels of lead in major United States cities increased about 5% per year over the last decade. The lead aerosol level above Fleet Street, London increased 68% from 1962 to 1971.

Lead is an element with no known beneficial function in human metabolism. Its toxicity has been known for 2000 years. The physiological effects are discussed in Chapter 6. The known cases of lead poisoning do not arise from lead aerosol inhalation, but rather from ingestion of lead from either paints or from the use of lead glaze on pottery.

TABLE 3.6. Typical lead levels

Region	Natural (Pre-auto)	Present
Air (μg/m$_3$)	0.0006	2–8 (urban)
Soil (ppm)	10–15	several hundred (urban)
Fresh water (μg/l)	0.5	10 to several hundred
Ocean water (μg/l)	0.015	0.2 (near shore)
Human blood (ppm)	0.01	0.25

There is no evidence that lead aerosols are toxic to man at present levels. However, it is possible that the effects of mild poisoning go undetected. The diagnosis of lead poisoning (plumbism) is difficult at low exposure because the symptoms are headaches and listlessness. Zoo mammals and snakes in New York suffer from lead poisoning, and the measured levels around the cages seem to point to the atmosphere as the culprit.

The principal concern about lead is the environmental impact on the ecosystem of the continued (and increasing) exposure to unnaturally high levels of lead. Plants can absorb lead from the soil, and the long-term effects are open to speculation.

TEL is present in gasoline to reduce engine "knock" and to boost the octane rating. Gasoline is a mixture of alkane hydrocarbons with six to ten carbon atoms. With some hydrocarbons, ignition in the engine occurs as an explosion rather than as a smooth burning. This not only reduces the efficiency of the engine but produces an audible "knock." Pure isooctane (equation 3.1) produces little knock, and for historical reasons has become a standard for fuel comparison. The "octane rating" measures quality of the gasoline. Regular gasoline is about 92 to 94 octane, whereas premium gasoline is 100 octane or higher.

Under economic conditions of the late 1960s, 88 octane gasoline was produced in refineries at a cost of about 10¢ per gallon. This could be boosted to 93 (or regular) by adding TEL at a cost of only 0.6¢ per gallon. High octane gas can be made without TEL; in fact the "Amoco" brand of the American Oil Company has been on the market for years. Increasing the octane number without TEL increases costs and could have other effects. For example, one method is to blend in more aromatic and olefin hydrocarbons. One study by the Bureau of Mines indicated that this particular type of nonleaded gas may create exhaust gases with increased photochemical reactivity, perhaps 25% higher than that of leaded gas.

Thus there has been considerable speculation, principally by the industries involved in TEL production, about the possibility of increased smog problems if a complete switch to lead-free gas with increased aromatics occurred. However, the Bureau of Mines study did not investigate the properties of nonleaded gasolines that used other approaches to octane boosting, such as an additional refining operation.

The trend toward low-leaded and nonleaded gasoline began in the early 1970s. The initial pressure arose because of public concern about the environmental impact of lead. Other factors soon became important, however. By 1973 most auto manufacturers had designed their engines to run on regular gasoline, reversing a long trend toward higher octane needs for auto engines. The manufacturers did this apparently because they had decided that reducing auto emissions to meet government specifications required techniques and devices, such as the catalytic converter (discussed in the following paragraphs) that operated best if lead were absent. A conversion to nonleaded gasoline would be more easily accomplished if most cars were burning lower octane (regular) gasoline. A conversion to the large-scale use of unleaded gasoline will have a number of effects. Additional refinery modifications will cost $1 billion to $4 billion at 1970 prices. Some auto maintenance problems may increase and some may decrease.

Engine Emission Controls

A new pre-1966 American automobile with no emission controls emitted at least 124 g/mile of CO, 17 g/mile of HC, and 5 g/mile of NO_x. On an annual basis this average car would produce 1240 kg of CO, 170 kg of HC, and 50 kg of NO_x. All of the CO and NO_x were emitted from the exhaust pipe, together with 60% of the HC. The remaining 40% of the HC was evenly divided between tank evaporation losses and "crankcase blow-by," gases escaping from the pistons into the engine crankcase from where they were vented to the atmosphere.

Blow-by emissions are controlled by recycling the vented gases into the engine air intake system for recombustion, a process called positive crankcase ventilation. Legislation requiring this became effective in California with the 1963 models and nationally with the 1968 models.

Evaporative emissions from the gasoline tank and from the carburetor are controlled by using a well-sealed gasoline tank and carburetor and an absorbant such as activated charcoal. When the engine is off, the hydrocarbons are absorbed by charcoal; this avoids a pressure buildup in the system. When the engine is running, air is circulated through the filter to remove the hydrocarbons from the charcoal and carry them to the carburetor for combustion. This cleans the filter and readies it for the next engine shutoff. Evaporative controls were required beginning with 1971 model autos.

An attack on the exhaust emissions may take several approaches. Changes may be made in the operation of the engine, or some device may be used to change the character of the exhaust gases before they are emitted. In practice both of these techniques are used. The exhaust conversion devices attempt to change the emissions to less noxious forms such as CO_2, O_2, H_2O, and N_2. These devices, such as the catalytic converter (discussed in the following paragraphs), did not come into use until 1975.

Initial control efforts (1969 to 1972) focused on CO and HC because it was believed control of these could be most easily accomplished and might provide the most immediate relief to some of the urban air pollution problems. General Motors, Ford, and American Motors originally used an air injection system in which air was injected into the exhaust close to the engine where the exhaust gases are hottest and most easily ignited. This would further the oxidation into CO_2 and H_2O without the extraction of much usable power. Chrysler developed a "clean air package" in which emissions were reduced through engine adjustment and operation modifications, such as the use of leaner and more precisely controlled air to fuel ratios. Unfortunately these approaches resulted in increased NO_x emissions, as could be predicted from Figure 3.9.

The yearly standards and performance data for vehicles are given in Table 3.7. The emission reductions have in general lagged behind United States

TABLE 3.7. Exhaust emission in grams per mile (except for buses) as set by law or as attained by various operational vehicles and engines.

Automobiles	HC	CO	NO$_x$
Actual uncontrolled emissions on pre-1966 cars (based on 1972 test procedures)	17.0	124.0	5.4
1966 California standards	3.4	34.0	—
1970 U.S. standards	2.2	23.0	—
U.S. Standards for 1975–79	1.5	15.0	2.0[a]
U.S. Standards for 1980 (proposed)	0.9	9.0	2.0
U.S. Standards for 1982 (proposed)[b]	0.41	3.4	0.4
1970 model U.S. autos (random tests) 10,000 cars)	2–4	5–25	—
DuPont thermal reactor (1970)	0.2	12.0	1.2
Diesel engine Mercedes (1973)	0.3	1.6	1.3
Diesel engine Peugeot (1973)	1.8	2.4	1.2
Stratified charge (1973)			
Honda (after 50,000 miles)	0.26	2.57	0.98
Ford Proco	0.37	0.93	0.33
Chrysler-Esso engines (1972)			
Manifold reactor	1.5	20.0	1.3
Catalytic reactor	1.7	12.0	1.0
Ethyl Corporation "lean reactor" car (1972)	0.7	10.4	2.5
Wankel engine (with controls) (1973)	4.0	23.0	2.2
Natural gas fueled internal combustion engine	1.5	6.0	1.5
Gas turbine	0.5–1.2	3.0–7.0	1.3–5.2

Buses	HC + NO$_x$	CO
1975 California standards (grams/brake horsepower hour)	5.0	25.0
Diesel (V8 and V6)	9–15	3–9
Experimental steam buses (San Francisco-Oakland)	1.3–3.1	1.6–2.7

[a] Becomes effective 1977.

[b] Originally set for 1975 model year by 1970 Clean Air Act, then delayed to model 1977 model year by Congress. In 1975, EPA proposed delaying these standards until 1982.

Note: Because of differences in test cycles and measurement methods the uncertainties in these numbers are rather large, and the numbers should not be considered accurate to within less than 50%.

government regulations. However, the progressively tighter controls have had a measurable effect on the amounts of the pollutants emitted by late model cars. The 1970 data in Table 3.7 were taken on about 10,000 randomly selected autos with an average mass of 1800 kg (4000 lb). The improvement in 1970 (and later) model autos is considerable over the pre-1966 cars. However, emissions often increase substantially after autos have been driven 20,000 miles or more, and most of the 1970 data were taken on cars that had been driven less than 10,000 miles.

To meet the stricter standards beginning in 1975, better control equipment was required, such as **thermal reactors** and **catalytic converters.** The thermal reactor replaces the engine manifold and permits additional gas reaction time at high temperatures to further the oxidation of HC and CO. Figure 3.14 shows the thermal reactor proposed by the Inter-Industry Emission Control Program (IIEC), which combines the test and research facilities of eleven major oil and auto companies. Some limitations of the thermal reactor are that expensive metals may be required to withstand the high temperatures and that a power decrease may occur. This device alone is not suitable for NO_x controls.

The catalytic converter is placed in the exhaust system, near the muffler, and operates at lower temperatures than the thermal reactor. Chemically, a cat-

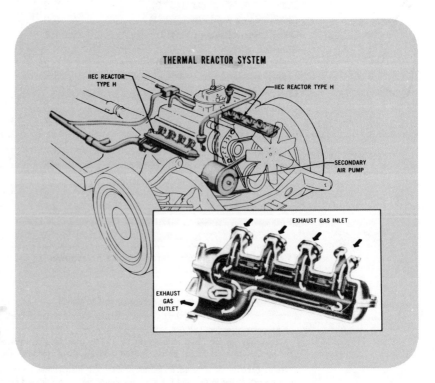

Figure 3.14. A thermal reactor system developed by IIEC. [Courtesy of Ford Motor Co.]

alyst is a substance that speeds up the rate of a chemical reaction but is not consumed in the process. For emissions control, the catalyst should speed the conversion of emissions to less harmful gases. Potential catalysts and details of their use are highly proprietary at present. Possible catalysts could be of platinum or various oxides of aluminum, iron, copper, and so forth, all of which promote the oxidation of CO and HC. Catalysts, which are important in many chemical manufacturing operations, were researched for auto emission use during the late 1950s and early 1960s and found wanting. They made a "comeback" and were installed on autos beginning with 1975 models because United States automakers decided they provided the best approach to reduced exhaust emissions. Apparently no one catalyst is known that will handle CO, HC, and NO_x simultaneously, so the concept of a dual-bed converter has evolved. In this device two catalysts are used. One converts CO and HC into CO_2 and H_2O, and the other converts NO_x into N_2 and O_2 (Figure 3.15).

Several problems may arise with the widespread use of catalytic converters. Research indicates that they may cause an efficient conversion of the small amount of sulfur in gasoline into a sulfuric acid mist that will be emitted through the exhaust. Sulfuric acid concentrations in the air along highways during rush hour might then exceed the levels that aggravate respiratory problems. These difficulties prompted the EPA in 1975 to propose delays in emission standards (see Table 3.7). Emissions of the catalyst metals in the form of aerosols may also create environmental problems. Because lead causes a reduction in the performance of the catalyst, unleaded gas is required for those autos equipped with catalytic converters. The installation cost of the catalytic converter is about $200 per auto.

NO_x emissions are difficult to control. One approach, in addition to the dual bed catalyst approach, is to use exhaust gasrecirculation, or EGR. The exhaust gases are returned to the carburetor for recombustion. This lowers combustion temperatures and reduces NO_x.

Table 3.7 lists the emission standards in recent years. The 1982 standards are those the 1970 Clean Air Act defined to be a 90% reduction of CO and HC below the 1970 emission standards. (The 1970 Act established these for the 1975 model year but Congress has on several occasions approved delays in their implementation.) The 1977 NO_x levels are to be 90% below the measured emissions on 1971 model automobiles. Another important facet of the 1970 act is that the autos must meet the standards of their model year for their "life," defined to be 5 years or 50,000 miles, whichever comes first.

No mass-produced American-manufactured automobile was available in 1974 that met the originally proposed 1975 (now 1982) standards. Prototype systems that meet that standard have been built, but it is a long step to mass production of such systems. Successful operation of these devices for 50,000 miles will probably require more attention to maintenance on the part of the owner.

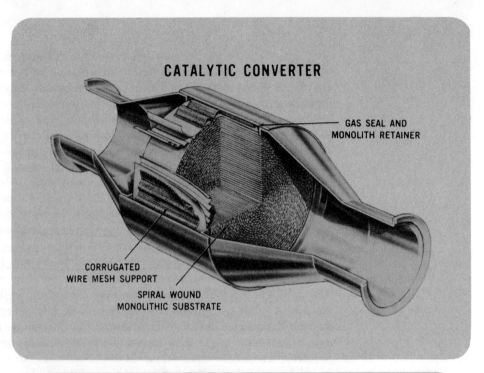

CATALYTIC CONVERTER

GAS SEAL AND
MONOLITH RETAINER

CORRUGATED
WIRE MESH SUPPORT

SPIRAL WOUND
MONOLITHIC SUBSTRATE

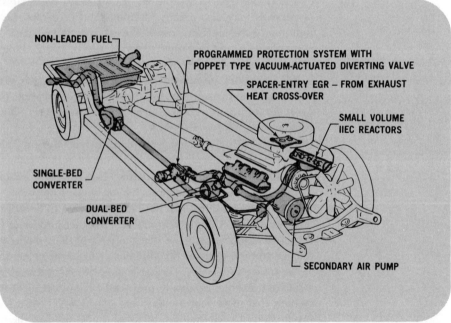

NON-LEADED FUEL

PROGRAMMED PROTECTION SYSTEM WITH
POPPET TYPE VACUUM-ACTUATED DIVERTING VALVE

SPACER-ENTRY EGR – FROM EXHAUST
HEAT CROSS-OVER

SMALL VOLUME
IIEC REACTORS

SINGLE-BED
CONVERTER

DUAL-BED
CONVERTER

SECONDARY AIR PUMP

Figure 3.15. One version of a dual bed catalytic converter system. This system also uses a thermal reactor and an exhaust gas recirculation (EGR) system. [Courtesy of Ford Motor Co.]

130

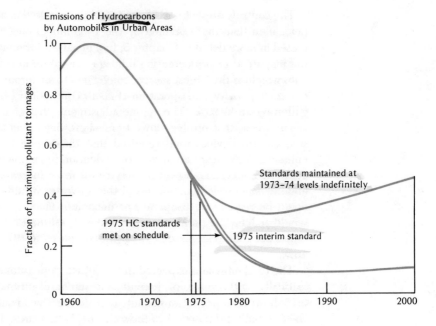

Emissions of Hydrocarbons
by Automobiles in Urban Areas

Fraction of maximum pollutant tonnages

Standards maintained at
1973–74 levels indefinitely

1975 HC standards
met on schedule

1975 interim standard

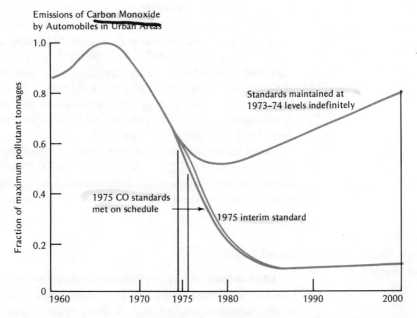

Emissions of Carbon Monoxide
by Automobiles in Urban Areas

Fraction of maximum pollutant tonnages

Standards maintained at
1973–74 levels indefinitely

1975 CO standards
met on schedule

1975 interim standard

Fig. 3.16. *Auto emission projections done in 1973. The 1975 standards referred to in the graphs are those established by the Clean Air Act of 1970. These were subsequently delayed until 1977, then until 1982 by Congress. [From the Environmental Protection Agency, 1973.]*

The controls present on 1973 automobiles resulted in about 9% less mileage per gallon than that obtained by the 1968 model uncontrolled cars. (As discussed in more detail in Chapter 5, fuel penalties for emission controls are not the major source of decreasing mileage per gallon in recent years. Increased auto weight is the largest source; conversion to an automatic transmission may cost a 6% penalty, and operation of an air conditioner can reduce mileage per gallon by up to 20%.) The engine adjustments that give maximum efficiency are not those that produce lowest emissions. Beginning about 1970, the emission control devices used required that the engine be adjusted for lowest emissions; this meant an automatic reduction in engine efficiency. The use of catalytic converters, however, may result in an increase in engine efficiency and mileage per gallon. Success of the converter would mean that the engine could be readjusted to operate at maximum efficiency while the converter would reduce the pollutant level. Industry estimates are that the efficiency (miles per gallon) increase in 1975 cars compared to 1974 models is as much as 15%.

The total emissions expected in the future from automobiles depend rather critically on the values of the emissions standards ultimately assigned. Figure 3.16 shows EPA projections of auto emissions with and without the adoption of the originally proposed 1975 (now the 1977) standards. This figure also shows the improvements since 1968, the year the auto emission controls began to be implemented. It is also obvious from the figure that the total emissions from automobiles will not continue to decrease throughout this century. A minimum in total emissions will be reached during the mid-1980s when the emissions from each vehicle will have reached their "practical" lowest value. Then as the number of cars increases, the total air pollution from automobiles will increase. The difference in total tonnage of pollutants created by autos operating with 1973–74 controls and those operating with proposed 1975 (now the 1977) controls is impressive.

United States auto manufacturers lobbied intensively against the 1975 standards proposal by the 1970 Clean Air Act, and their efforts were a factor in the decision to successively delay these standards until 1982. (The impact of emission controls on the economy and on energy consumption is discussed in Chapters 4 and 5.)

Other Internal Combustion Engines

Diesel Engines

Diesel engines, which are also internal combustion engines, are present in about 1% of United States vehicles (principally trucks) and consume 3% to 5% of the transportation fuel used. A diesel engine uses no carburetor; instead, it compresses only air in the cylinders. The compression heats the air, and diesel fuel (not gasoline) is injected into the cylinder. No spark plug is used; ignition occurs when the fuel meets the hot air. Because combustion occurs with an excess of air, the diesel emits less CO than the gas engine. The diesel emits HC and NO_x values that may approach the values emitted by the gasoline engine. From Table 3.7, it can be seen that the diesel engines used in the Mercedes and Peugeot meet most of the standards proposed for the 1980s.

The diesel has special smoke and odor problems, although a properly maintained engine will not emit the dark plumes often seen. Barium additives in the diesel fuel can reduce smoke emissions by about 50%; three fourths of the barium is exhausted as barium sulfate which is insoluble in water and is apparently harmless to humans, although its extensive use might reveal problems. Diesels have the advantage that they do not emit lead aerosols, since TEL is not present in diesel fuel.

Wankel Engine

Interest in the rotary piston Wankel engine has heightened since its marketing in the Mazda automobile in the United States in the early 1970s. The Wankel, although novel, is still a four-cycle gasoline-fueled, internal combustion engine, so the type—though not necessarily the quantity—of the emissions should be similar to those of other internal combustion engines. The uncontrolled Wankel engine emits more HC and CO but less NO_x than the conventional gas engine. The Mazda Wankel uses a thermal reactor to reduce the emission levels.

The rotary engine is half as heavy and one third the size of the conventional engine. It also has fewer moving parts and runs on lower octane fuel. These advantages plus the expected lower cost of maintenance may make the Wankel more popular in the future. A decided disadvantage is that the efficiency of the Wankel is less (by 20% to 30%) than conventional internal combustion engines. As fuel prices increase, the Wankel may lose out to more efficient engines.

Stratified Charge Engine

A modified internal combustion engine that has appealingly low emissions yet maintains good engine efficiency is the stratified charge engine. In a conventional internal combustion engine, the mixture in the cylinder is homogeneous and rich. The mixture must be rich in order to ignite. Unfortunately, as we saw in the discussion of Figure 3.9, a rich mixture invariably creates large CO and HC emissions. The stratified charge concept modifies the combustion process. One small portion of the cylinder has a rich mixture and the rest of the cylinder is filled with a lean mixture. Combustion is initiated in the rich portion, and the fire spreads into the lean mixture, which has much lower emissions.

The concept dates from the 1920s, but the Japanese automakers have apparently pursued this approach more actively than the United States auto makers. Several technical approaches are used to achieve a small rich portion and a larger lean portion in the engine. Honda of Japan uses a small precombustion chamber for ignition of a rich mixture that is then forced into the cylinder containing the lean mixture. Both Texaco and Ford have developed independent versions of the engine that rely on the injection of fuel into the cylinder with an air swirl that concentrates the fuel near the spark plug leaving a lean mixture in the rest of the cylinder. The Ford version, called PROCO (for Programmed Combustion) meets the 1982 standards when operated under careful test conditions (Table 3.7). However, as Ford is quick to point out, such performance may be difficult to achieve on mass-produced cars operating under varying conditions with only casual maintenance and upkeep.

An advantage the stratified engine possesses over nearly all other approaches to engine emission controls is that no penalty in terms of engine efficiency is expected to be paid for the controls.

Alternatives to the Internal Combustion Engine

Given the emission characteristics of the internal combustion engine, it is natural to inquire whether new fuels could be developed or whether alternate engines could be built that might satisfy transportation needs with less harm to the environment. The fuels options (principally hydrogen and methanol) are discussed in Chapter 5. We shall here first analyze the required properties of alternate engines and then consider how some of the alternatives might perform. Many of the alternatives, the steam car and the electric car, for example, were manufactured in the United States at some time and lost out in the competition to the gasoline engine. As we shall see, it was probably no accident that the gasoline engine won the competition. As a further point, even with the application of modern technology, these alternative engines will probably have a difficult struggle against an engine that has undergone considerable improvement and development over the last 70 years and thus has a considerable economic and practical advantage.

Power and Energy of Motive Sources

It is instructive to discuss engine performance in terms of power and energy. In Chapter 1, we saw that all things possess energy but that only certain forms of energy are useful. We defined power as the energy consumed divided by the time required for consumption. (You may find it helpful to review that portion of Chapter 1 as well as the material at the end of Chapter 1 on units of energy and power.)

An engine must provide sufficient power to move the vehicle forward and to overcome wind resistance and friction. That is, sufficient power must be available to accelerate the car to a given speed and to maintain that speed. In crude terms, power measures how *fast* an engine can move the vehicle. The energy rating of an engine measures how *far* the engine can move the vehicle at a given speed before running out of fuel. Power and energy thus measure quite different characteristics of a motive system.

Power and energy requirements are both proportional to the weight of the car. Suppose a 1-ton (2000 lb) car moves 60 miles at a constant speed of 60 mph. The engine must deliver a certain amount of power and energy to accomplish this. If the weight of the car were doubled, by perhaps adding 1 ton of lead to the back seat, the engine would have to deliver *twice* the energy and *twice* the power to once again move the car the same distance at the same speed. The energy comes from the gasoline. In this example, doubling the weight of the car resulted in a doubling of the energy (the amount of gas) consumed per mile. In other words, doubling the weight of the car cut the gas mileage (miles per gallon) in half.

Power and energy requirements depend on the speed of the car, but we shall concern ourselves only with the energy requirements. If a car travels at a steady speed, a very simple calculation from basic physics shows that the energy required depends upon the weight of the car, its speed, and the distance traveled. The dependence is given in equation 3.16.

$$\text{energy: (weight of car)} \times \text{(speed)} \times \text{(distance traveled)} \qquad (3.16)$$

This equation illustrates the fact, noted previously, that doubling the weight of the car means that the energy required will be doubled if speed and distance remain unchanged. Furthermore, this very simple model predicts that the gas mileage at 70 mph is *half* that at 35 mph because the energy required is directly proportional to the speed of the car. In practice, the dependence of energy and power upon weight, speed, and distance is more complex than we have assumed, but our simple model gives a good approximation to the results found experimentally.

Because engine weight is an important factor in a motive system, it is convenient to classify engines in terms of the power and energy they possess per unit weight. The **specific power** and **specific energy** of an engine are defined as:

$$\text{specific power (watts/lb)} = \frac{\text{power available (watts)}}{\text{engine weight (lb)}} \qquad (3.17)$$

$$\text{specific energy (watt-hr/lb)} = \frac{\text{energy available (watt-hr)}}{\text{engine weight (lb)}} \qquad (3.18)$$

Figure 3.17 shows motive power sources classified according to specific power and specific energy. The figure is based on a typical subcompact car weight of 2000 lb of which 500 is the engine (and the fuel if the engine is not electric). Note that the internal combustion engine and the gas turbine engine have the largest values of both specific power and energy. That is, *per pound* of engine weight, these can provide higher power (speed) for a longer time than the other sources. Fuel cells, an electrical source, have very large specific energy ratings but insufficient power to drive at high speeds. Lead acid batteries have relatively large power characteristics but are capable of only short-range operation because of small specific energy values.

As an example, driving a small car with a weight of 2000 lb up a slight grade at a constant speed of 30 mph for 125 miles would require an engine with a specific power of about 100 watts/lb and a specific energy of about 90 watt-hr/lb. The data in Figure 3.17 indicate that only a few motive sources are

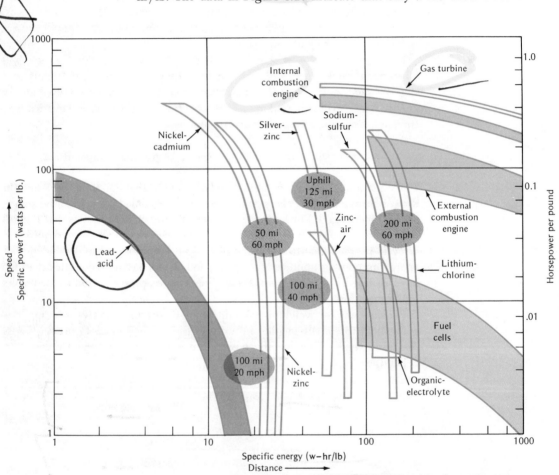

Figure 3.17. *Specific power and specific energy relationships for motive power sources. The area occupied by each source indicates the variation in attributes of different systems of the same type. The data assume a 2000-lb vehicle, a 500-lb motor, and steady driving. The distance and speed numbers in the body of the chart indicate the capacity of engines in that region. [From The Automobile and Air Pollution: A Program for Progress, U.S. Department of Commerce, October 1967.]*

capable of providing this performance. Those with specific power and energy in excess of the requirement are the internal and external combustion engines, the gas turbine engine, and two battery systems, sodium-sulfur and lithium-chlorine. (Both of these batteries are in the research phase.) The distances and speeds indicated in the center of Figure 3.17 indicate the capabilities of engines with given power and energy ratings.

Gas Turbines

Figure 3.17 indicates that gas turbines should be able to compete successfully with internal combustion engines. This engine's reliability has been established in its use in jet airplanes. Gas turbines are continuous combustion engines, the fuel being continuously sprayed into a combustion chamber where air is compressed to several times atmospheric pressure. Ignition creates an expanding gas that blows across turbine (or fan) blades, creating rotary motion. A power shaft connected to the turbine ultimately provides the forward motion by coupling to the rear wheels.

The gas turbine develops high specific power only at high turbine speeds, however. This creates a serious drawback for its use in city "stop and go" driving because the variable speed mechanism required between the turbine and the wheels is very expensive. This disadvantage is less serious for planes buses, and trucks, which tend to travel long distances at high speeds. Low gas mileage at low speeds and the high cost of the engine are further disadvantages for autos. The emissions expected are given in Table 3.7. HC and CO are low because of adequate air during combustion, but NO_x is about the same as for internal combustion engines.

Steam Engines

This external combustion engine has all the emission advantages of complete combustion indicated in equation 3.14, namely low CO and HC. Steam cars such as the Stanley, the White, and the Doble were common in the United States in the early 1900s. The almost universal use of steam engines for locomotives and for farm machinery during that period is a source of nostalgia for those whose childhood embraced that era.

In its simplest form the steam engine might be likened to burning fuel to boil water, then using the escaping steam as a power source—for example, to run a turbine or an airscrew. The steam is condensed, collected and returned to the heated area. In the steam engine fuel combustion takes place in the atmosphere at lower pressure and temperature than in the gasoline engine. As a power source for automobiles, the early steam engines were limited because of size and because of start-up times of several minutes. Also, there were dangers associated with what were, for the times, high pressures and temperatures created with steam. A further problem was that, if water were used as the fluid, it might freeze during cold periods.

The possibility of lower emissions and the advantage of running on lower quality fuel than gasoline have prompted several large-scale efforts on the part of private industry to develop a competitive steam-powered auto. The action has focused on the application of modern engineering technology and space age materials. Several prototype vehicles are in operation.

William P. Lear, an American industrialist who pioneered auto stereo tape players and the executive jet aircraft, formed the Lear Motors Corporation in 1968 to develop a steam power plant for use in modern automobiles. After four years of research costing several million dollars, the company settled on a design utilizing a new fluid, "Learium," as the driving fluid. This fluid is a mixture of an organic fluorocarbon and water. A fluorocarbon is a substituted hydrocarbon in which one or more fluorines substitute for hydrogen in the molecule.

Wallace Minto, an American inventor, has also developed an external combustion engine that uses a Freon as the fluid. Freons are a class of substituted hydrocarbons in which fluorine and chlorine replace various hydrogens. Some freons are used as the fluid in air conditioners. Minto's design has been licensed to Datsun, a Japanese automaker.

Beginning in 1973 the Department of Transportation sponsored in regular passenger service in California three experimental steam-powered buses to determine their performance and environmental characteristics. The bus engines, manufactured by Lear Motor Corporation, by Wm. Broebeck and Associates, and by Steam Power Systems, replaced the diesel engines in conventional buses. All the engines use water as the fluid and burned ∂2 diesel fuel. Transit system employees could tell little difference in bus performance. The early emission characteristics of some of these buses are in Table 3.7.

Electric Vehicles

The electric battery is a common source of electrical energy, and special purpose electric vehicles have been around as long as the gasoline engine. No engine is more attractive from the standpoint of emissions than the battery-electric motor vehicle. The vehicle emissions are nearly zero, although the power plant required to charge the batteries may be a major source of pollution. Furthermore, the electric motor is very quiet. Unfortunately the weakness of the present batteries is evident in Figure 3.17—a low value of specific energy. Ordinary lead acid batteries available today have 10 watt-hr of energy per pound, whereas the internal combustion engine has nearly 1000 watt-hr/lb. This implies that if a gasoline engine-equipped auto can go approximately 300 miles on a tankful of gas at 60 mph, an equivalent sized electric auto with batteries equal in weight to the gasoline engine would go approximately 3 miles at 60 mph. Of course, one could increase the weight of the batteries, but 2000 lb of batteries would be required to achieve a 40-mile range at 60 mph. The electric car is at present limited to short trips but could be used for city commuting where distances are short and the maximum speeds are low because of congestion. Experimental batteries of 100 watt-hr/lb are being tested (Figure 3.17) and 300 watt-hr/lb is suggested for the year 2000. Thus the potential for electric cars might improve.

Some other considerations would be involved if a massive changeover to electric vehicles were undertaken. The electrical power required to move all cars and trucks would have to be provided by electrical power plants; thus, an expansion in generating capacity would be necessary. The actual amount of increase would depend on several conditions. If the total car horsepower of American autos is converted to watt-hours of electricity, it amounts to about 75% of the 1968 United States peak power generating capacity. But that much increase would not be necessary for several reasons. American cars are ridiculously overpowered; on the average an American auto uses its full power capacity less than 1% of its operating time.

Because electric vehicles would be used under conditions such as commuting, where use of the maximum available engine power is often not even possible, the total power conversion mentioned above is not necessary. Furthermore, the batteries would be charged at night when demand for electrical power is low, considerably below the peak capacity. Thus a more realistic figure for the increase in capacity required might be 25%. This increase would, however, be in urban areas, which would feel the brunt of increased power plant emissions, although of course the auto emissions would be considerably reduced. If a conversion to electric cars occurred, there would probably be a change in economic factors such as a tax on electricity as well as on gasoline.

Fuel cells are another source of electricity. They are used in the space program and operate by converting chemical energy into electrical energy. For example, hydrogen gas and oxygen gas are supplied to the separate electrical connections, or electrodes, of a fuel cell. The electrodes are separated by an electrically conductive solution. When hydrogen and oxygen react to form water, electric current flows in the cell. The fuel is supplied continuously. Fuel cells have low specific power and at present their high cost is a limiting feature.

Mass Transit

Mass transit denotes the transportation of large numbers of people along well-defined corridors (usually away from streets) in urban areas. It includes the use of rail systems (trains, trolleys, subways) and buses. For urban environments, it offers many advantages over our present dependence on the automobile. Even with the present low ridership, mass transit offers air pollution emissions per passenger mile that are several times less than autos. In addition to reducing air pollution, the fuel consumption per passenger mile is one third or less than that of cars, so energy savings are substantial. (This is discussed further in Chapter 5.)

Furthermore, extensive use of mass transit in urban systems would reduce noise and street congestion. Major United States cities typically devote 40% of their land area to the automobile in the form of roads, parking, and service stations. A trend to mass transit would afford the opportunity to utilize this land in other, more aesthetically pleasing ways. In addition, the use of mass transit might stimulate the revitalization of the downtown shopping and living areas, whose deterioration is in large part due to the congestion, noise, and fumes of the auto.

Mass transit in the United States is at present an anachronism, a nineteenth-century scheme that has only been patched and repaired in this century. The San Francisco Bay Area Rapid Transit System (BART), partially completed in 1972, is the first major new mass transit system in the United States since 1907. For the most part, mass transit has not benefited from modern technology, and the existing systems are not conducive to increased ridership, although cities such as New York rely heavily on them.

Since World War II, a dramatic shift in passenger use from mass transit to cars has occurred. Although the total passenger miles has more than doubled since 1950, the percentage carried by mass transit has decreased by one fifth.

The auto has become a dominant factor in American life, in both transportation and economics. The automotive industry accounted for two thirds of the rubber, over half the lead, over one third of the zinc, and over one fifth of the steel consumed in the United States in 1965. Several factors have made the auto dominant over mass transit. Under present conditions of United States mass transit, the auto offers more flexibility. Convenience and time saving are often cited as automotive advantages, but these do not apply in many congested urban areas. (A horse-drawn vehicle could cross Manhattan Island in 1900 at a faster average rate than could autos in 1973).

The auto has thrived on enormous economic support relative to mass transit over the last 25 years. A major factor has been the heavy funding of highway construction, such as the Federal Interstate Highway system, through the Highway Trust Fund. This fund, which derives its monies from taxes on gasoline, tires, and auto accessories, provides about $5 billion per year for construction. The fund was created to finance the Interstate system in 1955, and the political and economic forces in favor of highways are significant and effective: the construction industry, petroleum industry, cement industry, auto industry, rubber industry, real estate interests, and chambers of commerce successfully promoted this (and other) highway programs. The Highway Trust Fund provides 90% of the funds for construction, leaving only 10% to be provided by local sources.

The particular significance of the trust fund concept is that these taxes do not have to compete in the United States budget with other needs, such as defense or health, because the highway use taxes can be used only for highway construction. It is a unique taxation scheme. On the other hand, funds for mass

TABLE 3.8. A typical federal transportation budget research funds for 1972 and earlier years, in percent

Highways	60
Air	21
Water	14
Mass transit	3
Railroads	2

transit have in the past had to compete in the budget with other needs, the amounts available were considerably less, and local sources had to provide half the funding. The railroads, meanwhile, found freight service more lucrative than passenger service, and the resulting decline in passenger facilities and service led to rider apathy. Mass transit is often a municipal service and therefore has not been sparked toward improvement by economic competition.

The federal transportation budget in the past has been distributed in approximately the manner shown in Table 3.8. Inasmuch as highways have for years received annually twenty times as much as mass transit and thirty times as much as railroads it is hardly surprising that we are a nation poor in mass transit.

Until 1973 the highway interests lobby successfully prevented the use of the Highway Trust Fund for any purpose other than highway construction. The 1973 highway bill included a timetable for the use of some of these funds for other transportation purposes including, by 1976, mass transit.

Various arguments have been used to justify the extensive road building program in the United States as initially proposed in the mid 1940s and early 1950s; the interstate system was to connect the perimeters of cities, but not enter their downtown areas. This approach was altered until the version passed by Congress included the inner city as well, and the 90/10 funding scheme was an inducement to extend the highways into city centers. Defense needs were also used to justify construction after World War II, although only a small amount of military traffic used highways during that war—most went by train. Any ideas that the system could be used by massive numbers of urban dwellers in an emergency can be dispelled by observing the morning traffic rush in any U.S. city. The safety of the system over other highways is also promoted. Although the interstates may be safer than other highways, if safety were the dominant concern, then mass transit would be preferable to the increased use of automobiles.

The fact that completed urban freeways are jammed upon completion is not necessarily an indication of prior "demand." Experience indicates that expressways encourage more auto traffic. Building urban highways that will not fill up may be a practical impossibility, and arguments stressing the use by motorists as an indication of "demand" are of little logical value unless the commuters are doing this even while offered the alternative of a modern mass transit system.

The realization that we now have an excellent highway system between cities, coupled with the problems created by highways in urban areas, has prompted a new objectivity with respect to federal highways in some cities. San Francisco and Boston have both canceled portions of urban highways in various stages of completion. In the past these and other cities have been eager for federal highway funds.

In view of our air pollution problem, our energy problem (Chapter 4), and the deterioration of our cities, more economic and planning emphasis should be placed on mass transit.

Requirements for Modern Mass Transit

A useful system should be convenient, economical, clean, safe, and quiet. The system would not be expected to replace all functions of the motor vehicle but should offer sufficient advantages to compete for ridership. Pricing policies are important, and some cities are introducing free mass transit in an effort to boost ridership. Such a plan reduces operating costs, but revenue stops. This approach considers mass transit as a public, tax-supported service, similar to police and fire protection. Seattle has recently initiated such a program. The use of mass transit systems can be encouraged through administrative procedures which make downtown auto use expensive or inconvenient, such as limited parking or use of some streets for pedestrians rather than autos.

Buses. In urban use, buses often have one fundamental limitation: a lack of speed due to auto congestion. This can be partially overcome by several techniques. An exclusive lane for buses, to improve transit time, has been used. Such a system on Shirley Highway in the Washington, D.C., metropolitan area provides a travel time of 14 minutes compared to a 37-minute auto trip. The market share of bus passengers has increased from 20% of all passengers (before express lanes) to over 50% at the current time. Over one third of the present users previously used cars. Another approach, where computer-controlled traffic lights are used, is to optimize bus flow patterns by, for example, holding lights green slightly longer when buses are detected approaching an intersection.

A home-to-work subscription bus service in Peoria, Illinois, designed to pick up workers at their home each morning, bring them to a large industrial plant, and return them at night was able to attract 28% of the potential riders. Over 70% of the bus riders had previously used cars to get to work.

Figure 3.18. *BART is a modern high speed rail system. This portion of the track is in Union City, California. [Courtesy Bay Area Rapid Transit District.]*

Rail Systems. A modern rail system utilizing space age technology is a pleasant surprise for those familiar with most of the present noisy, rough, poorly lighted trains in the United States. BART in the San Francisco Bay area is a prime example. Begun in 1951, the system used almost no federal funds. BART is an updated version of rail technology. The cars are carpeted, air conditioned, well illuminated, and quiet. Including stops, the train averages 50 mph and runs every 90 sec during peak periods. Fares are charged on a mileage basis, using computerized tickets. BART is 75 miles long, and its builders were motivated by the profits expected from the rejuvenated city that might result. As the only major new mass transit system in the United States since 1907, BART's impact is being carefully studied. In its first years of operation, it has had problems. The principal difficulties are centered in the computerized control system which has not functioned properly. A driver is required on each train because of concern about inadequacies of the control system.

BART had a major impact on the Bay area even before operation began. It was a partial contributor to changes in building codes in the 1950s and 1960s and to a subsequent building boom including high rise office buildings. The nature and character of San Francisco were changed and part of this is due to BART.

Thus mass transit systems, as well as highways, can have a major impact on cities.

Some of the more common controls for major pollutants in stationary sources and industrial plants are shown in Figure 3.19. A few standard approaches to air emission control in power plants exist. The major characteristics of power plant emissions are determined by the type of fuel used. In industrial air pollution problems, however, a wide variety of controls is necessary because the emissions have an industry-to-industry (almost a plant-to-plant) variation.

Control of Particulate Emissions

Rainfall and gravity are the primary mechanisms for natural removal of particulates from the air. Particle size has a drastic effect on the efficiency of these processes, however, because small aerosols settle very slowly.

Man-made emissions are reduced by two approaches: reducing particle formation, or, failing this, removal after formation. Prevention is obviously the preferred approach when possible. Sometimes this involves the use in power plants of low fly ash coal, a type of coal that burns with a reduced particle output. Particulate prevention techniques are limited, however, and a variety of processes and devices known as stack controls are used to remove particulates from effluent gases. All of these methods lose efficiency as particle size decreases. Few remove small aerosols (diameter less than a few tenths of a micron). The application of these methods to specific stationary fuel sources is discussed later in the chapter.

Electrostatic Precipitator. Frederick Cottrell invented one of the most famous and widely used particle removers, the electrostatic precipitator. Cottrell's device takes advantage of the fact that particles carry, or can be given, a small amount of positive or negative electric charge. A high voltage (up to 100,000 volts) created between metal plates in the stack is then used to attract the charged particles to the metal plates by electrostatic attraction. Removal of the particles is accomplished by washing the plates or by knocking the particles loose. The precipitator removes from 95% to 99.5% of the visible particles (Figure 3.20). Cottrell established the Research Corporation with funds from his invention. This organization provides research grants to educational and nonprofit institutions for a variety of projects, especially those of young scientists just beginning their careers.

Cyclones. Cyclone separators create a whirlwind inside the stack. The larger particles are thrown against the side and fall to the bottom to be collected. Cleaner air flows upward through the center tube. Cyclones are relatively inexpensive to buy and operate, but they are efficient only for larger particles.

Wet Scrubbers. These devices clean the gas by wetting with a shower of water or aqueous solution, which removes particles as well as chemically

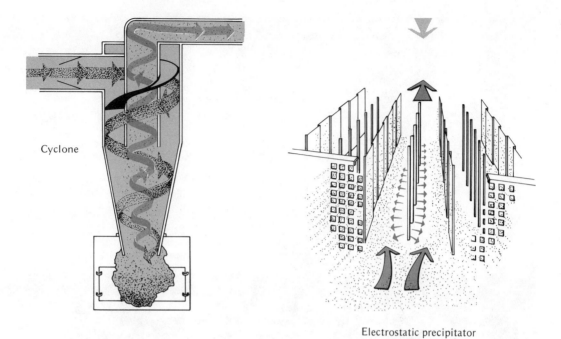

Cyclone

Electrostatic precipitator

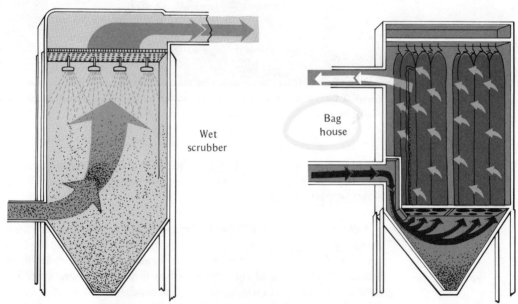

Wet
scrubber

Bag
house

Figure 3.19. *Stationary source and industrial air pollution control devices. [Courtesy of Manufacturing Chemists Association.]*

Figure 3.20.   *Operation of a power plant with and without an electrostatic precipitator.* [Courtesy of Eastman Kodak Company.]

reactive gases such as hydrochloric acid. Scrubbers capture particles smaller than those removed by cyclones but are more costly to operate than the other forms of particle removers.

Filters. Particles can be removed by filters in stacks, just as they are removed by filters in vacuum cleaners and air conditioners. However, the filter system must be designed for a specific plant. Consideration must be given to the volume and velocity of the gas as well as to its temperature, chemical reactivity, and clogging characteristics. One widely used technique is the bag house, an arrangement comparable to a series of giant vacuum cleaner bags.

Fig. 3.20.   *Continued.*

The air passes through the bags and the particles are removed (Figure 3.19). Installation costs of bag houses are high; operating expenses vary with the type of gases treated and their volume.

Control of Sulfur Oxides Emissions

Because of the drastic impact of SO_2, more methods have been investigated for its control than for any other pollutant. Despite this emphasis, the extent of control is often poor because the SO_x is often present in low concentration in the stack. Removing a dilute pollutant from a high volume gas flow is costly.

Use of low sulfur fuels has been the frequent initial choice of emission control techniques. The sulfur content in domestic coal fields varies from less than 1% to nearly 6%. Supplies of low sulfur coal are somewhat limited and often expensive because of their distance from the combustion site.

Removal of sulfur from coal can be accomplished, but the economics of the processes has prevented large scale efforts. A promising approach is coal gasification—conversion to a completely gaseous form, mostly methane (CH_4), with sulfur compounds being removed during the formation of the gas. If this could be accomplished on a massive scale, it would also extend the world natural gas reserves; consequently much research is presently concentrated on coal gasification (see Chapter 5).

Removal of SO_2 from effluent gases may in principal be accomplished by several chemical reactions. Two of the most important are the conversion to sulfuric acid (H_2SO_4) and the conversion to gypsum (calcium sulfate, $CaSO_4$).

Conversion of SO_2 to commercial grade H_2SO_4 is an economically attractive procedure because sulfuric acid is a valuable industrial chemical. This technique is effective only with gas streams containing very high concentrations (greater than 3.5 vol %) of SO_2. SO_2, O_2, and H_2O are combined to give sulfuric acid. This process is useful in some smelter operations. Coal-burning power plant emissions are often too dilute in SO_2 to use this procedure. Instead, the effluent gases are passed over pulverized limestone (calcium carbonate) or dolomite (calcium and magnesium carbonate) and the SO_2 is converted to gypsum (calcium sulfate) with a release of CO_2.

$$2 \text{ CaCO}_3 + 2 \text{ SO}_2 + \text{O}_2 \rightarrow 2 \text{ CaSO}_4 + 2 \text{ CO}_2 \qquad (3.12)$$
$$\text{limestone} \qquad\qquad\qquad \text{gypsum}$$

This process creates a difficulty, however: what to do with the gypsum. A typical large coal plant serving 100,000 people emits 500 tons of sulfur oxides per day. Application of the limestone process would yield 1000 tons per day of gypsum that must be disposed of! Gypsum is not water soluble, so the disposal problems are not minor.

In urban areas, other approaches to the reduction of sulfur oxides have been used. In downtown London in the 1950s, the sulfur oxides came principally from the domestic use of coal and only secondarily from power plants. To alleviate this problem, the "London Laws," as they came to be called, were formulated. Only low sulfur coal was allowed to be used in homes. This had an immediate effect on the street level SO_2 concentration. Power plants then used the higher sulfur content coal, but they were required to install tall stacks (several hundred feet) which served to dilute the SO_2 around the countryside and ultimately reduce the concentration at ground level.

Nitrogen Oxides Controls

As in the automobile, NO_x control is an extremely difficult problem. Emission reduction often emphasizes a preventative approach, that is, avoiding the formation of NO_x. This involves carrying on combustion at the lowest possible temperature and the lowest excess oxygen consistent with complete oxidation. This process is not always compatible with power generation and industrial operations because, by the second law of thermodynamics, lower temperatures mean lower efficiency. Two approaches often used are "staged combustion" and "flue gas recirculation." Staged combustion requires a combustion chamber design in which both high and low temperature combustion occur. Most of the fuel is initially burned in the high temperature stage with insufficient oxygen. The products (mostly HC, excess fuel, and CO) move to the cooler region where the remainder of the fuel is burned with excess air. The cooler temperatures in this region result in lower NO_x formation. Flue recirculation involves the return of stack gases to the combustion chamber. This results in a lowered flame temperature and reduced emissions.

A combination of these two techniques can lower NO_x emissions by 50% to 80%.

Summary

Most, though not all, of the air pollutants produced by man occur naturally and are integrated into the biogeochemical cycles which serve as a purification scheme. Although nature dwarfs man's output on a global basis, man dominates emissions in urban areas and overloads the capacity of the natural cycles. Air pollution causes health problems, destroys materials, and wastes resources. The costs to the public are comparable in size to the costs of air emission controls.

The automobile creates 50% of the total tonnage of air pollution, power plants are responsible for about 21%, and industry for about 17%. Total tonnage is not a good measure of impact; the effects of individual pollutants must be weighed. Carbon monoxide is produced in combustion. Its toxic properties arise from the fact that it can replace oxygen in hemoglobin in the bloodstream. Sulfur oxides, produced predominantly by fuel burning in stationary sources, cause severe respiratory problems, destroy vegetation, and weaken building materials. Particulates and aerosols, created in a variety of sizes by industry (50%) and power plants (24%), can be transported for hundreds of miles and deposited by rain. Particulates are responsible for haze, the most visible form of air pollution. Because many chemicals can form aerosols, the chemical action of particulates is varied; they often intensify the effects of other pollutants. Asbestos dust induces symptoms of respiratory ailments and lung cancer. Nitrogen oxides are produced during combustion when nitrogen in the air combines with oxygen; they cause vegetation damage and are the trigger for photochemical smog formation. Over half of the hydrocarbons are created by the automobile. Some hydrocarbons, such as benzo(a)pyrene, are carcinogenic and constitute a health hazard. Other hydrocarbons are important intermediates in the formation of smog. Photochemical oxidants, such as ozone, cause lung and eye irritation, materials damage, and vegetation damage.

Meteorological conditions, in particular thermal inversions, can increase the severity of air pollution problems by preventing the natural upward drafts of air. The result is a stagnant condition that holds the pollutants near the surface of the earth.

Man's increasing emissions might influence future temperatures and climates by affecting the heat balance of the earth. Carbon dioxide absorbs infrared radiation leading to the "greenhouse effect" and perhaps higher temperatures. Particulates lead to cooler temperatures on earth by reflecting incoming solar radition. Both effects may be offset, however, by small changes in cloud formation; this mechanism might occur naturally and counteract man's impact.

Photochemical smog formation is a complex series of chemical reactions. It is known that the reaction is triggered by the absorption of solar radiation by nitrogen dioxide, that hydrocarbons are involved, and that the end products include peroxyacyl nitrates, known eye irritants. Smog aggravates asthma, emphysema, and other respiratory diseases.

The nature of the internal combustion engine, which operates using incomplete combustion, dictates that the engine must always create carbon monoxide, hydrocarbons, and nitrogen oxides during operation. These can be reduced by a variety of emission controls including catalysts which convert emissions into the less harmful compounds carbon dioxide, water, and nitrogen. The stratified charge concept applied to the gasoline engine might make significant reductions in its emissions. Alternatives to the internal combustion gasoline engine might also produce relief from air pollution. Diesel engines used in automobiles have emissions lower than those of gasoline engines. Power and energy limitations mean that electric cars are probably most suitable for short trips, such as commuting. Conversion to electric cars would mean an increased load on power stations and would result in an increase in air pollution from that source. Steam cars and gas turbine engines have energy and power characteristics comparable to those of the internal combustion gasoline engine. Modern mass transit offers energy savings, noise reduction, and less congestion than an auto-based transportation system. Funding of highways through the Highway Trust Fund has been so dominant in the last 25 years that mass transit has not received the benefits of technological advances that are available for its improvement.

A variety of pollution controls are used to limit emissions from stationary sources. Filters, wet scrubbers, cyclones, and electrostatic precipitators can remove the larger particles. Sulfur oxide emissions can be reduced by burning low sulfur fuels or, in some instances, by chemical reactions in the stack. The latter are expensive and often create the additional problem of the disposal of solid sulfates.

Nitrogen oxide control is very difficult. Methods have focused on reducing the combustion temperature, sometimes by recycling exhaust gases into the combustion area.

Review and Study Questions

1. What is an air pollutant?
2. Explain the relationships between the biogeochemical cycles, nature's output of chemicals into the air, and man's output of air pollution.
3. What are the major air pollutants? List the two largest man-made sources of each (as classified in this chapter).
4. On the basis of *toxicity,* which pollutant accounts for the major impact of air pollution in the United States?
5. Why has global concentration of CO remained fixed, although man produces enough to double the global concentrations every 4 years?
6. Why are nitrogen oxide emissions referred to as NO_x?
7. Man's sulfur pollutant emissions constitute what fraction of the naturally occurring sulfur emissions?
8. What are the principal environmental effects of SO_x?
9. What are aerosols? In what respects, if any, do they differ from particles?
10. What is the environmental importance of NO_2?
11. What is a carcinogen? Give an example.
12. What hampers the identification of substances that are responsible for causing cancer in humans?
13. What are photochemical oxidants?
14. What air layer is nearest the earth?
15. How does temperature vary with altitude in the presence of a thermal inversion located 1500 ft above the surface of the earth?
16. Explain how the earth maintains an approximately constant average temperature.
17. What might act to counteract the effect of man's emissions and thus maintain a relatively constant temperature on earth?
18. What substances are responsible for most of the impact of London smog? Of photochemical smog?
19. Why is incomplete combustion an important process in the environment?
20. How does the air/fuel ratio in the engine affect automobile air pollutant emissions?

21. Why is TEL present in gasoline?
22. What is used to reduce hydrocarbon evaporation emissions from automobiles?
23. What is a catalyst?
24. How do Wankel engine emissions compare with those from conventional internal combustion engines?
25. How does the combustion in a stratified charge engine differ from that in a conventional internal combustion engine?
26. Distinguish between power and energy in reference to motive sources.
27. What is the potential of the gas turbine engine for automobiles?
28. How do steam engine air emissions differ from those of gasoline engines? What is the basis for any difference?
29. What major mass transit systems have been designed in the U.S. during this century?
30. What are the unique features of the Highway Trust Fund?
31. What emission control would be best suited to an industrial plant that emitted sulfuric acid fumes and solid particulates?
32. Give one chemical process that can remove SO_2 after combustion.

Questions Requiring Outside Reading

33. What are some quantitative relationships between health, air pollution, and dollars? (L. B. Lave and E. P. Seskin, "Air Pollution and Human Health," *Science* **169** [August 21, 1970], 723.)
34. Carbon monoxide has several other suspected effects in addition to those discussed in this chapter. What are they? (L. D. Bodkin, "Carbon Monoxide and Smog," *Environment* **12** [May 1970], 4.)
35. What are the essentials of the sulfur cycle; how does nature's production compare to that of man? (W. W. Kellogg, R. D. Cadle, E. R. Allen, A. L. Lazarus and E. A. Martell, *Science* **175** [February 11, 1972], 587.)
36. What aspects of molecular structure affect the odor a compound presents in the human olfactory system? (J. E. Moore, J. W. Johnston, Jr., and M. Rubin, "The Stereochemical Theory of Odor," *Scientific American* [February 1964].)
37. Is the particulate and aerosol level higher inside than outside the home? (V. J. Schaefer, V. A. Mohnen, and V. R. Viers, "Air Quality of American Homes," *Science* **175** [January 14, 1972], 173.)
38. Can severe air pollution problems exist on a regional (several state) basis? (Virginia Brodine, "Episode 104," *Environment* **13** [January 1971], 2.)
39. What is the attitude of industry regarding the regulation of asbestos emissions? (Paul Brodeur, "Annals of Industry," *New Yorker*, October 29, November 5, 12, 19, and 26, 1973. Bruce Porter, "An Asbestos Town Struggles with a Killer" *Saturday Review of the Society*. March 1973, 26.)

40. LaPorte, Indiana, is downwind from the industrial centers of Chicago and Gary, Indiana. Do the industrial air pollutants affect the weather of La-Porte? (S. A. Changnon, Jr., "The LaPorte Weather Anomaly—Fact or Fiction?," *Bulletin of the American Meteorological Society* 49 [1968], 4.)

41. What evidence is cited to estimate the temperature changes that occurred to initiate the onset of glacial periods? (R. A. Bryson, "A Perspective on Climatic Change," *Science* 184 [May 17, 1974], 753.)

42. What diseases are linked to cigarette smoking and on what data are these based? (A. Ochsmer, "The Health Menace of Tobacco," *American Scientist* 59 [March–April 1971], 246.)

43. Are auto emissions in grams/mile from small cars, such as Volkswagens, smaller than those from larger cars? (W. S. Craig, "Not a Question of Size," *Environment* 12 [June 1970]. 2.)

44. How do motorcycle emissions in the U.S. compare (in total tonnage) with those from autos? (*Federal Register* 39, No. 12 [January 17, 1974], 2108–21.)

45. How would a conversion to electric cars and trains affect the oil industry? (Bruce Netschert, "The Economic Impact of Electric Vehicles: A Scenario," *Bulletin of the Atomic Scientists* 26 [May 1970], 29.)

46. One alternative to the internal combustion engine is a partially electric "flywheel car." What is it; what are its prospects? (R. F. Post and S. F. Post, "Flywheels," *Scientific American* 12 [December, 1973], 17.)

47. What lead levels have been found near zoos in which animals have been reported to suffer from lead poisoning? (R. J. Bazell, "Lead Poisoning: Zoo Animals May Be the First Victims," *Science* 173 [July 9, 1971], 130.)

Suggested Outside Reading

"The Biosphere," *Scientific American* (September 1970). The entire issue is devoted to a discussion of the biogeochemical cycles and the impact of man through his food, energy, and materials production.

BRODINE, VIRGINIA. *Air Pollution.* Harcourt Brace Jovanovich, New York, 1973. A general treatment of the problems based on a series by the author in *Environment.* Health aspects are discussed in "Point of Damage," reprinted from the May 1972 issue of *Environment.*

BROECKER, W. S. "Man's Oxygen Reserves," *Science* 168 (June 26, 1970), 1537. Discusses the abundance of oxygen and changes that might occur due to activities of man. Many other problems will arise before we face an oxygen shortage.

CHOW, T. J., AND J. L. EARL. "Lead Aerosols in the Atmosphere: Increasing Concentrations," *Science* 169 (August 7, 1970), 577. Gives data on lead aerosol concentration in urban areas, in rural areas, and in noninhabited areas of the U.S.

"A Citizen's Guide to Clean Air," January 1972. Available from the Conservation Foundation, 1717 Massachusetts Avenue, N.W., Washington, D.C. 20036. A general discussion of air pollution problems and the implications of the 1970 Clean Air Act.

DEMAREE, ALLAN T. "Cars and Cities on a Collision Course," *Fortune,* February 1970, p. 124. Analyzes the change cars have wrought on cities and discusses some alternatives to present practices.

EDEL, M. "Autos, Energy and Air Pollution," *Environment* 15 (October 1973), 10. A look at how the auto achieved prominence in the cities. The energy demand of automobiles relative to other forms of transportation is discussed.

Environmental Quality. Annual Reports of Council on Environmental Quality, 1970–, U.S. Superintendent of Documents. Useful discussions of environmental problems, the programs designed to alleviate the problems, and the success of the programs.

LEAVITT, HELEN. *Super Highway—Super Hoax.* Ballantine Books, New York, 1970. A scathing analysis of the history of the Interstate Highway system. The book traces the history of the legislative hearings that established the fund, emphasizing changes in philosophy that occurred.

MAUGH, T. H., II. "Carbon Monoxide: Natural Sources Dwarf Man's Output," *Science* 177 (July 28, 1972), 338. A discussion of the carbon monoxide balance on earth.

NEWMAN, JOHN. "A Ride for Everyone," *Environment* 16 (June 1974), 11. A proposal for personalized mass transit using existing, but presently uncoordinated, forms of transportation. The author claims this scheme is cheaper than an auto-based system.

ROOS, DANIEL. "Doorstep Transit," *Environment* 16 (June 1974), 19. An analysis of "Dial-A-Ride" personalized mass transit.

Study of Critical Environmental Problems: Man's Impact on the Global Environment. MIT Press, Cambridge, Mass., 1970. An excellent discussion of some of the important problems in man-environment relationships.

WOUK, VICTOR. "Electric Cars: The Battery Problem," *Bulletin of the Atomic Scientists* (April 1971), 26. An analysis of the advantages and limitations of electric vehicles.

ZWIRLING, S. "BART," *Environment* 15 (December 1973), 14. The impact of BART on the skyline and the congestion of San Francisco.

4 Energy

CON·SUME . . . *v.t.* 1. to destroy. . . 2. to use up; spend wastefully; squander. . . *v.i.* to waste or burn away.

Energy is the ultimate currency of civilization.
Athelstan Spilhaus

Introduction

Man's existence is completely dependent on energy. Energy from food is obviously the single most important necessity for life, but if one looks beyond this to other factors in human existence, the second most important contributor is undoubtedly energy in its other forms as utilized to increase food production and to provide heating, cooling, electricity, and transportation. Man's pursuit of energy and energy sources has permitted civilizations to flourish and expand (and war with each other) and has thus profoundly affected the course of history. A significant characteristic of the growth of energy consumption in the developed nations is that it always tends to increase several times faster than the growth of the population. Consequently, as the total population has increased, the energy doubling times (see Chapter 2) have become progressively shorter. As noted in Chapter 2, short doubling times (10 to 20 years for example), if maintained for several generations, have serious consequences in terms of resource depletion. More than half of man's energy consumption has occurred in the last 25 years. The United States, for example, is expected to consume more energy between 1970 and the year 2000 than it consumed in its entire history prior to 1970. Over this span, the United States annual consumption is expected to double, and the world annual demand is expected to triple.

Providing sufficient energy to satisfy this almost insatiable use is not done without a price. Finding, obtaining, processing, and transporting increasingly larger amounts of fuel extract a toll from the environment in terms of strip mining or drilling efforts and the energy expended in processing and transporting the fuel. Environmental disruption is inevitable when energy is provided. Furthermore, as discussed in the preceding chapter, over three quarters of man-made air pollution is created by a need for power. Water pollution is created by the mining and transportation of man's energy sources, especially oil and coal. Almost all energy conversion schemes inevitably discharge heat to the environment, creating changes in the surrounding ecosystem.

Energy use, energy growth, and energy policies embrace crucial environmental concerns, economic considerations, and national security decisions. We shall now examine the use of energy, its role in the future, and some alternatives to present practices and policies.

155

A Primer on Energy and Power

In Chapter 1 we discussed energy, work, power, and thermodynamics. We will review these briefly here. For a more detailed review, restudy those portions of Chapter 1 devoted to these topics.

Energy may exist in many different forms, but two general classifications are made: potential energy and kinetic energy. **Potential energy** is energy held in abeyance until its release is desired. A weight poised above the ground, the water behind a dam, and the chemical energy of fuels are examples of potential energy. **Kinetic energy** is the energy of motion; it is difficult to store and must be used when available.

Work is energy that accomplishes some useful purpose. The scientific definition of work includes the requirement of the movement of an object or objects. As we saw in Chapter 1, the conversion of energy from one form to another is possible and in fact is often necessary if work is to be obtained from an energy source. In nearly all types of energy conversion, the conversion efficiencies are less than 100% by the second law of thermodynamics. That is, some of the initial form of energy is converted to heat during conversion.

Depending on the application, power may be defined as the energy consumed per second or the useful energy produced per second. Thus, energy and power are related as

$$\text{energy} = \text{power} \times \text{time} \tag{4.1}$$

$$\text{power} = \frac{\text{energy flow or work done}}{\text{time required}} \tag{4.2}$$

Many units may be used to measure energy. The Btu (British thermal unit) is the amount of heat required to raise the temperature of 1 lb of water 1°F, and the cal (defined calorie) is the amount of heat required to raise the temperature of 1 g of water 1°C; 1 Btu is equivalent to 252 cal.

Since power is energy divided by time, power may be measured in Btu per second, Btu per hour, calories per second, etc., but a more commonly used measure of power is the watt, commonly tabulated as thousands of watts or kilowatts (kw) (see Table 1.6 for a review of metric units and prefixes). One kilo-watt of power corresponds to an energy flow of 3412 Btu/hr. Because of the prevalence of the watt and the kilowatt as power units, it has become common to measure energy, especially electrical energy, in terms of kilowatt-hours (kw-hr). This is an energy unit because it corresponds to power multiplied by time as in equation 4.1. The numerical relationships between the units are given in Table 4.1.

To develop some intuitive feeling for the sizes of these units, let us consider some examples. A typical American daily food intake is about 3000 kcal

TABLE 4.1a. Energy and power units and conversions

ENERGY

1 British thermal unit (Btu) = 252 calories (cal) = 0.000293 kilowatt-hour (kw-hr)
1 kw-hr = 3412 Btu = 859,800 cal
Quadrillion Btu = Q = 1,000,000,000,000,000 Btu = 10^{15} Btu

POWER

1 kw = 3,412 Btu/hr = 859,800 cal/hr
1 kw = 1.341 horsepower (hp)

TABLE 4.1b. Energy equivalents (very approximate)

Source	Energy 10^6 Btu
1 gal gasoline	0.126
1 ton bituminous coal	25.
1 ton lignite coal	13.4
1 thousand cubic feet (ft³) natural gas	1.03
1 barrel (bbl) oil (42 gal)	5.8
1 lb uranium oxide (U_3O_8-fission)	200.
1 lb uranium oxide (U_3O_8-breeder)	20,000.

SOME OTHER APPROXIMATE RELATIONSHIPS

1 cord seasoned hardwood = 1 ton coal = 200 gal fuel oil

Source: United States Office of Emergency Preparedness

(12,000 Btu) per day. Heating a cup of water from room temperature to the boiling point requires 12,600 cal (50 Btu or 0.0147 kw-hr). A man can accomplish about 430,000 cal (1700 Btu or 0.5 kw-hr) of work (useful energy) in an 8-hr day. Table 4.2 illustrates the number of hours a man would have to work to generate an amount of energy equivalent to that consumed in some common processes and devices.

A man would have to work nearly 2 hr to generate sufficient energy to light a 100-watt bulb for 1 hr. He would have to work 10 hr to provide the energy to make one beverage bottle, and over 6 days to provide the energy to operate his air conditioner for 1 hr! A gallon of gasoline possesses 126,000 Btu, and Hoover dam generates 100 billion (1×10^9) Btu daily. The energy consumption in the United States in 1970 was 67 quadrillion (1×10^{15}) Btu. (To avoid carrying so many zeroes when discussing quadrillion (1,000,000,000,000,000), it is customary to refer to it as Q or to use scientific notation, 10^{15}, which means the numeral 1 followed by 15 zeroes; see Table 4.1.)

A feeling for power units can be illustrated by the fact that lifting a book from the floor to a desk top requires about 1 watt. Running at top speed up a

TABLE 4.2. Relative energy consumption by common devices

Device	Number of Work Hours Required for a Man to Generate an Equivalent Amount of Energy[a]
100-watt light bulb burning 1 hr	1.7
Color TV on for 1 hr	3.5
Manufacture of one pop bottle	10.
Total electric energy in home (1 hr, no air conditioning)	24.
Room air conditioner (operate 1 hr)	50.
Electric dryer (operate 1 hr)	70.
Small lawn mower (operate 1 hr)	160.
Motor bike (operate 1 hr)	400.
Riding lawn mower (operate 1 hr)	650.
Small car, 45 mph for 1 hr	1000.
Large car, 70 mph for 1 hr	4500.

[a] One man does about .06 kw-hr of work in 1 hr.

Source: Citizens Workshop Handbook, U.S. Atomic Energy Commission, 1973.

flight of stairs requires about 1 kw. This is in contrast to the human heart, which averages only 0.01 watts output. The burning of 3000 kcal of food every 24 hr in the human body means it has an average power rating of 150 watts, about equal to that of a light bulb.

Some nonhuman sources of power are those in a bird's song (0.0001 watts), a flashlight (4 or 5 watts), an automobile engine (200,000 watts), and a train engine (10 million watts).

In this chapter we shall concentrate on energy use and energy resources. One important aspect of this is that in the past, energy has been cheap. For example, in most United States cities one can purchase from a utility company for less than 2¢ an amount of electrical energy equivalent to the work one man can do in 8 hr. It is not surprising that we waste energy when it is this inexpensive.

Calculating Energy Available from Fuels

A competent scientist can use thermodynamics to calculate the *maximum* energy available (usually as chemical energy or heat) in a given fuel. This is calculated per unit weight, as, for example, the amount of energy available per pound of coal or per gram of uranium isotope. Table 4.1 is the result of such calculations. Once this total available energy is known, it is possible to compare types of fuels as regards not only their relative merits for certain applications but also how long given reserves of each might last. Furthermore, this same scientist could use this energy information and thermodynamics to determine what fraction of the fuel's energy will flow as heat to the environment when conversion to some other form of energy is attempted. Thus estimates of fuel resources and lifetimes depend crucially upon thermodynamics.

Energy Users

Obviously the highly developed nations are the largest users of energy on a per capita basis. The total economic output of a nation, the Gross National Product (GNP) affords a classification of per capita energy use. Figure 4.1 presents data of commercial energy use per capita as a function of GNP for some of the world's nations. The United States citizen is by far the world's largest energy user. The United States with 6% of the world's population accounts for one third to one half of the world's resource consumption, much of which is done in the name of energy. In contrast, India with 15% of the world's population consumes about 1.5% of the world's energy. In Figure 4.1, the United States total is 30% larger than that of the second leading consumer (Canada), 50% larger than that of the United Kingdom, and 2.5 times as large

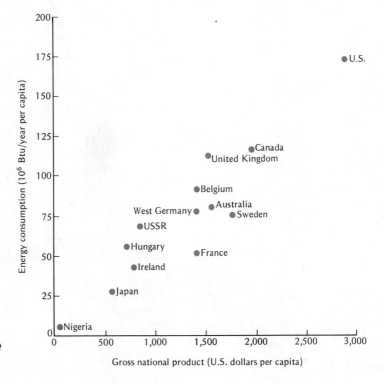

Figure 4.1. Commercial energy use per capita for various nations. [After C. Starr, "Energy and Power," in Energy and Power. A Scientific American Book. Copyright 1972.]

as that of the Soviet Union. As a further comparison, the European Economic Community of ten nations (including the United Kingdom, West Germany, France, and the Scandinavian countries among others) in 1971 had a 20% larger population than the United States, produced 10% more vehicles and 10% more steel than the United States, but consumed only about half as much energy.

What is socially and politically disturbing is that India and the other under-developed nations cannot under present conditions hope to better themselves economically unless they also greatly increase their energy consumption, which will certainly increase environmental and resource depletion and will probably increase political disagreements over energy resources.

Energy Growth Rates

As noted earlier, our energy consumption grows at a much faster rate than does population. Over the last quarter of a century, the United States energy consumption increased an average of 3.25% per year. This growth rate has

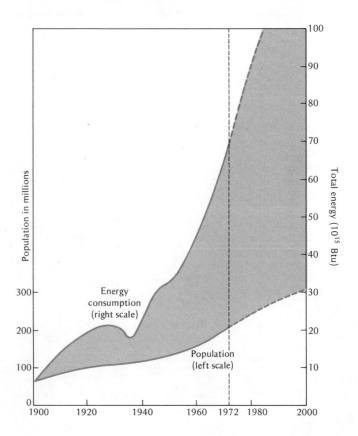

Figure 4.2. Energy consumption and population growth in the United States. Prior to 1900, energy and population grew at the same rate.

increased in recent years, so that in the early 1970s, total energy had an annual growth rate of 4.5% per year, which predicts a doubling time of about 17 years. This growth rate is more than five times the natural population growth rate in the United States. Figure 4.2 compares the growth of energy consumption with that of population in the United States since 1870. While all types of energy consumption display increasing growth rates, electricity has grown fastest (Figure 4.3), with annual rates of 8% to 12% per year in the late 1960s and early 1970s.

In the period covered in Figure 4.3, the United States was fortunate to have had access to ample supplies of natural resources to provide those annual increases in consumption. Energy has traditionally been an abundant and cheap commodity with respect to other goods and services. Such a situation is not conducive to careful use of energy, and it is not surprising that virtually all

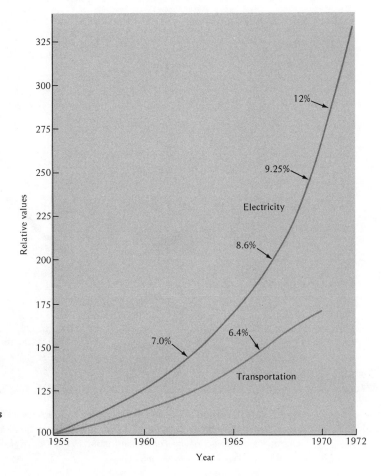

Figure 4.3. *Relative growth rates of electrical and transportation energy. The average annual percentage increases for various regions of the graph are labeled. The 1955 values of each curve are arbitrarily assigned the value 100.*

sectors of our society have grown to rely more heavily on energy—to the point where a sudden disruption in its availability would create chaos.

"Power corrupts" was coined in application to political processes, but an analogy holds for energy use. Energy is similar to an addictive drug: the more we use the more we need and demand. Unfortunately our increasing demand will inevitably encounter constraints in the form of resource depletion, environmental problems, and spiraling costs. Some of these difficulties began to become evident in the late 1960s. In the next section, we shall focus upon our use of resources.

Energy Sources and Reserves

Annual United States energy consumption has multiplied by 30 times since 1850. The total energy consumed and the sources used are given in Figure 4.4. In Figure 4.5 the same data are plotted on a percentage basis to illustrate the

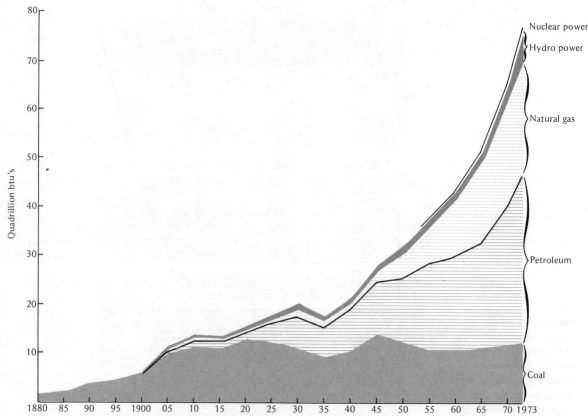

Figure 4.4 *Energy consumption in the United States since 1890. The energy from wood, an important source prior to 1900, is not shown [From Exploring Energy Choices, Energy Policy Project of the Ford Foundation, 1974. Copyright 1974 by the Ford Foundation. Reprinted with permission.]*

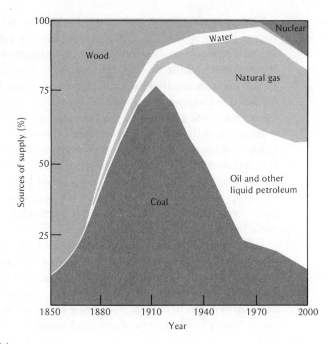

(a)

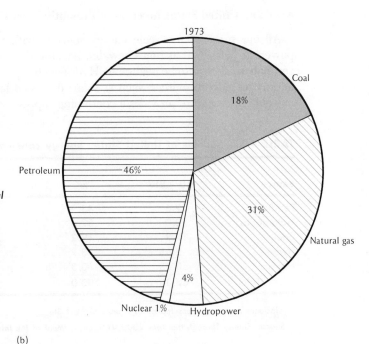

Figure 4.5. (a) Energy sources by percentage in the United States. The 1973 to 2000 figures are estimates that assume present sources will provide all our power. (b) Fossil fuels are the source of 95% of our total energy needs at present. [Data from U.S. Bureau of Mines. Figure from Exploring Energy Choices, Energy Policy Project of the Ford Foundation, 1974. Copyright 1974 by the Ford Foundation. Reprinted with permission.]

(b)

relative contribution of each energy source each year. Wood was the major energy source in the mid-1800s, but the development of electrical generators led to large-scale coal use near the end of the nineteenth century. Since 1900 the *percentage* of the total energy load carried by coal has decreased, although Figure 4.4 indicates that the amount used each year has always increased and is expected to increase annually in the future. The *percentage* changes since 1900 from coal to oil and natural gas have occurred for a variety of reasons. The emergence of the automobile increased the use of oil in the form of gasoline. Cross-country pipelines afforded a cheap means of gas transportation and this, coupled with the fact that the gas price was regulated at a very low level, led to increased use. The changeover in railroad locomotives from steam engines to diesels changed their fuel needs from coal to oil. In the late 1960s the air pollution emissions of coal, especially particulates and sulfur oxides, placed some bias against its use.

It is important to note that fossil fuels (coal, oil, natural gas) supply over 95% of our energy, and there is every indication that these will be our major energy sources during the remainder of this century, as indicated in Figure 4.5 and Table 4.3. Near the end of this century, the major changeover is predicted to be a massive shift to nuclear power as a source of electrical energy. Even if this prediction is fulfilled, 70% of our total energy will still be derived from fossil fuels.

Available United States Reserves of Presently Used Fuels

All our present large-scale energy sources, with the exception of hydro power, are classified as **nonrenewable** because their creation, if not zero, is certainly so slow as to be negligible. Hydroelectric power and several possible alternative energy sources (such as solar) discussed later are classified as **renewable** since nature provides a continuing supply.

TABLE 4.3. **Sources of United States energy consumption**

| | 1970 | | 2000 | |
Source	10^{15} Btu	%	10^{15} Btu	%
Petroleum	29.6	43.9	71.4	37.2
Natural gas	22.0	32.7	34.0	17.7
Coal	12.9	19.2	31.4	16.4
Hydro	2.65	3.9	6.0	3.1
Nuclear	0.23	0.3	49.2	25.6
Total	67.38[a]	100.0	192.0	100.0

[a] Includes nonenergy consumption of 4 units of 10^{15} Btu.

Source: Energy Through the Year 2000, *U.S. Department of the Interior.*

As noted in Chapter 2, many factors complicate the estimation of the size of nonrenewable energy resources. Economic considerations and the fact that most of the detailed exploration data are proprietary are major factors. In addition, one must be able to estimate both the rate of discovery and the size of future finds. This is commonly done by comparing geologic surveys of new areas with those where finds were located in the past. To arrive at definitive conclusions concerning the lifetimes of our energy sources, it is essential that we carefully define terms.

The term **energy resource** is generally applied to all of the energy of a given energy source. This includes all known deposits plus probable finds (based on geological evidence) plus very speculative finds (in unexplored or undrilled areas including offshore regions). All deposits are included, even those whose exploitation would require more favorable economic conditions or better technology than now exists. The fact that a resource exists does not, of course, guarantee that any or all of it can actually be utilized. Economics may preclude its use entirely. If on the other hand economics dictates that the extraction of fuel is feasible, existing technology may recover only a small fraction of the entire deposit; the rest remains in the ground. Obviously only that which can actually be recovered can be considered a viable source of energy. Current average recovery rates for crude oil are about 31%, with newer wells recovering as much as 45% of the field resource. Some oil fields no longer in production yielded no more than 15% to 20% recovery before they were abandoned. Present technology recovers about 50% of the coal in a vein and about 82% of the natural gas in a well.

For long-range planning, the maximum energy available to us can be estimated by considering **recoverable resources,** those portions of the total estimated energy resources that are ultimately recoverable using the best present technology. Table 4.4 provides an estimate of the total recoverable resources in the United States. The uranium estimates are for deposits that will produce the fuel (U_3O_8) for a price of $15 per pound or less; the 1972 price was $8 per pound. Nuclear power is discussed later in the chapter, but for our present consideration it is sufficient to note that all presently used fission plants (light water reactors) use a rare isotope of uranium (U^{235}), whereas breeder plants (coming in the mid-1980s) use a much more abundant isotope (U^{238}).

We shall use the term **energy reserve** to denote that part of the total recoverable resources that have definitely been identified and that are known to be recoverable under existing economic and operating conditions.

A **proved reserve** is the recoverable energy available *immediately* to satisfy present and near term future needs. Table 4.4 lists the proved United States energy reserves in 10^{15} Btu as of 1970. This category is extremely important, as it denotes a domestic supply we *know* exists and we are *certain* we can obtain immediately with existing technology. Coal is obviously our most abundant proved reserve, with the other energy sources possessing only a small fraction

TABLE 4.4. Proved United States domestic reserves and estimated total recoverable domestic resources in 10^{15} Btu

	Proved	Percent of All Proved Sources	Total Recoverable[a]
Coal (assumed 50% recovery)	4200	86.0	33,700
Gas (assumed 82% recovery)	275	5.6	1,650
Crude oil (assumed 42% recovery)	202	4.1	1,392
Uranium (less than $15/lb of U_3O_8)	208[b]	4.3	600[b]
	4885	100.0	37,342

[a] The total recoverable resources include known, probable, and speculative future finds.

[b] As used in present U-235 fission plants. If used in breeder plants these numbers are 100 times larger.

Source: Report of the Cornell Workshop on the Major Issues of a National Energy Research and Development Program, *Cornell University, December 1973.*

of its potential. As a reserve, coal is exceeded only by nuclear breeder plants using fuel at costs about twice as large as the present prices.

Estimating the lifetimes of these resources is extremely risky. Accurate predictions must include demand projections, technological advances, lifestyle changes, and possible conversion to other types of power sources, such as those discussed later in the chapter. We can, of course, make very crude predictions if we estimate the demand curve. In 1970 the United States energy consumption was 67×10^{15} Btu, and many forecasters predict that it will increase at 3.8% annually until the end of the century to reach a consumption of 192×10^{15} Btu in the year 2000. We can provide a crude estimate of the lifetimes by assuming that this growth rate is correct and that the distribution will be that given in Table 4.3. The columns in Table 4.5 labeled 1970 are estimates of the lifetimes if we continuously used the fuels at the energy rate we used them in 1970 (no growth assumed). The columns in the table labeled

TABLE 4.5. Lifetimes of United States proved reserves and recoverable resources (years) if used at 1970 and at projected 2000 rates

	Coal		Oil		Gas		Uranium[a]	
	1970	2000	1970	2000	1970	2000	1970	2000
Proved energy reserves	326	134	7	3	13	6	904	4
Total recoverable resources	2600	1070	47	20	75	49	2600	12

[a] As used in present light water reactors. The use of nuclear power is expected to increase rapidly over this 30-year interval.

2000 list the lifetimes if we used the fuels continuously at the rate we are expected to use them in 2000. The short lifetimes indicated in Table 4.5 are indicative of the continuing energy squeeze in which we are involved. Oil and gas lifetimes by any of these estimates are less than a century. The projected rapid expansion of nuclear power will rapidly deplete the uranium reserves, so that the fuel for the present light water fission reactors will be nearly exhausted by the end of this century. In anticipation of this, the nuclear industry is pushing for the development of breeder reactors (Chapter 5) which use a more abundant uranium isotope. (All of these estimates are for *domestic* sources.)

The Role of Imports

It is important to note that the United States energy growth shown in Figure 4.4 cannot at present be accommodated by domestic supplies, even though the domestic reserves are ample for the short term (the next 10 years, for example). Importation is used to provide the additional energy, about 15% in 1972. This source is expected to become even more important as shown in Figure 4.6. By the year 2000, nearly one third of our supply will be imported. Reliance on such imports is risky because they can be stopped either immediately, as occurred during the 1973 Mideast War, or gradually, as might occur with increasing nationalism and industrialization on the part of the suppliers.

The present United States reliance on imports arises from several factors. Of major significance is of course the rapid growth in energy demand. Further-

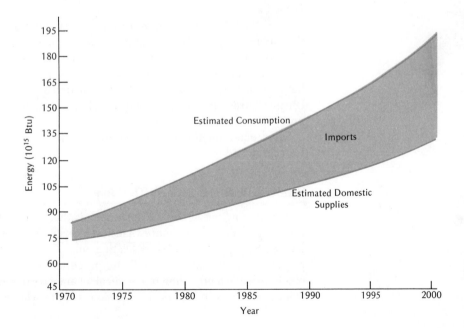

Figure 4.6. *Imported and domestic sources of energy. [From United States Energy until the Year 2000, U.S. Department of the Interior, December 1972.]*

more, there has been in recent years a general decrease in the amount of fuel resources, especially oil and gas, discovered in the United States. In the 1956 peak year, more than 58,000 wells were drilled, whereas in 1972 about half that many were tried. The reasons are in part economic; it has been more profitable in the last 20 years for oil companies to develop foreign sources. The oil depletion allowance was originally established to provide an economic incentive for exploration and speculative drilling. It allows 22% of the taxable income to be deducted for tax purposes. Allowing the oil depletion allowance to apply also to foreign operations encouraged the development of foreign sources. Import quotas that existed until mid-1973, designed to reduce the flow of cheaper imported oil into the United States and keep domestic prices high, also encouraged the oil companies to pursue the higher profits available abroad. This trend was further encouraged in the 1950s when royalties paid to foreign operators for oil became classified as tax credits and thus became tax incentives rather than direct expenses. Thus foreign drilling and exploration operations as well as foreign refinery construction were encouraged at the expense of domestic production. These policies were not created at random by government. A major force behind these decisions came from the oil companies.

Oil production peaked in the United States in 1971 and has declined since. Natural gas is an ideal fuel, but it is the fossil fuel that is in shortest supply. It was priced so low in the past that companies did not drill for it when they could drill for oil instead. When found in oil drilling operations it was often burned at the wellhead.

Since the late 1960s environmental concerns have acted to limit coal use at the expense of gas and oil and at the same time have limited drilling operations in certain regions, for example, offshore. The effects of these environmental controls were presumably small in the early 1970s because of the lag time between drilling operations and the production of useful energy. Finally, it has become harder to find new deposits simply because the most geologically probable areas have already been located.

Our reliance on imports is bound to result in increased political importance for energy and energy-related matters in the future.

Petrochemicals and Refinery Decisions

A further aspect of the energy crunch is that we are dependent upon fossil fuels, especially oil, for more than just energy. The refining of oil produces lubricants, asphalt, fuels, and heating oils. Some options are possible that require decisions during the refining process. Gasoline and heating oils (fuel oil) have sufficiently similar chemical characteristics that their production from crude oil is an "either-or" decision. That is, whether one gets gasoline or heating oil must be decided in advance of refining. Such decisions during the fall and winter usually favor fuel oil at the expense of gasoline and can therefore result in winter shortages of gasoline.

Petrochemicals are chemical products derived from petroleum and other fossil fuels. Included in this important classification are plastics, polymers, paints, coatings, nylon, tires, antifreeze, aspirin, and polyester fabrics (Figure 4.7). Decisions must be made on whether to use certain portions of the fossil fuels for the energy industry or for petrochemicals. In 1972, for example, about 5% of the crude petroleum was used for petrochemicals rather than for energy. Increased prices for crude oil obviously increase the prices of plastic products and synthetic fabrics such as double-knit clothing. Furthermore oil shortages increase the pressure to use petroleum for energy sources rather than diverting portions of it to petrochemicals.

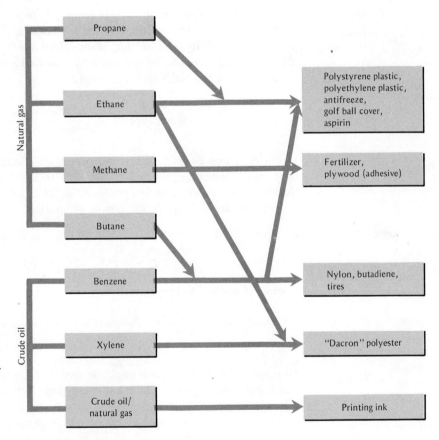

Figure 4.7. *Petroleum sources and products. [From Du Pont Context,* **3,** *(2) 1974. Published by the Du Pont Company. Reprinted with permission.]*

Energy Crisis

Whether an energy crisis existed in the early 1970s or exists now is for our purposes almost an academic question. Continued expansion of fossil fuel use will ultimately lead to fuel exhaustion. Local "crises" can of course be created by factors other than a shortage of proved reserves. Transportation problems, unequal distribution, importation changes, and hoarding can all result in apparent shortages in a given region.

The salient fact that emerges is that continuing our incessant growth for an extensive time with the present technology, resources, and policies is impossible. We wish to address some possible alternatives (Chapter 5), but we must begin by surveying energy use in more detail. This we do in the next section.

Energy Flow in the United States

Table 4.3 and Figures 4.4 and 4.5 indicate our consumption of energy. It is instructive to see how we utilize this energy. As noted previously, energy can never be converted completely into its intended applications such as transportation, electricity, and heating. Some (often most) of the energy is inevitably lost as waste heat to the environment. We shall find it useful to discuss the **efficiency** (ϵ) of the conversion of energy.

$$\text{efficiency, } \epsilon = \frac{\text{useful work obtained}}{\text{total energy consumed}} \tag{4.3}$$

Efficiency may be given as a fraction (less than 1) as in equation 4.3 or it may be given as a percentage by multiplying ϵ by 100 as discussed in Chapter 1. We shall use both expressions for efficiency in what follows. Equation 4.3 is a general definition. If applied to power generation,

$$\epsilon_{gen} = \frac{\text{electrical energy produced}}{\text{total energy consumed}} \tag{4.4}$$

whereas if applied to transportation,

$$\epsilon_{trans} = \frac{\text{kinetic energy (forward motion of vehicle)}}{\text{total energy consumed}} \tag{4.5}$$

More specialized applications might be used. For example, the efficiency with which light bulbs convert electrical energy into light might be defined as

$$\epsilon_{light} = \frac{\text{energy of visible radiation}}{\text{total electrical energy consumed}} \tag{4.6}$$

The second law of thermodynamics states that in any of these applications the efficiency is always less than 100%. Thus the ϵ values in equations 4.3 to 4.6 are all less than 1. That this is so is a fundamental law of nature. It does *not* arise because of frictional losses, poor design, or poor engineering. As noted in Chapter 1, the second law predicts the *maximum* ϵ that could be obtained under ideal conditions. The additional (and inevitable) effects of friction, poor design, or poor engineering mean that the actual efficiency will be even less than that predicted by the second law.

The heat that is lost to the environment is commonly termed "waste heat" because it can only rarely be utilized and because it may in fact cause problems. According to the first law, the total energy consumed must be numerically exactly equal to the sum of the useful work obtained plus the waste heat produced.

The flow of energy in the United States in 1971 is shown in Figure 4.8. The line-widths are proportional to the energy flow in Btu. At the left side is given

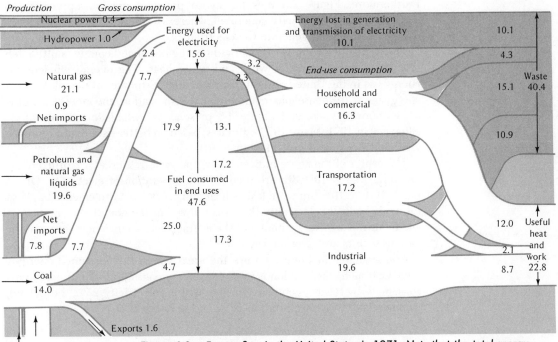

Figure 4.8. *Energy flow in the United States in 1971. Note that the total energy consumed on the left-hand side (63.2 x 10^{15} btu excluding nonenergy uses of fossil fuels) equals the sum of the useful energy plus the waste heat on the right-hand side as must be true by the first law of thermodynamics. The efficiency estimates for the various sectors (except for electrical energy generation) are very approximate. [Courtesy Professor Earl Cook, Texas A & M University.]*

TABLE 4.6. Average efficiency of energy use in various sectors of the United States

Electricity generation	30–36%
Transportation	12–15%
Household/commercial	65–75%
Industrial	40–50%

Note: Wide variations may occur. Data are merely representative.

the total energy consumed from each of the fuel sources including the effects of net imports and exports. Figure 4.8 uses data slightly different from what one might predict from the data presented earlier in Table 4.3 because oil used for petrochemicals is included in Table 4.3, but not in the data for Figure 4.8. Furthermore, Figure 4.8 uses for hydro power a number approximately equivalent to the electricity generated, whereas Table 4.5 lists the total hydro energy consumed to generate the electricity.

The right side of Figure 4.8 shows how the consumed energy is ultimately distributed between useful work and waste heat. The overall efficiency of energy use in the United States is less than 40%! Estimates of the efficiency of energy use in the household/commercial, transportation, and industrial sectors are extremely crude, and Figure 4.8 presents an *optimistic* estimate of the efficiency in 1971. Many 1973 estimates, based on better data than Figure 4.8, yield an efficiency of much less than 40%.

The center of the figure indicates how the various sources are utilized. For example, 15.6×10^{15} Btu are used for the generation of electricity. Of this input, 10.1×10^{15} Btu were lost as heat and 5.5×10^{15} Btu emerged as "useful work," namely electricity that was utilized in the household/commercial and industrial sectors. Estimates of the average actual efficiencies of use in the various sectors are given in Table 4.6

Also available in Figure 4.8 are the areas where the consumed energy is used. One such classification is to divide energy consumption into percentage consumed by each sector as in the figure: electricity 25%, household/commercial 21%, transportation 27%, and industry 27%.

Efficiency of Various Processes

A deeper appreciation of our energy use can be obtained by studying the efficiency of utilization in some individual applications. Equation 4.3 is a general definition of efficiency, and we shall examine its application here.

Electrical Power Generation

Electricity can be obtained by either of two processes: direction conversion or dynamic conversion. These differ in that **dynamic conversion** requires a mechanical device to convert some type of motion (usually rotation) into electricity, whereas **direct conversion** involves "no moving parts."

A battery, which converts the energy of a chemical reaction into electricity, is an example of **direct conversion** (Figure 4.9). Fuel cells, as commonly used in manned space missions, are also examples of direct conversion. In the more common fuel cells, hydrogen gas and oxygen gas are the fuel. When these combine to form water, electricity is produced. A third example is a photovoltaic cell (solar cell) such as that used in light meters on cameras. Such a cell produces electricity in the presence of light. Each of these devices is used in specialized cases for power generation, usually when the amount of power required is small. In all cases some limitation, usually cost, has prevented their use for large-scale installations. The best efficiencies attained with some of

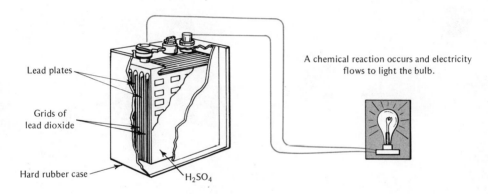

Lead plates

Grids of
lead dioxide

Hard rubber case

H_2SO_4

A chemical reaction occurs and electricity
flows to light the bulb.

Lead Acid Battery (Car Battery)

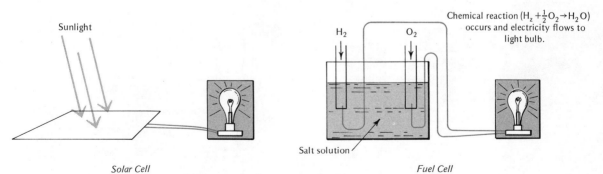

Sunlight

Chemical reaction ($H_2 + \frac{1}{2}O_2 \rightarrow H_2O$)
occurs and electricity flows to
light bulb.

H_2 O_2

Salt solution

Solar Cell   *Fuel Cell*

Figure 4.9.   *Direct conversion of electricity.*

TABLE 4.7. Efficiency of direct conversion

Device	Energy Conversion Scheme	Efficiency (%)
Dry cell battery	chemical to electrical	90
Lead storage battery	chemical to electrical	75
Fuel cell	chemical to electrical	60
Solar cell	light to electrical	10

Source: Conservation of Energy, Committee on Interior and Insular Affairs, U.S. Senate, 1972.

these devices are shown in Table 4.7. Batteries are the most efficient devices listed. Lead storage batteries convert 75% of their chemical energy into electricity, but it should be noted that they must be recharged, usually by means of a dynamic conversion scheme.

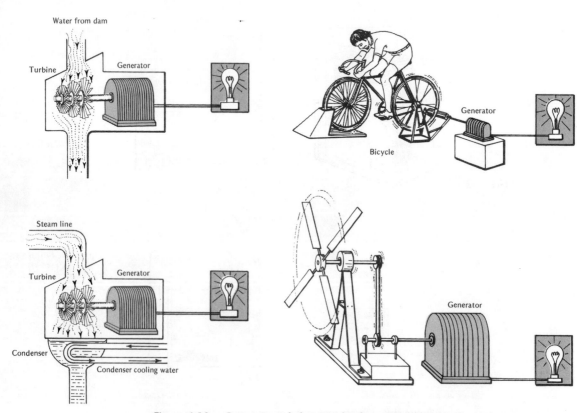

Figure 4.10. *Generation of electricity by* **dynamic conversion.** *In each case mechanical motion is involved; a source of rotary motion turns the generator that produces electricity.*

Dynamic conversion is used in all commercial power plants. The basis for dynamic conversion is the generator, a device in which the rotation of shaft produces electricity, as shown in Figure 4.10. The shaft rotation may be driven by any of a number of schemes: a bicycle, a windmill, or a gasoline engine, for example. In power plants, the rotation is accomplished by a turbine, a fanlike machine that rotates when water flows across it (hydroelectric dam) or when

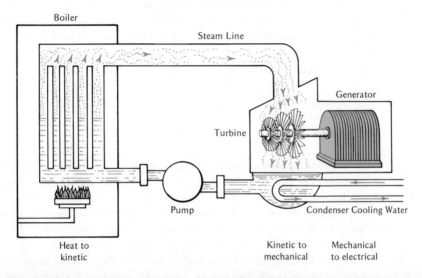

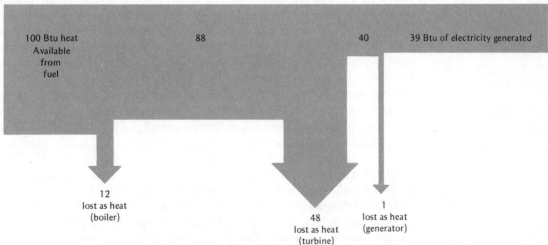

Figure 4.11. *A steam power plant for electrical power generation. A source of heat produces steam in the boiler, and the steam drives a turbine. The approximate heat loss at each stage is shown. The overall efficiency of this plant is 39%, about the best obtainable with present technology.*

hot steam is blown across it (power plant). Steam power plants in the United States convert chemical energy into electrical energy with an average efficiency of about 32%.

Figure 4.11 demonstrates the basis of a modern fossil fuel plant with the presently best available conversion at each step. The boiler converts heat into steam (thermal energy) with an efficiency of about 88%, and the turbine converts 46% of the thermal energy into rotational energy. The generator is a very efficient device, converting 99% of the mechanical energy it receives into electricity. The overall efficiency of conversion is $39/100 = 0.39$ or 39%, about the best possible with a modern fossil fuel plant. Reflected in the figure is the fact that the individual efficiencies of each stage must be multiplied to obtain the overall efficiency.

$$\text{overall efficiency} = \epsilon_{\text{boiler}} \times \epsilon_{\text{turbine}} \times \epsilon_{\text{gen}} \tag{4.7}$$

(converting heat
to electricity) $= 0.88 \times 0.40 \times 0.99 = 0.39$, or 39%

Power plant efficiency has improved markedly since 1900 when it was 5%. At present, the limiting step is the turbine. Unfortunately the prospects for *major* improvements in this stage are limited. The second law implies a maximum possible efficiency of 53% for the turbine using present materials and steam temperatures; this would yield an overall efficiency of 46%, just slightly better than that obtained in a modern fossil fuel plant.

The heat lost (Figure 4.11) is discharged to the environment, and has been somewhat erroneously labeled as "thermal pollution." Heat is not usually considered a pollutant because living systems depend on energy conversion and heat flow. However, the term "thermal pollution" is indicative of the possibility that the impact of heat production may be severe in the environment and that minimizing this impact is at present a significant consideration. We discuss the effects of the waste heat later in this chapter.

We shall use in this text 32%, the average value in the United States, for the efficiency of conversion of heat to electricity. Even this figure, which is for the complete power plant, does not accurately reflect our utilization of the fossil fuels. The transmission line system that conducts electricity from the plant to a home or a factory has an average efficiency of about 93%, so that 7% of the electricity conducted by the line is lost as waste heat. This means that the efficiency with which we convert fossil fuels to electricity at the wall socket (delivered efficiency) in our home is about $0.32 \times 0.93 = 0.30$, or 30%. Furthermore the devices plugged into the socket do not have 100% efficiency, as noted in the next section, so that losses occur in their use also.

We use a wide variety of electrical devices of varying efficiency. The efficiency of an individual electrical device is defined analogously to equation 4.6, namely useful work produced divided by electrical power consumed. In light-

TABLE 4.8.   Overall efficiency with which fossil fuel energy is converted to useful energy by electrical devices

Device	Conversion	Efficiency (%)
Incandescent bulb	fossil fuel into electricity, then into light	1.7
Fluorescent lamp	fossil fuel into electricity, then into light	6.6
All lighting		
(U.S. average)	fossil fuel into electricity, then into light	4.2
Small electric motor	fossil fuel into electicity, then into rotational energy	20.0

ing, for example, an incandescent bulb converts only 5% of the electrical energy it receives into (visible) radiant energy. The other 95% is given off as heat. A fluorescent lamp is four times more efficient, converting 20% of its electrical consumption into light. It is important to note that the *total* efficiency with which we convert fossil fuel sources into light must include the efficiency of power generation, that is,

$$\epsilon_{lighting} = \epsilon_{elec.\ gen.} \times \epsilon_{bulb} \tag{4.8}$$
$$= 0.32 \times 0.05 = 0.0160$$

or only 1.6% for incandescent bulbs. Over 98% of the energy of a fossil fuel is lost as heat if the fuel is used for lighting with incandescent lamps. A similar calculation for the fluorescent light shows an overall lighting efficiency of 6.6%.

Heating

The efficiency of our use of fuel sources for space heating also varies widely. Central furnaces provide the most effective use of fuels, with an upper limit on efficiency of 85% (gas) or 65% (oil), as noted in Table 4.9. However, the efficiency varies markedly with the details of design, care in installation,

TABLE 4.9.   The efficiency of converting fossil fuel energy into useful heat for space heating

Device	Process	Efficiency (%)
Central gas furnace	gas burned in furnace	85
Central oil furnace	oil burned in furnace	65
Electric resistance heating	fuel generates electricity which heats	33
Fireplace	fuel burned in fireplace	20

amount of home insulation, and degree of maintenance of the system (filter cleaning, and so on). An average value for central furnace use in the United States is probably 50% to 60%. This average value means that nearly two homes with central furnaces can be heated with the fossil fuels required to heat one electrically. The most effective use of our fossil fuels for heating is obviously central furnaces, rather than the costly (in the energy sense) use of electrical heat. We will expand this view when we consider energy conservation.

Electrical Power: Assets and Liabilities of Present Sources

Electrical power use is the fastest growing component of United States energy. Since World War II the doubling time for energy growth has been 10 years, and if continued it would mean that United States consumption in the year 2000 would be eight times larger than the consumption in 1970. Most experts feel that the 10 year doubling time cannot be sustained much longer, however, and that by 2000 the annual consumption will be "only" five times the 1970 value. The present and projected consumption (energy input) for power generation is shown in Figure 4.12. Of the energy devoted to electrical power 80% comes from fossil fuel sources, 16% from hydroelectric plants, and 4% from nuclear plants. A drastic shift toward nuclear power is expected in the next 30 years until it provides 60% of the electrical energy in the year 2000 (Figure 4.12). It is important to note that, because of the explosive growth in electricity usage, the amount of fossil fuels consumed each year will continue to *increase*, despite this projected change to nuclear sources.

Under present economic and technological conditions the three types of power plants discussed above (fossil, hydro, nuclear) are the only competitive choices available if one wishes to invest 0.5 to 1 billion dollars and be *certain* of attaining within a decade a large power plant. The largest nonhydro power plants have power ratings of a little more than 1000 Mw, about enough electrical power for a typical city of 400,000 in the United States. It will be instructive to consider the advantages and disadvantages of each of the present types of plants before discussing in the next chapter some more speculative proposals for large-scale electrical power plants.

Fossil Fuel Plants

Of the energy input for electricity, 80% comes from fossil fuel plants. Power plants are a major source of air pollution, ranking as the leader in SO_x emissions and as the second largest producer of particulate emissions. These pollutants are characteristic of coal combustion, which provides about half of the energy used for fossil fuel generation of electricity. Coal was formed from vegetation deposited during mineral-rich, swampy conditions and is a complex mixture containing predominantly carbon (50% to 95%). Many elements and

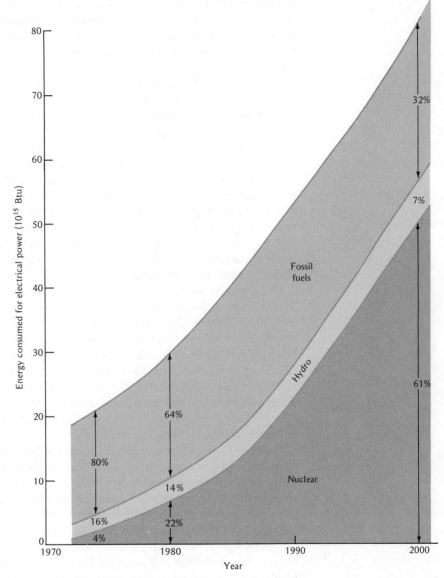

Figure 4.12. *Energy inputs for electrical power generation. The percentages show the approximate amount furnished by each source during that year. [From United States Energy Until the Year 2000, U.S. Department of the Interior, December 1972.]*

minerals are present in coal because of the variety of conditions that existed when the vegetation was deposited, and coals from different regions of the United States have widely varying chemical contents. One very troublesome element present in all deposits of coal is sulfur, which forms SO_2 upon combustion. The sulfur content of coal in the United States varies from 0.6% to more than 5%. Another characteristic of coal, one that is sometimes used to classify the various types, is the **inorganic ash** content, the residue left after combustion. This inorganic ash or "fly ash" is composed of minerals, and leads to particulate emissions. The fly ash content in United States coal varies from 3% to 12%.

The largest coal-burning plants in the world are in the "Four Corners" region of the Southwest, an area once famed for clear skies and 100-mile vistas. A brief look at the problems encountered with these plants will serve to illustrate what can be expected with future large coal plants. A consortium of power companies, including Southern California Edison and Arizona Public Service, were instrumental in the planning during the late 1950s for these massive plants. One major factor in the selection was the large coal deposits in the region which could be strip-mined to provide fuel. An additional incentive was the small population in this area of the United States. A sparse population meant fewer worries about air pollution controls, although it was expected that the particulates would be rapidly dispersed. Furthermore, many people living in the Southwest welcomed the plants, believing they meant jobs, population growth, and "progress."

The proposed plants were not greeted with universal enthusiasm, however. The location of the plants and the coal reserves involved Indian lands, and a controversy erupted in the tribes over whether lands for the plants should be leased. Many of the Hopi and Navajo who follow traditional tribal ways believed that 35 years (the lease lifetime) of violence to their land was not worth any price, even if the land could be reclaimed later. To them the land is holy, and Black Mesa, the site of much of the strip mining, is holier than most. Other Indians supported the plant, feeling that the estimated $100 million in jobs and lease payments was a fair trade for their ancestral lands. The Indians ultimately sold the rights.

The Four Corners plant (Figure 4.13) was the first to be completed. This one plant consists of five generating units. Units 1–3 were completed in 1964 and Units 4 and 5 were operating by 1970. At this time, the total generating capacity of 2.2 giga watts (Gw) made this the largest coal-burning plant in the world. (A gigawatt, Gw, is 1 billion (1×10^9) watts; see Table 1.6.) It consumes 25,000 tons of coal per day from the adjacent Navajo strip mine. The specifications for the original units called for controls to reduce particulate emissions by 87%, which did not represent the best available technology at that time. Electrostatic precipitators were not installed. Only about 78% reduction was actually achieved on Units 1–3, and the plant emitted an estimated 250 tons of particulates per day, which was more than the particulate tonnage emitted by the entire city of Los Angeles. A 1970 report from the U.S. Department of Health, Education, and Welfare (HEW) indicated that, because of the high ash quality of the coal and the inadequate pollution controls, the Four Corners

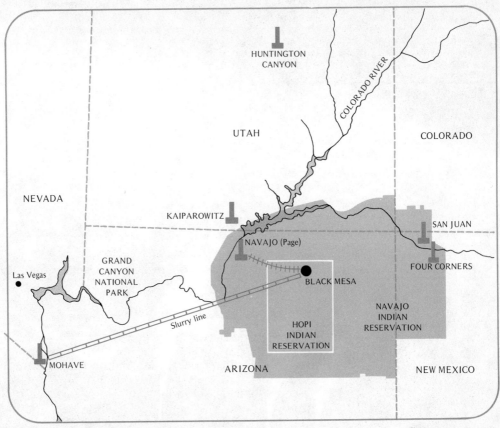

Figure 4.13. *The Four Corners region of the United States. Indian lands are indicated by the shaded area.*

plant emitted five times as much particulate waste as other plants of comparable size.

The impact on visibility in the area was striking. Particulate emissions were noted over a thousand square mile area, and the plume from the plants became a convenient beacon for pilots of high-flying aircraft. Spurred by public pressure, the state of New Mexico established new emission standards that required 99.2% reduction of particulates. To meet this requirement, the utilities installed electrostatic precipitators, wet scrubbers, and bag house filters (see Chapter 3) at Four Corners. These conversions, which were to be completed during 1973, have the capability (on paper) to reduce particulate emissions from 250 tons per day to about 30. Included in the 30, however, are most of the very hazardous smallest particles.

Figure 4.14a. *Smoke pours out of the coal-burning Four Corners plant. [Courtesy of Terrence Moore.]*

The long-range impact of strip mining is unclear. More than 100 square miles (64,000 acres) has been leased at the Black Mesa site alone (Figure 4.14b). The agreement calls for the mines to be backfilled and reseeded when the mining is completed. Whether the fragile desert plants will become reestablished is unclear. While a few strip-mine reclamation projects have received much favorable publicity, reclamation on a scale this large has yet to be

Figure 4.14b. *Strip mining, used to provide the coal for many fossil fuel plants, has the potential for extensive environmental harm. This aerial view is of a Peabody strip coal mining operation near Nucla, Colorado. [EPA-DOCUMERICA-Bill Gillette.]*

demonstrated in actual mining operations at a cost companies are willing to pay.

Water to cool the plants is also a focus of concern. The water comes from the Colorado River and its tributaries in quantities sufficiently large that it will compete with agricultural uses and with future development options of the area. The dams constructed on the Colorado and the increasing use of that river's waters for irrigation have already resulted in major changes in the condition of the river as it flows to Mexico. The flow has been greatly reduced in recent decades, and the salt concentration, or salinity, has increased. Increasing United States appropriations of the Colorado water has increased tensions with Mexico because of the adverse impact on Mexican agriculture along the lower reaches of the river.

The Four Corners plant was merely the first massive site. Five other large plants are now in operation or near completion in this part of the Southwest.

The San Juan plant is 10 miles from the Four Corners units (Figure 4.13). The Mohave plant is fed with coal mined in the Black Mesa and transported to the plant by means of a pipeline: pulverized coal is mixed with water, and the resulting "slurry" is pumped in the pipe as a fluid. A special railroad was constructed to carry coal to the Navajo plant, which is slightly larger than Four Corners. Kaiparowitz, the largest plant of all, will consume 50,000 tons of coal per day.

The air pollution, mining, and water problems of the area will increase markedly as these plants begin to operate. Included in the impact area are six national parks, 28 national monuments, and the Lake Powell and Lake Meade recreational areas.

Eventually, the generating capacity of all six plants will be nearly 14 Gw, half again what Consolidated Edison supplied to New York City in 1972 and almost equal to the total output of the Tennessee Valley Authority. But most of this power is not used in the area that suffers the impact. Many of the Navajo hogans in the area are not wired for electricity. Nearly half is transmitted to Los Angeles. In fact, the power from just three of the plants would provide all the electricity needed in the entire four states of Colorado, Arizona, New Mexico, and Utah. It is, of course, an advantage for Los Angeles, which has sufficient air pollution problems already, to obtain its power from such a remote site.

These problems encountered in coal plants notwithstanding, an increase in the rate of construction of these plants is predicted because of the enormous reserves of coal relative to oil or natural gas.

Hydroelectric Power

Water power is a renewable energy source, offering power until the climate changes. In a hydroelectric plant, the potential energy of the water, attained by virtue of its elevation behind the dam, is converted to kinetic energy and run across a turbine that operates a generator to produce electricity. The construction of a dam and the creation of a large lake result in ecosystem changes. The water flow below the dam is altered. Standing bodies of water, such as a lake, often support aquatic life different from that of flowing rivers.

Most of the major rivers in the United States have been tapped for hydroelectric power, and the capacity to expand production is not large. The Federal Power Commission recently estimated that we presently obtain about one-fourth of the maximum hydroelectric power that might be available in the United States.

Although the water supply might be available forever, the lifetime of a hydroelectric plant is considerably shorter. The natural filling of the lake by silting means that unless the silt is removed, the lifetime of most plants is about one century.

Nuclear Power   The fossil fuels yield energy from combustion. Their utilization relies upon the heat released when atoms rearrange to form new compounds, an example being the combustion of methane (CH_4) to yield CO_2 and H_2O. Such reactions involve only the outer electrons of an atom. The center of the atom, the nucleus, is sufficiently stable that it is not affected by combustion or other chemical reactions. This is true even for hydrogen, which has only one electron.

Since 1939 man has learned how to cause nuclear reactions, that is, cause changes in the structure of the nucleus. Because the particles that compose the nucleus are bound together so tightly, an amount of energy much greater than that commonly involved in chemical reactions is required to cause these nuclear changes. The nuclear reactions involve either the splitting (**nuclear fission**) of the nucleus into two smaller nuclei or the combination of two nuclei into a third (nuclear **fusion**). In both fission and fusion reactions, an enormous quantity of energy is released per nucleus because a small amount of nuclear mass is converted into energy. Einstein's famous equation,

$$E = mc^2 \tag{4.9}$$

where m is the mass converted and c is the speed of light, correctly predicts that nuclear reactions release about a million times more energy per atom than do combustion reactions. One pound of uranium, for example, can generate as much electricity as 1400 tons of coal. It is this enormously greater energy that has focused interest on nuclear reactions.

If the energy is released too quickly, nuclear reactions become explosions, a characteristic used in bombs and warheads. The atomic bombs used against Japan in World War II utilized fission reactions, whereas the hydrogen bomb, developed shortly after that war, uses a fusion reaction. For use as electrical power sources, nuclear reactions are of course initiated under conditions where the rate of the nuclear reaction is controlled to achieve a relatively slow energy release.

Fission reactions are the basis for all presently used nuclear power plants. The heat released from the nuclear reaction is used to form steam, and dynamic conversion is then used to generate electricity in a manner quite similar to that used in fossil fuel plants. Although fossil fuel and nuclear plants have some similarities, such as dynamic conversion, each has certain environmental advantages (and disadvantages) relative to the other. As we shall see, arguments for or against nuclear power often involve individual value judgments applied to the various advantages or disadvantages that nuclear power may possess relative to other types of power plants. To gain a better comprehension of the environmental effects of nuclear power, we must investigate nuclear structure, radioactivity, and nuclear reactions.

Nuclear Structure and Isotopes

In our discussion of elementary chemistry in Chapter 1, we discussed the essentials of nuclear structure. (You may wish to review that material at this time.) We noted that the **atomic number, Z,** of an element is the number of protons in the nucleus, and the **mass number** of an element is equal to the sum of the number of protons plus the number of neutrons in the nucleus. The atomic number is unique because it identifies the element. However, the mass number of an element is not unique: a given element may have *more* than one mass number depending on the number of neutrons contained in the nucleus. Atoms of the same element possessing different mass numbers are called **isotopes.** We need to investigate isotopes to obtain a basis of understanding for nuclear reactions.

Because the mass can vary, we must denote isotopes of an element by writing the mass number as a superscript to the chemical symbol, as in Table 4.10. Thus, ^{235}U and ^{238}U are isotopes of uranium with mass numbers 235 and 238 respectively. They are commonly called "U-235" and "U-238."

The element hydrogen (Z = 1) exists as three isotopes and is unusual in that for this element alone the isotopes are sometimes given different symbols: ^1H, hydrogen; ^2H, heavy hydrogen (sometimes called deuterium, D); ^3H, tritium (sometimes indicated as T).

Isotopes differ only in the number of neutrons in the nucleus. The bonding in molecules is not changed and isotopes of the same element exhibit similar chemical behavior. For example, water is the name given the compound H_2O, but the isotopes of hydrogen also combine similarly with oxygen. Heavy water, D_2O, is well known, and the compounds HDO, T_2O, and DTO also exist.

Of the 92 elements that occur naturally on earth, only 20 exist as a single isotope. The rest of the elements have two or more isotopes. Tin has the greatest number, ten in all. Isotopes of an element often occur in grossly unequal amounts. As in Table 4.10, 99.98% of the naturally occurring atoms of hydrogen has a mass number of 1; 0.02% has a mass of 2 (deuterium); and mass

TABLE 4.10. Some common isotopes, their nuclear structures and abundances

| Name | Nuclear Composition | | Mass No. | Atomic No. | Symbol | Abundance (%) |
	Protons	Neutrons				
Hydrogen	1	0	1	1	^1H	99.98
Deuterium	1	1	2	1	D (or ^2H)	0.02
Tritium	1	2	3	1	T (or ^3H)	trace
Carbon-12	6	6	12	6	^{12}C	98.9
Carbon-13	6	7	13	6	^{13}C	1.1
Uranium-235	92	143	235	92	^{235}U	0.7
Uranium-238	92	146	238	92	^{238}U	99.3

3 (tritium) is found in only trace amounts. Abundances of other isotopes are given in the table. The uranium abundances will be of importance in our discussions of nuclear power.

Since man has learned how to initiate nuclear reactions, he has been able to create isotopes. Over 1000 man-made isotopes are known, many of which are significant in medicine, chemical research, biological studies, and nuclear physics.

Radioactivity

Some nuclei possess a natural tendency to decompose or change into another nucleus. They do this spontaneously, that is, without assistance by or interference from man. This decomposition occurs with the emission from the nucleus of a particle or an energy ray or both. For example, the following reaction occurs:

$$^{238}U \longrightarrow {}^{234}Th + \alpha \text{ particle} \tag{4.10}$$

Uranium of mass 238 emits a particle (called an alpha particle) and changes into a nucleus of thorium of mass 234 (Figure 4.15). Such a process is called **radioactive decay,** and the element undergoing decay is said to be **radioactive** or to be a **radioisotope.** Natural radioactivity was first observed by Henri Becquerel in 1896 in a uranium compound. Over 50 naturally occurring radioactive isotopes are known and nearly 1000 more have been created by man.

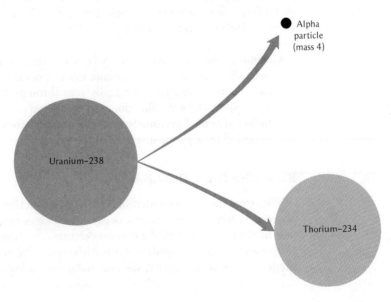

Figure 4.15. *Radioactive decay of uranium by α emission. The uranium-238 (^{238}U) decays to thorium-234 (^{234}Th).*

Types of Radioactive Decay. Three major types of radiation are emitted by radioisotopes: alpha particles (α), beta particles (β), and gamma rays (γ). The types of emissions provide the names applied to each of the radioactive processes: in equation 4.10, the ^{238}U underwent **alpha emission.** The emitted species (α, β, and γ) are ejected from the nucleus with considerable energy. They can interact with molecules in the air, in water, or in living systems and cause ionization, the removal of electrons from the molecules. Consequently, α, β, and γ are often collectively termed "ionizing radiation." Radiation damage arises from the subsequent reactions of the ionized molecules with nearby matter.

Alpha Particles. When a nucleus undergoes α decay, it loses a particle composed of 2 protons and 2 neutrons. The alpha particle has a charge of $+2$ and a mass of 4; it is identical to a helium (He) nucleus. Of the radiation particles, α is the most massive and biologically the most destructive. Fortunately, however, α has the least penetrating power. It can be stopped or "shielded" by a centimeter or so of plastic or of paper. Often α radiation does not pass through the skin, but havoc can result if an α-emitting nucleus is ingested into a biological system, because the positively charged nucleus attracts electrons so strongly that it rips them off surrounding molecules.

Beta Particles. When β emission occurs, a high-speed electron is ejected from the nucleus. Within the nucleus a complex process has occurred in which a neutron has changed to a proton with the attendant release of a β. This particle can penetrate more than a centimeter of flesh but can be shielded with metal foil a few millimeters thick. Inside the tissue, the β particle can also cause ionization of surrounding molecules.

Gamma Rays. Radioactive decay by γ emission is quite different from the two preceding processes. The gamma ray is a packet of electromagnetic energy, a powerful x ray that can easily pass through the human body causing ionization along the way. Shielding of γ rays often requires several feet of lead.

A radioactive nucleus usually loses just one particle (α or β), and this is often accompanied by a γ emission as well.

Radioactive Decay Half-Lives

Whatever the type of radiation emitted, radioactive decay is a spontaneous process. When a given nucleus decays, it changes into another nucleus. The new nucleus may be that of a different element, such as the change of uranium to thorium given in equation 4.10. Although we can never predict when a single nucleus might decay, we can make very accurate predictions about a statistically large sample of a radioisotope, say, 1,000,000,000,000,000,000,000 or 10^{21} nuclei, about the number in a gram of metal.

We might compare the radioactive decay of such a large collection of nuclei to the random breaking of a large collection of balloons as they blow about a field. The bursting of a balloon corresponds to spontaneous decay. If each balloon has each hour a 50/50 chance of breaking, we still cannot predict when an individual one might break; it might survive for days. If we begin with 32,000, however, we would expect after 1 hr that half will have burst, leaving 16,000. After another hour, half of these will have burst leaving 8000. Half of these will break during hour 3, and half of those remaining will burst in hour 4, leaving only 2000. The process continues hour after hour, but 1 hr is the time required for half of any initially observed sample to disappear. For this experiment, 1 hr is the half-life of the balloons.

The **half-life** of a radioisotope is the time required for the sample to decay to half of its initial size. Each isotope has a characteristic half-life, and the lengths range from thousandths of a second to billions of years. The length of the half-life is crucial to an understanding of radioisotopes and to an evaluation of their environmental impact. A radioisotope with a short half-life—say, an hour or a few days—will decay very rapidly to another nucleus. After 3 half-lives the isotope will have 1/8 of its starting concentration, after 10 half-lives it will be 1/1024 as prevalent, and after 20 half-lives it will be reduced by 1/1,048,576. The balloons in the preceding example would have all been broken after 15 hr (15 half-lives), which you can easily verify by simple multiplication by 2. (Actually a few might remain. Remember that our observations are valid only for statistically large numbers and might not hold for the five or six balloons left at the end).

Most of the naturally occurring radioisotopes that exist today must have very long half-lives (hundreds or thousands of years). Any shorter-lived isotopes that might have been created when the earth was formed would have decayed over the past 5 billion or more years to amounts too insignificant to be located.

The nuclei resulting from radioactive decay may or may not be radioactive. For example, $^{230}\text{Th} \rightarrow \, ^{226}\text{Ra} + \alpha$ describes the decay of thorium-230 into radium-226. This has a half-life of 83 thousand years. Radium is formed by this reaction and is always found in thorium deposits. Radium, however, is also radioactive and decays, with a half-life of 1600 years, to radon-222 by alpha emission. The scheme does not end there, because radon is also radioactive, and the following series of decays occurs:

$$
\begin{aligned}
&\text{radon-222} \;\;\rightarrow\; \text{polonium-218} \rightarrow \text{lead-214} \rightarrow \\
&\text{bismuth-214} \rightarrow \text{polonium-214} \rightarrow \text{lead-210} \rightarrow \\
&\text{bismuth-210} \rightarrow \text{polonium-210} \rightarrow \text{lead-206}
\end{aligned}
\tag{4.11}
$$

This involves three isotopes of lead, two of bismuth, and two of polonium. This decay of thorium thus results in lead-206 which is a stable (nonradioactive) nucleus. The half-lives in each step range from fractions of a second to 20 years.

Nuclear Transformations

By bombarding heavier nuclei (such as uranium) with neutrons, two types of nuclear reactions may be initiated. The nucleus may absorb ("capture") the neutron and become a different isotope, or the nucleus may fission (split) approximately in half to form two other elements (fission fragments) releasing energy in the process. The fission fragments are often radioactive.

Fission. Nuclear fission, the splitting of the atom, is to be distinguished from radioactive decay. Decay is a spontaneous process, not forced or hastened by external forces or acts, whereas fission is initiated by intentional neutron bombardment. Furthermore the energy released in radioactive decay is miniscule compared with the energy released in fission. The fission of ^{235}U nucleus can yield many different fission products. One of the more important possibilities is in Figure 4.16. This reaction, together with many others, occurs during fission in a reaction. Depending on which reaction occurs, either two or three neutrons are released when a single neutron strikes the uranium nucleus. The

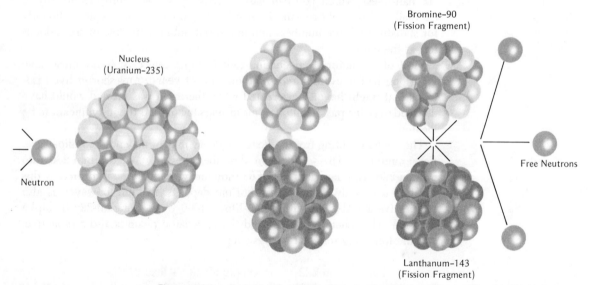

Figure 4.16. *Fission of the ^{235}U nucleus. Bombardment with a neutron results in fission. Bromine-90 (^{90}Br) and lanthanum-143 (^{143}La) represent only one set of possible fission fragments; ^{90}Br is radioactive and ultimately becomes ^{90}Sr.*

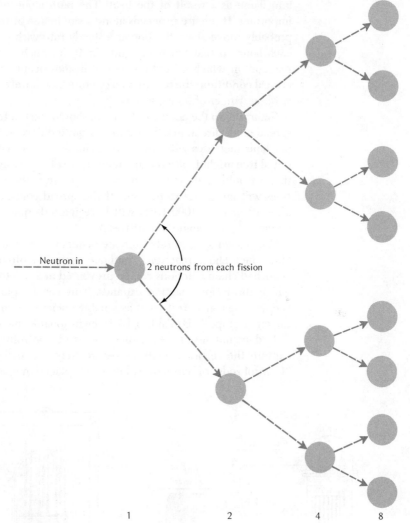

Figure 4.17. An
expanding branched
chain reaction in which
the number of neutrons
doubles with every fission.
Two neutrons are
released (arrows) when
each nucleus (solid circles)
fissions. The released
neutrons then strike two
other nuclei which fission,
each releasing two more
neutrons. Such a reaction,
if continued, will result in
an explosion.

average for all possible products is 2.4 neutrons released per ^{235}U nucleus.
(Many of the fission products, including those in Figure 4.16, are radioactive,
and their treatment and disposal represent a vexing problem which we discuss
later.)

The released neutrons can also cause fission should they strike other ^{235}U
nuclei. This process can lead to a "chain reaction" in which the fission process
expands throughout the sample. A simple analogy for a fission reaction is the
spread of a forest fire. If one tree is ignited (fission), adjacent trees may burst

into flame as a result of the heat. The path along which the fire spreads is important. If the fire proceeds along a single line of trees, for example, it will probably move along the line at a steady rate with perhaps one tree igniting each hour, so that ten trees ignite in 10 hr. Such a steady rate in a nuclear reaction, in which a fixed number of fissions occurs per second, is called the **critical condition**: the reactor can operate indefinitely at this steady fission rate with no danger of an explosion.

Returning to the analogy, if the fire should start in the center of a forest and spread in all directions, the picture is quite different: the rate of tree ignition per hour increases with time. For example, 4 trees equally spaced about the initial tree might ignite the first hour. Each of these might then ignite 2 trees so that 8 would ignite the second hour. If these 8 also ignite two each, then 16 trees will ignite the third hour. If the spread continues in this fashion, then after 10 hr over 4000 trees will have ignited—quite a change from the fire moving along a single line of trees!

When such a rapid rate increase occurs in a fission reaction, it is termed a "branched chain reaction," and an explosion often results (Figure 4.17). Branched chain reactions can be prevented in a reactor by slowing the rate at which the fission reaction expands. This can be partially accomplished by keeping the concentration of fissionable nuclei very small. In present reactors, the rare isotope ^{235}U (Table 4.10) is the fissionable nucleus. The more abundant ^{238}U does not fission. A further means of controlling the fission rate is to capture the released neutrons before they can strike a fissionable nucleus. "Control rods" of boron or cadmium are placed in reactors to accomplish this.

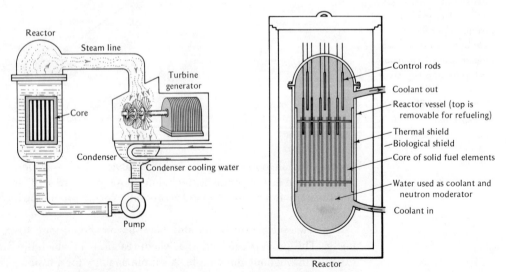

Figure 4.18. *A nuclear power plant and its reactor. The core, in which the reaction occurs, generates steam that turns the turbine.*

The nuclei of these metals have a marked tendency to capture neutrons, and the rate of the reaction can be controlled by varying the depth of insertion of the control rods into the reactor.

Nuclear Fission Power Plants

Figure 4.18 shows the essential components of a boiling water reactor nuclear power plant. Another common type of reactor, the pressurized water reactor, is a bit more complex but operates in essentially the same manner. The heart of the plant is the core that contains the fuel, uranium dioxide (UO_2), enriched in ^{235}U to about 3%. Here the nuclear energy is converted to heat that is then transferred to the circulating water to make steam. The steam passes across a turbine, then into the cooling tubes where it is condensed to water. The water also serves another purpose, that of a "moderator" to lower the velocity of the neutrons. As released in the fission reaction, the neutrons are traveling too fast to cause fission efficiently in other nuclei they might encounter. The reduction of their velocity by water increases the efficiency of the fission process.

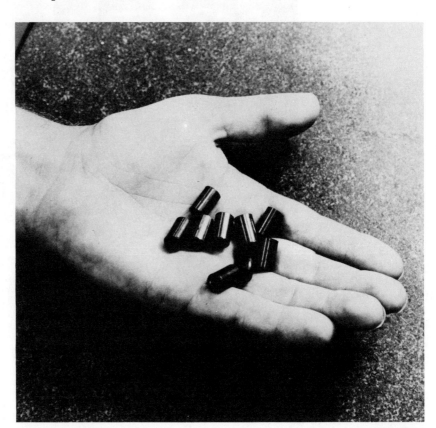

Figure 4.19. Fuel pellets and a fuel assembly of a reactor core. (a) Pellets of uranium dioxide that will be fabricated into fuel assemblies. When these assemblies are placed in the core of a nuclear reactor, each pellet can produce more energy than a ton of coal.

Figure 4.19. (b) Lowering a fuel assembly into a reactor core. [Courtesy of the Atomic Energy Commission.]

A 500 Mw nuclear plant (about half the size of plants now being constructed) would have a core of about 70 tons of uranium dioxide in the form of ceramic pellets. The pellets are packed into 23,000 "fuel rods": thin-walled tubes of stainless steel each ½ in. in diameter and 12 ft long (Figure 4.19). Inserted between the tubes are the control rods which can be lowered into the

Figure 4.20a. *The containment vessel of the Vermont Yankee Plant during construction at Vernon, Vermont. [Courtesy of the Atomic Energy Commission.]*

core to slow or to stop the reaction. The neutrons pass from fuel rod to fuel rod through the water. As the reaction proceeds, the fission products increase. These tend to absorb neutrons and decrease the reaction rate. The fuel rods are used until about 6% of the ^{235}U has been consumed. More extensive reaction would result in excessively high radioactivity in the fuel elements and a possible weakening of the metal holding the fuel. About one third of the fuel elements are replaced at the end of each year of operation.

The core is enclosed in a reactor vessel with steel walls nearly a foot thick. Finally, the entire reactor is covered by a dome-shaped, concrete vessel about 150 ft high with reinforced concrete walls 3½ ft thick over a steel liner (Figure

Figure 4.20b. The Trojan nuclear power plant on the Columbia River near Prescott, Washington, in 1972. The enormous cooling tower in the background dwarfs the multistory buildings on the bank of the river. [EPA-DOCUMEMERICA-Gene Daniels.]

4.20a). The plants are designed to withstand earthquakes and any external explosions short of a direct hit by a nuclear warhead. (See Figure 4.20b.)

The efficiency with which such a plant converts heat to electricity is about 32%, so that more than two thirds of the heat released by fission goes to the environment. Some of the newer types of nuclear power reactors, such as the high temperature gas reactor, achieve efficiencies of nearly 40%, comparable to the best fossil fuel plants.

All presently operating nuclear power plants use ^{235}U fission. Because this is a rare isotope of a not especially abundant element and because of the expected increase in nuclear plant construction, serious ^{235}U shortages may develop in the next 20 years for plants being planned and constructed now. The ^{235}U fission plants do not represent a long-term solution to the energy problem. Rather, they represent a stopgap measure to serve until power from a more abundant fuel source is available.

Breeder Reactions and Power Plants. One possible means of avoiding the ^{235}U shortage is to find some other more abundant fissionable nucleus to use as fuel. It is possible to cause a nuclear transformation of the abundant uranium isotope 238 (over 99% abundance) into the element plutonium (Pu) which will fission. A representation of how this proceeds is shown in Figure 4.21. ^{238}U absorbs a neutron and undergoes a quick series of radioactive decays to become ^{239}Pu. This isotope will fission if struck by a high energy neutron. The three neutrons released during fission serve dual purposes: they can create more ^{239}Pu if they are captured by ^{238}U, and they can also cause fission if they strike a plutonium nucleus. A detailed analysis indicates that such a reaction can create more plutonium than it uses, hence the term "breeder" is used to describe such a nuclear reaction.

A breeder reactor can create fissionable fuel (^{239}Pu) for other reactors while generating electricity. Higher energy (that is, higher speed) neutrons are needed for the breeder reaction than for the ^{235}U fission reaction. Conse-

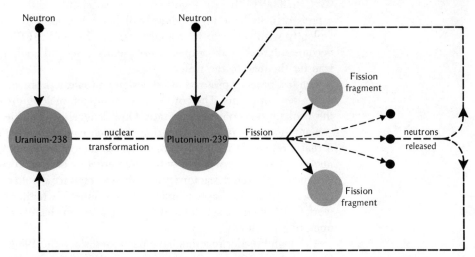

Figure 4.21. *A schematic representation of a breeder reaction. Absorption of a neutron converts ^{238}U into plutonium. Neutron bombardment of ^{239}Pu causes it to fission. The three neutrons released during fission can hit ^{238}U and thus create more plutonium or can hit ^{239}Pu and thus cause more fission.*

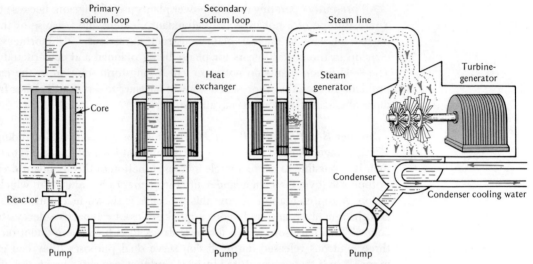

Figure 4.22a. *A schematic representation of the liquid metal fast breeder reactor (LMFBR). Liquid sodium is used to transfer heat to steam line.*

quently, in the design of breeder reactor power plants, no moderator (water) is used. Instead, as in Figure 4.22a, liquid sodium, which will not slow the neutrons, is used to transfer the heat. Such a design is called a liquid metal fast breeder reactor (LMFBR). Sodium is a silvery metal that exists as a liquid from 98°C to 900°C. It is a highly reactive substance, and great care must be exercised in its handling. Sodium will burn if exposed to air and will react violently with water to form hydrogen gas. In the LMFBR, the sodium will become intensely radioactive; consequently a second cooling loop is used to transfer the heat to the water.

The fast breeder presents more design and safety problems than do present fission reactor plants. The core is more compact, making greater demands on the coolant and coolant flow rate. One danger in all nuclear plants is that coolant loss might result in a heating of the core until it melts. The danger then is that a release of radioactive materials to the environment might occur. All nuclear plants are equipped with an emergency backup core cooling system designed to flood the reactor with coolant if a failure should occur. As we shall discuss later in the chapter (under Reactor Safety), whether such systems will work as designed is a subject of debate because none has ever been tested on an operating reactor.

Although breeder reactors have been used since the 1950s for research and for prototype power generation, construction of the first large-scale LMFBR for commercial power generation began in 1974. The plant, to be completed in 1985, is in Oak Ridge, Tennessee. A substantial portion of the development costs, which have escalated sharply since construction began, are being borne

Figure 4.22b. *A reactor vessel head used in a fast breeder test program at Hanford, Washington. [Du Pont Context, 3 (2), (1974), p. 20. Published by the Du Pont Company. Reprinted with permission.]*

by the United States government. The Soviet Union will have two large breeder reactors operating for power generation before 1985. The efficiency of the LMFBR at Oak Ridge is expected to be about 38%.

Environmental Hazards of Nuclear Power

The second law of thermodynamics implies that any form of energy production will result in some form of environmental disruption. Nuclear power plants present a number of environmental problems, some of which are common to fossil fuel plants and some of which are unique to nuclear plants.

Disposal of Waste Heat: "Thermal Pollution." The efficiency of fission plants is about 33%. Because of its lower efficiency, a nuclear plant releases about 40% more waste heat per kilowatt than a fossil fuel plant of the same electrical capacity. This waste heat must be transferred to the environment, which can be done by cooling the plant with air, using a cooling tower, or by cooling with water, using a river, a pond, or an ocean.

Cooling towers are huge "radiators," often 100 ft in diameter and 400 ft tall (Figure 4.20b). The hot water from the turbine is pumped to the top of the tower and drained over cooling fins. This method of cooling is expensive, often adding 10% to the cost of electricity (see the cost comparisons between nuclear and fossil fuel plants later in this chapter). Furthermore, in drier climates the water loss through evaporation from the use of towers exceeds the loss from other methods of plant cooling. If maintaining adequate ground water supplies is a serious concern, as it often is in the western and southwestern portions of the United States, then cooling towers should not be used. In addition to these problems, the massive size of the towers is somewhat of an aesthetic problem.

If water cooling is used, the total water requirements are impressively large. A 1000 Mw nuclear plant requires about 7500 gal of cooling water per second, and it is usually heated about 10°C (22°F). Obtaining sufficient water is often difficult. Most rivers in the United States have a flow too small during the summer to handle the heat loss, so cooling ponds are often used. The water is pumped from one end of the pond into the condensers and then returned to the opposite end of the pond. A 2000-acre pond (about 3 square miles) a few feet deep is required for a 1000 Mw nuclear plant. Warming the water increases the evaporation rate. In the drier portions of the United States, the evaporation losses from such a pond would be about equal to the water used by a city of 60,000 persons. Because of the possibility of the buildup of radioisotopes in the aquatic system, cooling ponds are often used only for cooling; recreational use, such as fishing and boating, is in many instances prohibited.

An alternative to the cooling pond is "once through" cooling by diverting the flow of a river through the condensers. For such a process, the critical concern is what fraction of the river should be used for this purpose. Increasing the temperature of a water body will inevitably affect the aquatic ecosystem. These effects have in fact led to the use of the term "thermal pollution" to describe the heat disposal problems.

A wide variety of chemical and biological effects may occur as the temperature is increased. The rate of a chemical reaction increases, often doubling for every 10°C (22°F) increase. As the temperature is raised the dissolved oxygen content of water *decreases*, but the oxygen consumption requirements of aquatic organisms *increases*. Algal plant growth and bacterial action also increase with temperature. Water temperature is closely related to the metabolic rates of organisms, especially those of cold-blooded species such as fishes and snakes. Spawning, feeding, and migration are also influenced by temperature changes.

Not many fish survive at the temperature (32°C or 90°F or more) of the thermal discharges, and no fish native to the United States survives temperatures greater than about 35°C (95°F). Brook trout, for example, thrives best at water temperatures of 4.4° to 10°C (40° to 50°F). At higher temperatures, say, 15.6°C (60°F), the fish swims slower, is less active, and feeds less. However, the fish's metabolic rate is faster at this higher temperature, so it actually needs more food than at 4.4°C (40°F). Above 25°C (77°F) the trout cannot survive.

A generalization is that "trash" fish tend to survive warmer water temperatures better than do "game" fish. Carp, alewives, sunfish, and perch survive at temperatures higher than do bass, trout, and salmon.

The effects of thermal discharges can be intensified by the fact that power plants do not run continuously. Nuclear plants in particular are plagued by intentional power reductions or complete plant shutdowns. The variations in plant operation create sudden temperature changes which may occur more rapidly than most species can adjust to or escape from the higher temperatures.

We should emphasize that, although the heating of an aquatic ecosystem causes changes, it does not necessarily follow that all of them are harmful. Recreational fishing in the area, for example, may improve. Many aquatic biologists believe, however, that in the *long* term the damages from increased algal growth and eutrophication (see Chapter 6, Water Pollution) will exceed the benefits.

From a practical point of view, thermal effects in a *small* portion of a waterway can be tolerated. The disturbance in that area, in many (though not all) cases, would constitute a negligible change in the ecosystem of the entire water body. (The present standards for running water coolants, for example, require that the surface temperature increase 300 ft downstream from a plant must not exceed 3°F.)

A dominating feature of thermal discharges from power plants is that the rapid growth in electrical generating capacity will threaten *major* portions of many rivers and lakes, even those as large as Lake Michigan. As an indication of the impact of this rapid growth, Figure 4.23 shows the power plant capacity (both nuclear and fossil fuel) in the United States in 1970 and the capacity projected for 1990. If this growth rate continues, an amount of water equivalent to half the entire annual freshwater runoff will be needed to cool power plants in the United States by the year 2000.

A number of proposals have been made to utilize this heat. Enough heat is discharged by power plants to heat all homes in the United States. Space

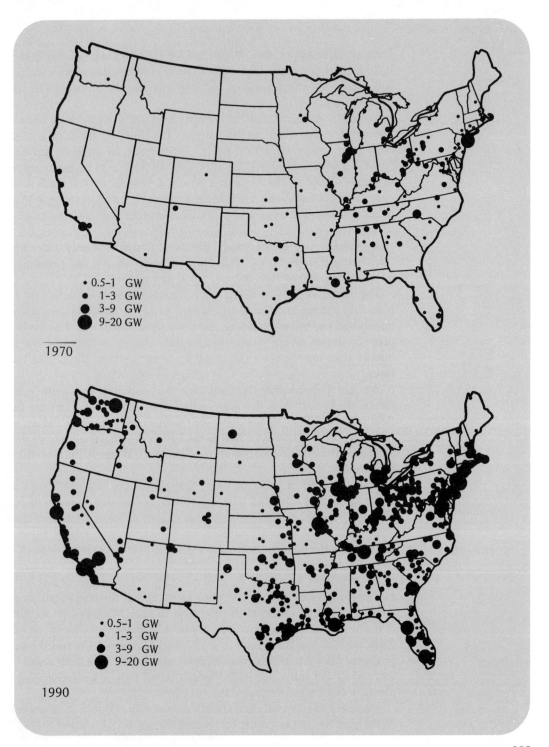

0.5–1 GW
1–3 GW
3–9 GW
9–20 GW

1970

0.5–1 GW
1–3 GW
3–9 GW
9–20 GW

1990

202

Figure 4.23.   *Major steam generating centers of all types 1970 and 1990. [From Rolf Eliassen, ''Power Generation and the Environment,'' Bulletin of the Atomic Scientists, September 1971. Copyright © 1971 by the Educational Foundation for Nuclear Science. Reprinted with permission of the author and the publisher.]*

heating using power plant waste heat is limited by the fact that heated water or air must be pumped to the homes from the plants, and our present policy of locating plants away from urban centers increases the costs of such schemes. The fact that aquatic plant growth is faster in warmer water has led to the use, notably in Japan, of the warmed water for food growth (aquaculture).

Radioactivity. Radiation exposure created by the use of nuclear power is an environmental problem of some consequence. The mining of uranium and the preparation of the fuel elements create exposure problems for workers in those areas. Some leakage of radioisotopes invariably occurs even during normal plant operation, and accidents with reactors could result in the release of enormous amounts. Processing the spent fuel rods after use in the reactor creates the possibility for more exposure. Finally, radioactive wastes created during the fission reaction must be kept out of the air and water for many generations.

We shall begin by considering the effects of radioactivity on living systems, then focus our discussion on nuclear power plants.

Radiation Damage. Earlier in the chapter we discussed three types of ionizing radiation, α, β, and γ. All three are capable of creating radiation damage, genetic transformations, and death in organisms. A living organism can be irradiated in either of two ways: from external exposure or from ingestion of the radioisotope. Of these two, the latter is potentially the more harmful because the nucleus may remain in the body, perhaps substituting for an essential element of life, an example being strontium-90 (^{90}Sr) replacing calcium in bones. The organism is then subjected to continuous radiation effects for an extended period. In contrast, external radiation is transitory. Furthermore, α particles, which are the most destructive, often fail to penetrate the skin.

Radiation damage occurs by disruption of molecules in cells, and the effects are classified as either somatic or genetic. Damage that affects the individual organism but is *not* transmitted to future generations is **somatic**; examples are radiation sickness, cataracts, cancer, and death. These arise from damage to cells that are not involved in reproduction.

Genetic effects relate to the disruption of the cells involved in reproduction. Such effects may result in abnormal offspring in future generations. Each of the 50 trillion cells in the human body contains 23 pairs of structures called chromosomes which in turn contain many genes. Growth and healing of the cell depend on the fact that the cell can divide and reproduce. The chromosomes and genes provide the "genetic code" that determines the many characteristics of an individual cell and ultimately, therefore, determines the characteristics of an individual. Each chromosome reproduces itself during division, so that the daughter cells also possess 23 pairs with the same genetic code.

Chromosomes and genes are composed of chemical compounds arranged in a very carefully structured order. When a change in the carefully ordered chromosomal chemical structure occurs or when ionization of a chromosomal structure occurs, an imperfect reproduction may result—color blindness, for example, or the possession of two joints on each finger instead of three. Such a "malfunction" is called a **mutation.** Radiation can cause mutations by ionizing chemical entities in the chromosomes. In addition radiation damage to a more abundant chemical in the cell, such as water, can result in an alteration of reproduction because the ionized water can react with and alter the other portions of the cell including the chromosomes.

An important aspect of a radiation-induced genetic mutation is that it is frequently a recessive trait. Genes are classified as dominant or recessive, and the traits of offspring are determined by the parental genes. For example, brown eye genes are dominant whereas blue eye genes are recessive. If one parent has the brown eye genes (dominant) and the other has a blue eye gene (recessive), the offspring will have brown eyes but will continue to have and transmit both the recessive blue eye trait and the dominant brown eye trait. If in later generations both adults possess the recessive blue eye gene, their offspring may have blue eyes. Thus, the radiation-induced mutations may not appear in the first generation but could have a significant impact in subsequent generations.

Measurement of Radiation. Several units are used to measure radiation exposure and dosage. We shall define only those relevant to our discussion. The **rad** is used to measure radiation dosage, that is, the total energy of radiation that has impacted upon or has been absorbed by a body. (One rad is equivalent to 100 ergs of radiation energy per gram of absorbing material.) The rad can be used to measure the radiation received by any substance—metals, plastics, vegetation, or living tissues, for example. It does not discriminate between α, β, or γ; it merely reflects the total energy received from radiation of all types. For biological systems it is often important to know, in addition to the rads received, how much of each type of radiation has been absorbed. Because of its much heavier mass and much more effective ionization properties, an α does, on the average, about ten times more biological damage than a β or a γ. Consequently 1 rad of α creates ten times more harm in living tissue than 1 rad of β or 1 rad of γ. To measure the effects of radiation in *biological* systems more accurately, the **rem** is used as a measure of radiation dosage.

$$\text{radiation dosage in rems} = (\text{rads}) \times (\text{relative biological effect}) \quad (4.12)$$

The **relative biological effect** is merely a number that equals 10 for α, 1 for β, and 1 for γ. That is, the absorption of 2.5 rads of beta is a dosage of 2.5 rems, but the absorption of 2.5 rads of alpha is 25 rems dosage. In our discussion we shall use only rems. For smaller exposures it is conventional to measure dosage in **millirems** (or thousandths of a rem), which is abbreviated as **mrem**.

Effect of Radiation on Living Organisms. Our knowledge of the effects of radiation comes from a variety of sources. In World War I, radium was painted on dials to make them glow in the dark. The painting was done by hand, and the painters ingested sufficient radium to cause serious consequences in later years. Extensive studies have been done on the Japanese exposed to the atomic blasts of 1945. Miners and workers in industries using radioisotopes are exposed to higher amounts of radiation than most people and are hence studied for possible effects. Various medical procedures in the last 50 years have used high x-ray dosages to treat or attempt to treat diseases, and the effects of the exposure on those individuals have been recorded.

Death from radiation occurs only at heavy exposures of several hundred rems. A single dosage of 100 or 200 rems will cause sickness and nausea (due to the killed cells) but will only rarely cause death.

Genetic effects of radiation appear at much lower levels, but it is difficult to obtain a quantitative relationship between low-level exposure and genetic mutations because of the effects of the natural background radiation to which we are all exposed. Table 4.11 gives the average yearly dosage in the United States. The natural background arises from cosmic radiation (that from outer space) incident on the earth and from natural radioactivity of the elements in the earth's crust. This natural level varies from state to state, those states with higher elevation receiving more radiation from cosmic sources. Colorado has the highest figure—about 250 rems/year; Louisiana, the lowest—about 100. Of

TABLE 4.11. Whole body yearly radiation dosage in the United States

Radiation Source	1970	2000
Natural	130.	130.
Man-made		
Medical procedures	74.	88.
Radioactive fallout	4.	5.
Miscellaneous	3.5	2.
Nuclear power	0.05	0.5
Other AEC activities	0.015	0.012
Total	211.	225.

the approximately 80 mrem from man-made sources, over 90% arises from medical practices, especially x-rays. Many smaller man-made sources exist.

The natural background and man-made sources of radioactivity undoubtedly result in genetic and somatic effects in living organisms although there is controversy about the exact number that might be caused. We obviously can exert regulatory control only on the man-made sources listed in Table 4.11. Guidelines for maximum radiation exposure, which have existed for over 50 years, reflect efforts to balance societal good against predicted genetic and somatic risks. The medical exposure levels, principally from x-rays (Table 4.11) are an example. A single chest x-ray gives about 100 mrem exposure and a dental x-ray gives about 20. The Health Physics Department of the Oak Ridge National Laboratory estimates that deaths owing to x-ray-induced diseases (such as bone cancer, thyroid, tumor, leukemia, and other maladies) range to 3000 annually. Children born to women who have had pelvic x-rays during pregnancy have a higher (though still very small) risk of developing leukemia than children whose mothers have not had such exposures.

Genetic transformations due to x-rays may create an even greater number of deaths. However, the benefits of x-rays in medicine in the hands of competent physicians are assumed to offset this liability. That is, many thousands of lives are saved each year by the use of x-rays. (X-rays should not, however, be regarded as "routine." Estimates have been made that as many as 30% of the x-rays taken annually are unnecessary and hence present needless exposure.) Should a limit on x-ray exposure be instituted? The EPA, which now has jurisdiction in this area, prefers to leave the responsibility to the discretion of physicians.

The present guidelines for maximum exposure of the general population are 170 mrem per year *in addition to* natural and medical exposure. This level was established in the late 1950s. Historically the allowed exposure levels have decreased as knowledge of radiation effects has increased. In the late 1960s and early 1970s considerable controversy existed as to whether the 170 mrem standard should be lowered. Those most strongly in favor of lowering the standard were represented by Drs. John Gofman and Arthur Tamplin of the Livermore Laboratory which was at the time operated by the Atomic Energy Commission (AEC). They were supported by Dr. Linus Pauling, a winner of two Nobel Prizes, one for chemistry and one for peace.

Gofman and Tamplin were involved in research on low level radiation and predicted that if the 170 mrem additional dosage *were* actually received by each member of the United States population, 30,000 extra deaths from cancer would result each year, an increase of roughly 10% in the current death rate from cancer. Pauling's estimates were even larger. These men urged that the additional exposure maximum be reduced tenfold to 17 mrem per year. The AEC strongly challenged the assertion of its employees, Gofman and Tamplin, arguing that their projections were overestimations.

In late 1972 a special committee of the National Academy of Sciences, which had been asked to investigate the standards, reported that a more accurate estimate of the impact of exposing the entire United States population to the full 170 mrem above natural and medical sources would probably be closer to 6000 deaths. In addition, the Academy committee projected an added total of genetic effects and diseases of several thousands per year if this maximum allowed exposure actually occurred. The committee emphasized that such estimates were fraught with uncertainty. As noted in Table 4.11, the *average* exposure in addition to natural and medical sources is nowhere near 170 mrem, as the AEC was quick to point out. Gofman and Tamplin have responded by inquiring that, if the extra exposure levels are low, why not lower the standards? Why maintain the legal possibility of a considerable increase in exposure?

Emissions from the Nuclear Power Industry. The standard for normal operation of fission plants is that a level of 5 mrem per year shall not be exceeded at the outside boundary of the plant. This was established by the AEC and has been maintained by its successor, the Nuclear Regulatory Commission (NRC). Under emergency conditions this level can be increased to 17 mrem per year. The exposure from nuclear plants shown in Table 4.11 refers to an average exposure from nuclear plants to the *entire* population. It will increase as nuclear power proliferates. The radiation emitted by an individual plant decreases drastically with the distance from the plant. Those persons closest to the plant will receive larger doses than indicated in Table 4.11, but in no case should this exceed 17 mrem per year. The NRC further contends that most of the newer plants will operate well below the 5 mrem standard.

The emission standard of 5 mrem does not apply to fuel processing plants (or to breeder reactors). During the processing of spent fuel rods, radiation is emitted, a particular problem being release of gaseous tritium and krypton-85. Krypton, which does not dissolve in water and forms almost no compounds, has a half-life of 9 years. Measurements outside the one processing plant in the United States indicated levels of 75 mrem per year.

Fossil fuel plants also emit radioactive isotopes. Coal and oil contain radioisotopes, especially radium-226 (^{226}Ra) and radium-228 (^{228}Ra) which are released upon combustion. The total radiation from the operation of a large coal plant, such as the Four Corners plant, may be comparable to emissions from a large nuclear power plant operating properly within the AEC standard.

The hazard of an additional 5 mrem per year, even if absorbed by the entire United States population (which probably will not happen in the near future), is almost impossible to establish. On a statistical basis, to establish (to within 95% confidence) the anticipated effect of this 5 mrem increase above the natural background would require experiments on 8 billion mice.

One other aspect of radiation from plants should not be overlooked: the released radioisotopes may concentrate (magnify) in the food chain. (This biological magnification also occurs with other substances, such as metals and DDT as discussed in Chapters 1, 6, and 8.) Organisms can concentrate radio-isotopes to levels much higher than those found in their surroundings. For example, iodine-131 (^{131}I) concentrates in the thyroid gland, strontium-90 (^{90}Sr) in the bones, and cesium-197 (^{197}Cs) in the muscles. If minute organisms should concentrate the isotopes they ingest, they pass these on to organisms higher up the chain (see Chapter 1). At the top of the chain, the radioisotope levels may be 20,000 times larger than those present in the water.

Radioactive Waste Treatment and Storage. The operation of a nuclear reactor creates many radioactive fission fragments and other radioactive elements. In the present scheme of nuclear power, when the "burned" fuel rods are removed (approximately yearly), the spent rods are stored for several months at the reactor until the most intensely radioactive elements, those with half-lives of a few days, have decayed to levels that make handling more feasible. The rods are then transported to a reprocessing facility. Here those isotopes useful in research, in medicine, and in nuclear weapons are removed to within the limits of chemical technology. (As we noted in Chapter 1, no chemical impurity is ever completely separated from a mixture.) Uranium is also removed because the "burning" of the fuel rods uses only a few percent of the ^{235}U initially present.

After processing, a number of radioisotopes that have no use or that cannot be efficiently separated, are left as "waste." These radioisotopes must be stored to ensure that they are not disseminated in the environment. To estimate the storage time, we might assume that the storage should last at least ten half-lives of the longest-lived species. At the end of such a storage period, the longest-lived species would be 2000 times lower in concentration. Plutonium-239 (^{239}Pu) the longest-lived species, has a half-life of over 24,000 years; hence the waste storage must function for 200,000 years! The transportation and storage of these wastes are some of the most vexing problems of nuclear power and have been a focus of the pro and con arguments about nuclear power. Use of fission plants means a commitment to waste storage in perpetuity (or at least for hundreds of generations). This time span is sufficiently long to bring into question whether societal order is sufficiently stable to ensure care for the duration. Other questions also arise, such as whether the storage procedure will survive another ice age that might occur during the time of storage.

Transportation of the wastes will be by a special shipping cask (Figure 4.24). Dr. Alvin Weinberg, former director of the AEC's Oak Ridge National Laboratory, estimates that by the year 2000 there will be 7000 to 12,000 annual shipments of spent fuel from reactors to chemical processing plants with nearly 100 in transit at all times. On the basis of present statistics, if these were shipped by rail, about ten per year would be involved in derailing accidents.

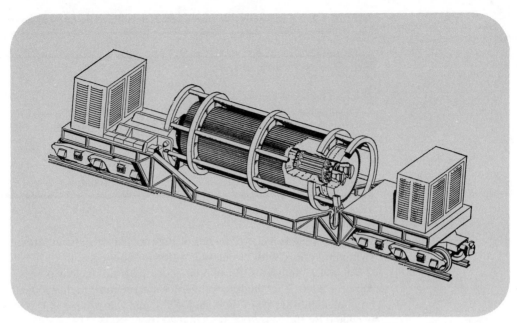

Figure 4.24. *A shipping cask used to transport fuel elements for nuclear reactors.*
[*From A. W. Weinberg, "Social Institutions and Nuclear Energy," Science,* **177** *(July 7, 1972). Copyright 1972 by the American Association for the Advancement of Science. Reprinted with permission.*]

The casks are designed to withstand these and other common accidents; they must be able to withstand a 30-ft fall onto a hard surface, a 30-min fire followed by quenching in water, and a fall onto a steel spike.

The necessity for waste storage and disposal has existed since the 1940s. At present almost 100 million gal of radioactive wastes are stored in liquid form in large tanks hear Hanford, Washington, and near Aiken, South Carolina. These were generated by military applications of nuclear energy but are chemically similar to the wastes expected from power facilities. The results with the liquid storage method have not been encouraging. Because of the effects of the radioactivity, these tanks seldom last more than 20 years. Over 16 tank failures have occurred, one that allowed 150,000 gal to drain into the soil before the leak was discovered. The danger of such leaks is that they will seep to an underground river and create a radioactive hazard that will last for generations. The AEC denied that any harm resulted from the leaks but acknowledged that a better method must be found.

The present procedure is to convert the wastes to a solid form for holding until a more permanent solution is decided upon. By boiling off the liquids and casting the radioactive residue in a ceramic material, a solid bulk is created

TABLE 4.12.   Radioactive waste in solid form from the nuclear power industry

	1980	2000
Fuel reprocessed (metric tons/year)	3,000	19,000
Solidified high-level waste		
Annual volume (ft³)	9,700	58,000
Accumulated volume (ft³)	44,000	770,000
Low-level wastes		
Annual volume (ft³)	1,000,000	1,000,000
Accumulated volume (ft³)	11,000,000	31,000,000

that avoids the leakage problems of liquids. The solid forms are held in large swimming pools, with circulating water for cooling.

Table 4.12 indicates the amount expected from nuclear power industry by the year 2000. The "high level" wastes are the most highly radioactive. Principal elements are ^{90}Sr, ^{137}Cs, and ^{239}Pu, and the volume of this classification expected by the year 2000 is a cube 90 ft on a side. The elements in the "low level" category are less hazardous and might be stored under less stringent conditions.

Because of radioactive decay, the nature of the wastes changes with time. In simplified terms, the problem can be described as a 1000-year storage of several radioisotopes followed by extended care of ^{239}Pu. Radioisotopes of cesium and strontium are present and have half-lives near 30 years; these would be reduced by 1/1,000,000 after 1000 years. Left in the waste will be ^{239}Pu with a 24,000-year half-life. Plutonium is one of the most toxic substances ever studied. Less than a millionth of a gram (about the size of a pollen grain) can cause cancer. Several tons of ^{239}Pu and chemically similar elements (called the actinides) will be present in the wastes by the end of this century.

The technical problem is thus: How are these wastes to be stored "permanently"?

The only place where the wastes will never interfere with the earth is far outer space or in the sun where the high temperature fusion reaction would make for excellent waste disposal. However, sending the wastes into outer space would raise the cost of electricity by as much as 30%. In addition to this, a hazard would exist if the launch had to be aborted.

The traditional American approach to waste disposal has been to toss wastes into the nearest uninhabited area or body of water. This tradition fortunately is on the decline, but a consideration of the disposal of nuclear wastes in the traditional American manner gives insight into the magnitude of radiation expected. We might consider dispersing the pollutants evenly in the entire atmosphere or in the oceans while attempting to stay within the present inhalation and ingestion standards for radioactive isotopes. The entire atmosphere of the earth is 1000 times too small to permit the dispersal within the allowed concentrations. Uniform dispersion of the wastes in the oceans would raise the radioactivity of the oceans to almost the maximum permitted level. Further dumping in future years would exceed the standard.

Another proposal has been the creation of holding areas, perhaps as pyramids in deserts, with the attendant formation of a "priesthood" whose duty is to care for the wastes, monitoring any possible leakage, and transmitting the information about treatment from generation to generation.

The NRC and its predecessor, the AEC, have searched for a solution that is not dependent on the maintenance of social order. Since the storage period is typical of geological time scales, one solution might be to place the wastes in a geologically stable formation, such as salt beds. Some of the bedded salt under the United States was laid down in Permian times and has been undisturbed (except for man's intrusions) for 200 million years. The proposal has been made that the solidified wastes be sealed in stainless steel cylinders about 6 in. in diameter and 10 ft long and buried in 1000 ft-deep holes in the salt. Because of the heat released during radioactive decay, the cans will reach temperatures of several hundred degrees centigrade and the salt, which has a consistency similar to hard wax, will undergo "plastic flow," a slow gradual tight packing around the containers almost as if the salt melted and resolidified. Even if the metal and ceramic disintegrate in a few years, the radioactive elements would hopefully remain encased in the salt. A major liability of such a scheme is the possibility that water might enter the salt bed, dissolve the salt, and carry the wastes to ground water supplies. Such a liability might result from geological changes, from earthquakes, or from man's activities.

After 15 years of discussion and some research, the AEC in the early 1970s announced a tentative plan using salt bed storage at a site near Lyons, Kansas. Salt mines are found near Lyons, and thus good rail connections existed. Questions concerning the suitability of the site came from Dr. William Hambleton of the Kansas Geologic Survey who felt, among other objections, that the AEC had insufficient data on the salt beds to justify the assumption that water was not likely to enter them. In hearings that followed, it developed that water intrusion into the beds was certainly a possibility. Over 100,000 oil wells had been drilled through these salt formations in Kansas alone. Thirty of these oil wells are near the site. Also, some solution mining, involving the pumping of water into the beds, had been done in the area. These problems and the rising political pressure in Kansas against the site forced the AEC in 1973 to look for another salt disposal site, perhaps in New Mexico (Figure 4.25).

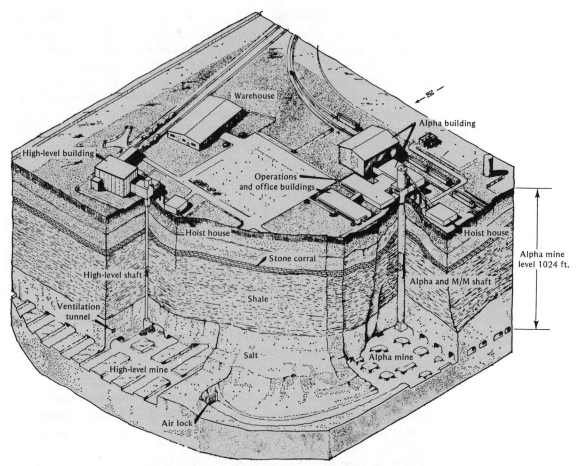

Figure 4.25. A model of the proposed federal repository for radioactive wastes. This proposal uses salt bed storage. [From A. W. Weinberg, "Social Institutions and Nuclear Energy," Science, **177** (July 7, 1972). Copyright 1972 by the American Association for the Advancement of Science. Reprinted with permission.]

Another suggested approach to waste disposal is to bury the wastes in the center of a polar ice cap. The proposed sites are 1000 miles from any ocean and are surrounded by mountain ranges. Transportation to the sites would be by large icebreakers specially designed for flotation in case of accidents. One advantage of such sites is that future generations are not likely to happen upon the wastes by accident. Only a technically advanced civilization, which presumably knows about radiation, would be digging in the polar caps. At present the Antarctic treaty of 1959 prevents radioactive waste disposal there, so the treaty would have to be altered if this procedure were tried.

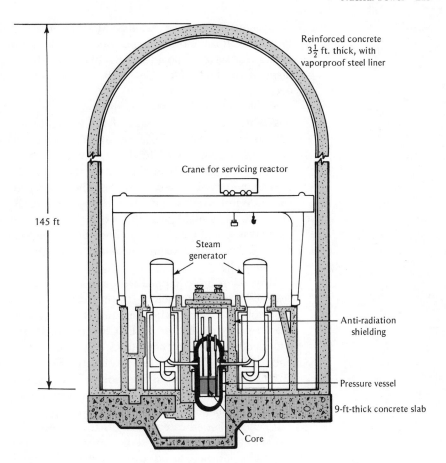

Figure 4.26. *Structure of a nuclear power plant. The outer shell is the containment structure. [After New York Times Co. Copyright 1971.]*

Some scientists feel permanent storage should not be attempted at this time. Rather, the wastes should be stored and tended above ground in anticipation that better chemical processing technology or waste disposal techniques might be available in a few centuries. Others involved in waste disposal problems point out that since after 1000 years the wastes are mostly plutonium, we are offering future generations this valuable energy source in exchange for several centuries of tending the storage bins.

Reactor Safety. Often paramount in the public mind, because of the association of nuclear power with bombs, is the worry about the safety of nuclear power plants. The reactors used in such plants are designed to minimize the chance of a drastic accident. Construction materials such as tubing and hardware are carefully checked for flaws. In the event an accident does occur, however, several "failsafe" devices are designed into the plant to shut down the reactor and cool the assembly. Many of these are backup schemes designed to activate if some preceding device has failed. In the event that all of these emergency systems should fail, the plant shell is designed to provide containment of the fuel under a majority of the postulated conditions that might occur.

Should an accident, such as the loss of coolant, occur there is almost no possibility of a fission reactor of the presently used variety exploding as a nuclear bomb. The ^{235}U concentration in these devices is too small. Furthermore the fission of ^{235}U tends to slow down if coolant is lost because the neutrons are no longer moderated by the water. Breeder reactors, however, present more trying safety problems. The breeder reaction tends to speed up if coolant is lost. Furthermore the ^{238}U is present in an abundance similar to that in bombs, and a slight possibility exists of an uncontrolled chain reaction if the meltdown should happen to produce the proper geometric rearrangement of the fuel. The liquid sodium coolant can present a hazard if it escapes the reactor shell.

All reactors are equipped with an emergency core cooling system (ECCS) designed to prevent melting of the reactor core by flooding it with excess coolant from as many as four separate storage vessels. A point of controversy in reactor safety is whether the ECCS will actually perform as intended. So far, under actual operating conditions ECCS has not been called on to function. But tests in 1972 raised the possibility that the coolant might bypass the reactor and thus fail to provide cooling. The possibility of a core meltdown has been described as the "China syndrome," wherein a core melts its way through the shell of the plant into the earth, heading presumably toward China.

It cannot be said categorically that a catastrophic failure of a reactor in a power plant is impossible, but elaborate measures are taken to make the probability of such an occurrence extremely slight. By 1976 more than 200 reactor years of commercial electrical reactor operation had been accumulated without loss of life owing to malfunction of the reactor. There have been accidents and fatalities with research reactors rather different from those now being promoted for use in power plants. It is realistic to assume that a catastrophic reactor accident might occur in the future because of human error, carelessness, mechanical failure, or an incredible series of events. If a major accident occurred that resulted in the release of radioactivity, its effects might be staggering.

The AEC recently assessed the accident risks of United States commercial power plants by means of computer simulation studies, and reported in 1974 that the risks were estimated to be smaller than many other man-made and natural risks. The study implies that even with 100 nuclear plants in operation (as is expected in the early 1980s) the chance of an accident severe enough to kill 100 persons is far less than the chance of an airplane crash of similar magnitude. The worst possible nuclear accident, which was estimated to lead to 2300 fatalities, $6 billion in property damage, and permanent contamination of 31 square miles of land around the reactor, has a probability rated less than the possibility that the same number of persons could be killed by a large meteor. The results of this study must be viewed within the limitations of computer modelling. The study could not, for example, predict the possible presence of fundamental design errors. Nor could it take into account the possibility that a plant might be sabotaged. The AEC contended, however, that acts of sabotage would require an extensive knowledge of plant design and operations and that good security procedures could make the possibility of sabotage unlikely.

Thus the study offers some assurance, but obviously not a guarantee, that the probability of a major accident is slight. If 100 or fewer deaths occur each century from nuclear power plant accidents, this rate would be lower than that in most other industries.

Some experts, among them Dr. Edward Teller of the AEC's Lawrence Laboratory, feel that nuclear reactors should be placed underground to minimize the danger should an accident occur.

Plutonium Proliferation. An increasing reliance on nuclear power means that by the year 2000, the United States and other nations will have inventories of hundreds of tons of ^{239}Pu, an essential ingredient of nuclear bombs and warheads. A single large fission reactor, even the ^{235}U variety, makes enough plutonium every year for several hundred nuclear bombs. As created in the reactor, the ^{239}Pu is contaminated with other plutonium isotopes that make the bomb less efficient, but it is still adequate for most purposes. Making nuclear weapons is now a technological problem principally related to obtaining plutonium. The techniques required to construct the device are well known to physicists in all nations and many nations have the capability to make nuclear weapons should they choose to devote a large amount of resources and money to the task. India's entry into the nuclear weapons field in 1974 probably came about by using plutonium from a research reactor sold to them in 1954 by Canada. Because of a legal loophole in the sales agreement, this was one of the few reactors in the world for which the plutonium inventory was not covered by international safeguards.

Shipments of nuclear fuel for the power industry use conventional civilian shipping methods, and the present record is not especially commendable. A number of accidental misroutings have occurred, one of which resulted in material being sent outside United States borders. In contrast, the shipping of nuclear materials used in military applications involves a special means of transportation, including the use of guards. The cost of this means of transportation, which would result in a substantial increase in the cost of electricity, has limited such procedures for the nuclear power industry.

Because of the large inventories expected and the poor handling of the fuels, the possibility of misroutings, theft, or hijacking of plutonium will increase in the future. This may lead to a profusion of nuclear weapons and increased worldwide tensions.

The Role of the AEC

The AEC was established following World War II to coordinate nuclear policy and developments in the United States. For the nuclear power industry, the AEC had a dual charge: to promote the use of nuclear power and, at the same time, to regulate the industry in terms of safety standards and radioactive emissions standards. The promotion function involved subsidizing the development and design costs of reactors and working closely with the industrial suppliers of the reactors. The two AEC functions could easily have conflicting goals, and critics charged that the AEC sacrificed its safety and environmental responsibilities at the expense of promoting nuclear power.

Such charges were stoutly denied by the AEC; nevertheless, in 1975 the AEC was dissolved, and its responsibilities were transferred to two newly created governmental agencies: the Nuclear Regulatory Commission (NRC) and the Energy Research and Development Administration (ERDA). The NRC inherited all of the AEC's regulatory powers for safety and handling of materials, and ERDA took over the responsibility for the development of nuclear reactors for electric power generation. ERDA also has the major responsibility for most of the research and development done in other energy areas: solar, geothermal, fossil, and energy conservation.

Comparison of Nuclear and Fossil Fuel Plants

A summarizing comparison of the operation of nuclear and fossil fuel plants is given in Table 4.13. The emissions from the fossil fuel plants are those that would occur without emissions controls. Radioactive emissions from the preparation and reprocessing of the nuclear fuel rods are not included. We have discussed the impact of emissions from fossil fuel plants in Chapter 3 and from nuclear plants in this chapter.

Any comparisons of the costs of electricity from nuclear and fossil fuel plants must be viewed carefully. We should distinguish between the generating cost (which is reflected in your electric bill) and the total cost of producing the power. The total cost must include the impact of the emissions, the mining, the cooling requirements, and the transmission of the power. Such a price is very difficult to establish. It is very easy to obtain a comparison of the generating costs, as in Table 4.14 which gives a comparison for 1000–Mw plants beginning operation in 1978. The estimated costs are obtained by averaging the construction costs (capital costs) over a 30-year plant life. Nuclear plants have much higher construction costs but much lower fuel costs than fossil fuel

TABLE 4.13. Comparison of 1000 MW power plants

	Coal	Oil	Gas	Nuclear
Annual fuel consumption	2.3×10^6 tons	460×10^6 barrels	6800×10^6 ft^3	2500 lb
Annual release of pollutants in millions of pounds (no controls)				
Oxides of sulfur	306	116	0.03	0
Oxides of nitrogen	46	48	27	0
Carbon monoxide	1.15	0.02		0
Hydrocarbons	0.46	1.47		0
Aldehydes	0.12	0.26	0.07	0
Fly ash (97.5% removed)	9.9	1.6	1.0	0
Annual release of radioactive nuclides[a]	^{225}Ra, ^{226}Ra	^{225}Ra, ^{226}Ra	0	^{138}Xe, ^{18}Kr
Overall efficiency		38%		32%
Heat energy				
Input to generate 1kw-hr of electricity (Btu)		8,600		11,500
Total waste heat (Btu)		5,200		8,100

[a] ^{225}Ra, radium-225; ^{226}Ra, radium-226; ^{138}Xe, xenon-138; ^{18}Kr, krypton-18

Note: The emissions from the fossil fuel plants are those that occur in the absence of air pollution emission controls.

Source: Andrew P. Hull, "Radiation in Perspective: Some Comparisons of the Environmental risks from Nuclear and Fossil Fueled Power Plants," Nuclear Safety, May–June 1971. "Problems in Disposal of Waste Heat from Steam-Electric Plants," Federal Power Commission, Bureau of Power, 1969.

TABLE 4.14. Estimated cost of power from 1000-Mw power plants in 1978

	Nuclear Plant		Coal Plant			
			No SO$_2$ Control		With SO$_2$ Control	
	Run of River Cooling	Cooling Towers	Run of River Cooling	Cooling Towers	Run of River Cooling	Cooling Towers
Capital cost (millions)	$365.	$382.	$297.	$311.	$344.	$358.
Fixed charges	7.8	8.2	6.4	6.6	7.4	7.7
Fuel cost	1.9	1.9	3.9[a]	3.9	3.9	3.9
Operation and maintenance costs	0.6	0.6	0.5	0.5	0.8	0.8
Total power cost mills/kw-hr[b]	10.3	10.7	10.8	11.0	12.1	12.4

[a] The coal cost is assumed to be $10.6/ton in 1978 (it was $8/ton in 1971).

[b] A mill is 0.1 cents.

Source: A. M. Weinberg, "Social Institutions and Nuclear Energy," Science 177 (July 7, 1972), 27.

plants. The difference in cost between the two types will vary from site to site, but for the site in the table the comparisons range from comparable to 20% less for the nuclear plant. The cost of the fossil fuel varies drastically with the distance of the plant from the mine, however, and this can alter the relative comparisons.

Many economic factors important in a comparison of the plant are not shown in the table. In general, nuclear plants are much more heavily subsidized by the government, principally through the AEC and, more recently, through ERDA, than are fossil fuel plants. The costs of enriching the nuclear fuel and processing and storing the nuclear wastes are borne in part by the government. Utilities must obtain insurance from insurance companies. These companies will fully insure fossil fuel plants but will provide only a small fraction of the coverage needed for nuclear plants. The United States government, through the Price-Anderson Act of 1957, provides liability coverage to a nuclear plant to a maximum coverage of $500 million. Finally, the costs of the environmental impact of both types of plants are not accurately reflected in the table.

Summary   The energy growth rate in the world is several times greater than the population growth rate, and providing this energy inevitably creates environmental disruption. Energy use is strongly correlated with the economic power (GNP) of a nation, and the United States is by far the greatest per capita consumer of energy. Annual energy growth in the United States is about 4.5% in recent years, nearly five times the natural population growth rate. Fossil fuels supply 95% of our total energy needs, and even with a massive shift to nuclear power to generate electricity, fossil fuels are expected to provide 70% of our needs in the year 2000. At present rates of use, lifetimes of oil and gas are less than a century, whereas coal will last a few centuries. Even uranium used in present fission reactors is expected to be in short supply by the year 2000. About 15% of our energy is imported, partly because of our enormous energy growth rate and partly because oil policies over the last 20 years have favored the development of foreign rather than domestic supplies.

Efficiency of energy use is critical to the lifetime of resources. The overall efficiency of energy use in the United States is less than 50%. Electrical power generation averages 32%, whereas automobile engines convert less than 25% of the energy in gasoline into mechanical energy. On the average, only about 4% of the energy in fuels emerges as light in electrical lighting, but fluorescent lamps are four times more efficient than incandescent bulbs. In terms of using fuels for heating, central furnaces are considerably more efficient than electrical heating.

Electrical power is provided principally by fossil fuel plants, hydroelectric dams, and nuclear plants. Fossil fuel plants are a major source of air pollution, their fuel supplies are limited, and the mining required causes enormous problems. Hydroelectric plant sites are becoming limited, so that extensive increases in this source should not be expected. Present nuclear plants use a rare uranium isotope (^{235}U) in a fission reaction. Disadvantages of nuclear power include mining, reactor safety and radiation problems, the necessity of 40% more cooling water than equivalent sized fossil fuel plants, and the lack of a reliable means of storing radioactive wastes for thousands of years. Because ^{235}U is in short supply, fission breeder plants using ^{238}U will have to be developed if nuclear power is to expand later in this century. An increase in nuclear power means that plutonium proliferation will be a problem. Plutonium, the most toxic substance ever studied, is produced in ton lots annually in nuclear reactors. Because it is an essential element in nuclear weapons, keeping close track of inventories is an essential but increasingly difficult task.

Estimates of future generating costs are that electricity from nuclear plants will be slightly cheaper, but these estimates vary widely with construction costs, with required safeguards, and with air pollution controls. Generation costs do not reflect the total cost of the power; environmental impact and government subsidies are not included.

Review and Study Questions

1. Define or describe: work, power, energy reserve, dynamic conversion, fusion, isotopes, fly ash.
2. Of what use is thermodynamics in energy studies?
3. Compare United States energy and population growth rates.
4. What form of energy consumption shows the most rapid growth rate in the United States?
5. If nuclear power expands until it supplies 50% of United States electrical demands by 2000, will coal consumption decrease relative to 1975 usage?
6. Define or describe: radioactive decay, critical condition in a nuclear reaction. LMFBR, rem, emergency core cooling system.
7. Fossil fuels supplied what portion of United States energy in 1970? What will they supply in 2000?
8. Give an example of a *renewable* energy source.
9. Which of each of the following has a longer lifetime if consumed continuously at the rate we expect to consume them in the year 2000? (Total domestic recoverable resources data) (a) coal or oil? (b) coal or uranium (^{235}U fuel priced at \$15/lb or less)? (c) gas or uranium (^{238}U in a breeder reactor, fuel priced at \$15/lb or less)?
10. What are the major reasons for the decrease in the number of oil wells drilled annually in the United States since 1956?
11. Name four products that are derived from petrochemicals.
12. Transportation accounts for what portion of the United States energy consumption?

13. Why is direct conversion not used to generate large amounts of electrical power?

14. Which particular component of a steam power plant is largely responsible for the low overall efficiency of conversion of heat to electricity?

15. If 100 Btu of oil are used to generate electricity to power an incandescent light, how many Btu of visible light are created? Account for the remaining Btu.

16. Trace the history and impact of the Four Corners coal plant.

17. How does radioactivity differ from fission?

18. How is a branched chain reaction prevented in a reactor?

19. Why are breeder reactors being developed?

20. What factors limit the use of cooling towers to handle the cooling of power plants?

21. Contrast somatic and genetic effects of radiation.

22. Which, on the average, probably creates a greater hazard for biological systems: absorbing 1 rad of gamma radiation or absorbing 1 rad of alpha radiation?

23. What are the allowed radiation levels at the boundary of a nuclear power plant?

24. Discuss how spent fuel rods are processed after removal from the reactor (assume a salt bed storage scheme is used for wastes).

25. The Lyons, Kansas, salt beds were ultimately discarded by the AEC as a storage site for nuclear wastes. Why?

26. What two purposes does water serve in a ^{235}U reactor?

27. Who provides insurance for nuclear plants? For fossil fuel plants?

Questions Requiring Outside Reading

28. How does United States energy consumption compare with that of other nations? Could we maintain our present standard of living with a lower consumption of energy? (A. B. Makhijani and A. J. Lichtenberg, "Energy and Well Being," *Environment* [June 1972], 10–18; M. Corr and D. MacLeod, "Getting It Together," *Environment* [November 1972], 2–9.)

29. How does the rate of production of energy from fuel sources change with depletion of the resource? (See M. King Hubbert in *Resources and Man* Suggested Outside Reading.)

30. What are the prospects for improving the efficiency of electrical energy transmissions? (See A. L. Hammond et al., *Energy and the Future*, Suggested Outside Reading.)

31. How have the efficiencies of the internal combustion engine, the electrical power plant, and electric light changed since 1920? (See Claude M. Summers in *Energy and Power*, Suggested Outside Reading.)

32. Home air conditioning is one of the most rapidly growing uses of electricity. The cooling efficiency of air conditioners (Btus of cooling per kW-hr of

electricity consumed) varies by as much as a factor of three for commercial units. Why? Which are the most efficient? Which give the best cooling per dollar? (See Hirst and Moyers, Suggested Outside Reading.)

33. The Alaskan pipeline, designed to transport oil from the north slope oil field to the southern port of Valdez, was a subject of considerable controversy before it was approved. What were the inherent problems in the pipeline as noted by the opponents of the pipeline? (R. O. Ramseier, "Oil on Ice," *Environment* [March 1974], 6–13; D. P. Morgan, "Oil By Rail," *Environment* [October 1972], 30–31; C. J. Cicchetti, "The Wrong Route," *Environment* [June 1973], 4–13; R. Moxness, "The Long Pipe," *Environment* [September 1970], 12–23.)

34. The uranium used as fuel in a ^{235}U fission power plant must be enriched to about 3% ^{235}U. How might this form a practical limitation on nuclear power in the 1980s? (V. V. Abajian and A. M. Fishman, "Supplying Enriched Uranium," *Physics Today* [August 1973], 23–29; *Science* [June 14, 1974], 1160–1161.)

35. How well were reactor safety research programs funded by the AEC? What are the opinions of the AEC researchers? (See Robert Gillette, Suggested Outside Reading; J. R. Emshwiller, *Wall Street Journal*, November 6, 1973, 1.)

36. What are the relative death rates caused by aviation, electric power generation, and natural disasters? How are each of these topics viewed in society? (See Chauncey Starr, Suggested Outside Reading.)

37. Have nuclear power plants in operation proved more reliable (fewer shutdowns for maintenance) than predicted? Than fossil fuel plants? (Thomas Ehrich, "Breakdowns and Errors in Operation Plague Nuclear Power Plants," *The Wall Street Journal*, May 3, 1973, p. 1.)

38. What nuclear power developments in the Soviet Union surprised a visiting group of United States scientists in 1971? (Robert Gillette, "Nuclear Power in the USSR," *Science* 173 [September 10, 1971], 1003–1006; Philip R. Pryde and Lucy Pryde, "Soviet Nuclear Power," *Environment* [April 1974], 26–34.)

39. By how much would nuclear power generating costs be increased if nuclear wastes were disposed of by rocket launchings into the sun? (See A. M. Weinberg, Suggested Outside Reading.)

40. What technical, safety, and economic factors might slow the development of the breeder reactor? (A. L. Hammond, *Science* [August 30, 1974], 768, and *Science* [December 21, 1973], 1236; Robert Gillette, *Science* 185 [May 10, 1974], 650, and *Science* [May 24, 1974], 877; B. Commoner, *Science* 185 [July 5, 1974], 9; K. A. Hub, *Science* [December 22, 1972], 1240.)

41. The British nuclear electric power program is older than that in the United States yet seems to have encountered less resistance on the part of the public. Why? (W. C. Patterson, "The British Atom," *Environment* [December 1972], 2–9.)

42. Critics of nuclear power have argued that the emergency core cooling systems may not work. What are the details of their argument? (I. A. Forbes, D. F. Ford, H. W. Kendall, and J. J. MacKenzie, "Cooling Water," *Environment* [January/February 1972], 40–47.)
43. What might be the impact on Lake Michigan of heat discharges from electric power plants? (P. F. Gustafson, "Nuclear Power and Thermal Pollution: Zion, Illinois," *Science and Public Affairs* [March 1970], 17–23.)

Suggested Outside Reading

ABRAHAMSON, DEAN E., *Environmental Cost of Electric Power*, Scientists' Institute for Public Information, 30 E. 69th St., New York, N.Y. 10021. Compares environmental costs of all types of electric power. Written for the layman.

Bulletin of the Atomic Scientists, (September 1970), 2–12. A series of articles on allowed radiation levels. Gofman, Tamplin, and Pauling state their positions, and the AEC responds.

———. (September; October; November 1971). An excellent series on energy, nuclear power, and alternatives to present systems. The articles present many diverse opinions on the nature of the crisis and on methods of resolving it.

———. (October; November 1974). A series of provocative articles, pro and con, on nuclear reactor safety, breeder reactors, plutonium, and the promotion of nuclear power.

CLARK, JOHN R. "Thermal Pollution and Aquatic Life," *Scientific American* (March 1969), 18–27. The impact of discharged heat on ecosystems is discussed, the relative merits of various cooling schemes are given, and the effect of increased temperature on aquatic organisms is explored.

Energy and Power, Scientific American (September 1971). The entire issue is devoted to energy. Eleven articles discuss virtually all topics, from energy use in hunting societies to energy in the universe to energy resources of the earth to the efficiency of energy conversion.

FORD, DANIEL F. AND H. W. KENDALL. "Nuclear Safety," *Environment* (September 1972), 2–9. A negative view of the safety of nuclear power reactors by two active in opposing nuclear power.

GILINSKY, VICTOR. "Bombs and Electricity," *Environment* (September 1972), 10–17. An article stressing the dangers of plutonium proliferation as the number of nuclear power plants increase. Included are data on the plutonium production capacity of various nations.

GILLETTE, ROBERT. "Nuclear Safety," *Science* (September 1, 1972), 771–776; (September 8, 1972), 867–871; (September 15, 1972), 971–975. An in-depth series in which the author interviews, sometimes clandestinely, administrators, scientists, and engineers within the AEC.

GOFMAN, JOHN W., and A. R. TAMPLIN. "Radiation: The Invisible Casualties," *Environment* (April 1970), 12–19. The case for lowering allowed radiation exposure standards by two former AEC research scientists.

HAMBLETON, W. W. "The Unsolved Problem of Nuclear Wastes," *Technology Review* (March–April 1972), 15. The problems of and proposed solutions for the storage of radioactive nuclear wastes. The author was a central figure in the Lyons, Kansas, controversy.

HAMMOND, A. L., W. D. METZ, and T. H. MAUGH, II. *Energy and the Future.* American Association for the Advancement of Science, Washington, D.C., 1973. A series of articles, most from *Science* magazine, on all aspects of energy: resources, efficiency of use, present and future power sources.

HIRST, ERIC AND JOHN C. MOYERS. "Efficiency of Energy Use in the United States," *Science* **179** (March 30, 1973), 1299–1304. An excellent article describing the efficiency with which energy is used. Transportation, space heating, space cooling, and the potential for energy conservation are discussed.

INGLIS, DAVID R. *Nuclear Energy: Its Physics and Its Social Challenge.* Addison-Wesley Publishing Co., Inc., Reading, Mass., 1973. Written for the layman, this excellent book treats nearly all technical and social aspects of nuclear power.

KUBO, A. S., AND DAVID J. ROSE. "Disposal of Nuclear Wastes," *Science* **182** (December 21, 1973), 1205–1211. A relatively technical article on the economic costs and technical limitations of the various options for nuclear waste disposal.

LEAR, JOHN. "Radioactive Ashes in Kansas Salt Cellar," *Saturday Review,* February 19, 1972, 39. An extended look at the controversy over the Lyons, Kansas, nuclear waste storage site. The role of Wm. Hambleton is a central theme.

Nation's Energy Future, The. U.S. Government Printing Office, December 1973. A report to President Nixon in response to his request for an integrated energy research and development program. The study was chaired by AEC Chairman Dixie Lee Ray, and it includes projections of energy use and supply, together with short- and long-term policy proposals.

NOVICK, SHELDON. "Nuclear Breeders," *Environment* (July/August 1974), 6–15. Breeder reactors, their hazards, costs and benefits are discussed. The shortcomings of the AEC's environmental impact statement are examined.

Resources and Man. W. H. Freeman and Co., San Francisco, 1969. A study and recommendations by the Committee on Resources and Man of the U.S. National Academy of Sciences/National Research Council. Especially good articles on food, mineral, and energy resources from the land and from the sea.

SAGAN, L. A. "Human Costs of Nuclear Power," *Science* (August 11, 1972), 487–493. An effort to assess the environmental costs of nuclear power. The author concludes our society is able to maintain low prices by evading environmental costs and by failing to pay for occupational injuries. At the same time, the price of nuclear energy is maintained at an artificially high level by a policy of reducing the possibility of exposure to radiation to a far greater extent than can be justified.

Science (April 19, 1974). The entire issue is devoted to energy, energy sources, and energy policy.

SEABORG, G. T., and J. L. BLOOM. "Fast Breeder Reactors," *Scientific American* (November 1970), 13–21. A general discussion of nuclear breeder reactors and power plants. The article emphasizes the advantages of breeders.

STARR, CHAUNCEY. "Social Benefit versus Technological Risk," *Science* **165** (September 19, 1969), 1232–1238. A general article that attempts a cost-benefit analysis of the risks inherent in nuclear power, other forms of energy production, and other human activities.

WEINBERG, A. M., and R. PHILIP HAMMOND. "Limits to the Use of Energy," *American Scientist* **58** (July–August 1970), 412–418. The fundamental limits to energy production from various sources are discussed along with the limitations of heat dissipation and radioactive waste disposal.

WEINBERG, A. M. "Prudence and Technology: A Technologist's Response to Predictions of Catastrophe," *BioScience* (April 1, 1971), 333–338. An address given by the director of the AEC's Oak Ridge National Laboratory. Weinberg presents his view that technology and especially nuclear power afford the best avenue to environmental improvement.

WOLFF, ANTHONY. "Showdown at Four Corners," *Saturday Review* (June 3, 1972), 29–41. A discussion of the development of coal-fired power plants in the Four Corners section of the U.S. The article covers the projects from the initial negotiations for leases through the 1971 air emission control requirements.

ZELLER, E. J., D. F. SAUNDERS, and E. E. ANGINO. "Putting Radioactive Wastes on Ice," *Bulletin of the Atomic Scientists* (January 1973), 4–9; and (April 1973), 2–3, 53–56. The first article proposes ice storage of radioactive wastes, the second consists of critical comments on the proposal.

5 Energy Options: Fuels, Sources, Conservation and Policy

It was characteristic of dwellers on earth they never looked ahead more than a million years and the amount of energy was ridiculously squandered.

Venusian announcer in
The Last Judgement by J. B. S. Haldane

To Save Energy—
In Winter: Lower people's body temperature to 68°F.
In Summer: Let birds fly around the house to keep the air circulating.
Anytime: Eat more carrots so we can see just as good with less light.
Dip everything that's made in stuff that glows in the dark.

Student Responses, Henry Elementary
School, St. Louis

In Chapter 4 we looked at the present status of energy and its attendant problems. The energy picture for the United States and for the word is bleak; fuel resources will dwindle, and energy demands, environmental disruption, and international tensions will increase. We should examine whether other courses are available. The world will of course be forced into changes as the fossil fuels decline. Do alternatives to energy production and use exist? What options are available? How soon can they be ready? What changes will they make in our world?

We shall examine the options that are available in the near future (a decade or less), the alternatives that might be available before the end of this century, and the policies that will dictate the path we will follow as the century progresses. Two significant future possibilities are the utilization of new fuels and the development of alternative energy production schemes. New fuels might be developed that substitute for fossil fuels in our present methods of energy use. These would probably result in only slight disruptions of our present economic and industrial base. Over a longer period, our entire energy production and use patterns may be changed if new devices for energy generation are developed. The impact of these changes may make profound alterations in our way of life.

In the sections that follow, we shall discuss new fuels, consider alternative power sources, and investigate the factors involved in energy growth and energy conservation.

New Fuels   **The Hydrogen Economy**

In its elemental form, H_2 gas, hydrogen is a superb fuel. Its major combustion product is H_2O. It burns cleanly with an almost invisible flame and, unlike fossil fuels, emits no pollutants such as sulfur oxides, carbon monoxide, or particulates. It does produce nitrogen oxides if burned in air, but even these can be controlled or, in fact, eliminated by burning hydrogen in oxygen. If the H_2 is obtained from water, it represents an *unlimited* fuel because it is recycled: H_2 obtained from water produces water when burned. In addition to its superb combustion properties, hydrogen has numerous other advantages as a fuel. It can be transported in pipelines, including those already in use for natural gas, which afford the cheapest form of energy transportation. At its destination, hydrogen can be used for heating or for the generation of electricity. In heating applications, very little venting is necessary because no carbon monoxide and no particulates are emitted. Consequently no chimney is needed, so building costs would be lower and heating efficiency would improve as much as 30%. Appliances such as stoves need not burn hydrogen in the conventional manner. They could use a catalytic heater similar to those used in camp heaters. These devices are flameless, and because they operate at low temperatures of 100 to 200°C (200 to 400°F), almost no nitrogen oxides are formed.

Hydrogen can be used in transportation. It is a fine fuel for turbojet engines; even internal combustion engines can burn hydrogen if the carburetor is modified.

Hydrogen also has a variety of nonfuel applications. It can be combined with carbon monoxide, carbon dioxide, and nitrogen oxides to provide fertilizer, foodstuffs, petrochemicals, and industrial materials. Furthermore it has many chemical applications in industries, for example, in metallurgical processing.

Although not abundant in nature in the form H_2, hydrogen can be synthesized from any of the fossil fuels or from water. For many applications, hydrogen can be stored in liquid form. It has a boiling point of −253°C (−424°F), which means it must be stored in vessels equipped with vacuum jackets. In scientific use such vessels are called Dewar flasks, or Dewars. Dewars are common; glass "thermos bottles" use the same principle of vacuum insulation. Large amounts of liquid hydrogen are stored and used in space flights and in scientific research. Handling this low temperature liquid has become routine.

Most of the hydrogen used today is made from natural gas, and this would of course have to be changed if we were to increase our reliance on hydrogen. The best source for extended hydrogen production is water, and the most efficient process for obtaining H_2 from water is **electrolysis.** In electrolysis, an electric current is passed through H_2O, and the water molecules are broken

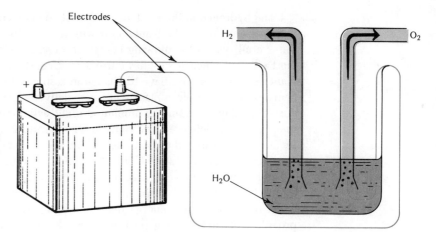

Figure 5.1. *The electrolysis of water to hydrogen and oxygen. As current flows between the electrodes in the water, H_2O is broken down into H_2 and O_2. H_2 evolves at one electrode and O_2 evolves at the other.*

down into the gases H_2 O_2. Ordinarily electrolysis can be done as shown schematically in Figure 5.1. Two electrodes representing the positive and negative poles of a battery are placed in the water. As current flows in the battery, the gases are generated at the electrodes and can be collected separately. Electrolysis has been known for over a century, and is used under a variety of conditions to produce other compounds.

Economics prevents the use of electrolysis as a source of commercial hydrogen today; in 1972 the price of electrolytic hydrogen was about 50% greater than that of hydrogen obtained from natural gas. However, large power plants devoted exclusively to hydrogen generation could greatly reduce the cost, and increasing costs of natural gas will make electrolytic hydrogen more competitive. In such a scheme, the plant would generate electricity that would immediately be used to create hydrogen by electrolysis. If done at the bottom of the ocean, this would result in hydrogen at high pressure ready for pipeline transport. The hydrogen, not electricity, would be piped to the destination because a pipeline can carry ten times more energy per hour than an overhead electrical power line. Although hydrogen would be a more expensive form of energy to *produce* than electricity, it would be a cheaper form of energy for the consumer to *buy* because of the low energy transportation costs offered by pipelines. By using power plants to generate hydrogen, important changes in our energy picture might appear. Because of the versatility of hydrogen, we would be transferring much of our energy demand for heating and transportation to electrical sources such as nuclear power (or perhaps solar power, discussed later in this chapter) which are capable of generating electrical energy. This would reduce the pressure on fossil fuels but increase the demand for electrical power sources.

Using hydrogen in the fashion just described in electrical power generation possesses an additional advantage: it will improve the operating efficiency of the overall system by allowing energy storage. An electrical power plant is most efficient when it supplies current to a constant load, that is, when the plant's output does not have to change in response to varying customer load demands. For example, a plant that has to deliver a large amount of electricity during the day and a smaller amount at night will not use its fuel as efficiently as it would if it delivered electricity at a fixed rate. When generating electricity into a power line, the plant must follow the customer demands because the electricity cannot be "stored" in anticipation of future increased demand.

Some utilities achieve storage by using the electricity available at night and during low use periods to pump water to a high altitude storage lake. The water represents stored energy that can be used as a hydroelectric system to provide electricity during peak use periods. In this fashion, the electricity demands on the power plant are kept constant. Hydrogen gas can be stored in underground tanks or in natural caverns such as those presently used to store natural gas. Thus a power plant could generate hydrogen at a constant rate, storing it for future heavier demand, and the resulting greater efficiency of the plant would provide a further reduction in cost.

The mechanics of the conversion to the use of hydrogen in industry, space heating, and power generation would be straightforward, although it would require changes in the gas-regulating equipment. Such a change is not without precedent; all equipment was changed when a conversion from manufactured gas to natural gas was made in the 1940s.

In transportation use, the principal problem is one of fuel storage. To provide sufficient energy per tank for vehicle range, the hydrogen would probably have to be stored as a liquid. In planes this presents no problem because Dewar storage could be used. The fact that liquid hydrogen provides more energy per pound than any other fuel is an enormous advantage for its use in airplanes.

Buses and large trucks might also adapt easily to use of liquid Dewars which could be exchanged at "filling" stations, but such a scheme might be too expensive for use in automobiles because of cost. There is perhaps an alternative to liquid storage for the internal combustion engine. Certain metal alloys (mixtures of magnesium, nickel, and copper) can store prodigious amounts of hydrogen by forming metal hydrides. Often the amount of hydrogen stored is equivalent to that contained in liquid hydrogen in the same volume. These interesting compounds are stable at room temperature but when heated (perhaps by engine exhaust) they release H_2.

Given the many advantages of hydrogen, it is worth inquiring about why it is not used more often. At present its use is limited by a concern for the hazard involved and by a lack of facilities for production on a massive scale. Worry about the safety of hydrogen is to a large extent exaggerated. All fuels, especially natural gas and gasoline, can explode under proper conditions, but the handling of these commonly used fuels has been developed to the point where the public can accept the hazard. In practice, the hazard of using hydrogen may not be much greater than that which presently exists with natural gas. Before natural gas became abundant immediately after World War II, Americans commonly used in their homes a manufactured gas that contained up to 50% H_2. No drastic problems occurred.

Proponents of hydrogen fuel feel that most of the worry about the danger of hydrogen comes from the publicity surrounding the burning of the blimp *Hindenburg,* a giant German airship filled with H_2 that burst into flames in New Jersey in 1937; it provided a spectacular fire, killed one third of the passengers, and received wide news coverage. The resulting "Hindenburg syndrome" relating to the dangers of H_2 may remain in the public mind, despite the fact that a similar fire could have occurred with natural gas, for

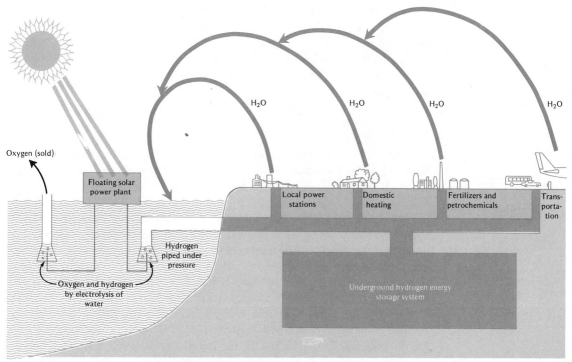

Figure 5.2.   *The hydrogen economy. The electricity from a floating solar power plant is used to generate hydrogen and oxygen by electrolysis. The oxygen is sold and the hydrogen is piped ashore for use. Pipelines afford cheap energy transportation, and the hydrogen can be used for power generation, for petrochemicals manufacture, and for transportation (principally buses and jet planes). As a fuel, hydrogen creates few air pollutants. When burned, hydrogen is converted to water and is thus a renewable fuel source.*

example. One group interested in promoting hydrogen as a universal fuel has formed an organization known as the H₂indenberg Society which meets regularly to exchange research ideas.

The advantages of hydrogen are so numerous that its use is bound to become more widespread. In fact, the enormous versatility of hydrogen, and the fact that it can be used in present domestic and industrial operations with little disruption, have led to the use of the term "hydrogen economy" to describe its impact. Hydrogen would make a significant impact in all sectors of the economy, just as do fossil fuels today (Figure 5.2).

If a conversion from fossil fuels to hydrogen occurred, the quantity of hydrogen required would be enormous. If we wanted to substitute hydrogen from electrolysis to replace just the natural gas presently consumed, we would need nearly four times our present United States electrical generation capacity just to produce hydrogen! This is equal to the output of about 1000 of the largest nuclear plants that would be required in addition to most of our present electrical demand. These figures are a further indication of the magnitude of our use of fossil fuels.

The use of alternative power sources such as the solar sea power plants (described later) may provide a stimulus for the development of the hydrogen economy.

Methyl Fuel

Methanol is an alcohol that shares many of the virtues of hydrogen. It can be made from many substances: fossil fuels, wood, farm wastes, and municipal wastes. It burns with a clean blue flame, emits no sulfur oxides or particulates, and can be used for heating, power generation, and transportation. Methanol used as a gasoline additive in present engines without modification increases engine power and engine economy, while decreasing engine emissions. A liquid at room temperature, methanol is easily transported and stored and has many uses in the chemical industry. Methanol's flexibility, its ease of synthesis from many materials, and its ease of handling compared to hydrogen are advantages that will probably lead to its increasing use.

Methanol has been used as a fuel for over a century. It was originally distilled from wood, and is also known as wood alcohol or methyl alcohol. It has the chemical formula CH_3OH, but for energy purposes, pure methanol is not needed. In practice, the first step in the production of methanol results in a mixture of alcohols: CH_3OH, C_2H_5OH (ethanol or grain alcohol), and C_3H_7OH (propyl alcohol). This mixture, usually called **methyl fuel,** is cheaper than pure methanol and has in fact superior combustion properties for most uses. Such commercial products as Sterno use methyl fuel as the combustion agent. We shall use the words *methanol* and *methyl fuel* interchangeably.

The internal combustion engine can operate without modification on gasoline-methyl fuel mixtures containing as much as 30% methyl fuel but a 15% mixture appears optimum: carbon monoxide emissions decrease by as much as 70%, fuel economy improves by about 10%, and acceleration improves by about 7% compared to operation on gasoline. Because methanol has an octane number of over 100, it could be used as an additive to replace tetraethyl lead. Its use would also stretch gasoline reserves.

An internal combustion engine can be modified at a small cost (about $100) to run on methyl fuel alone. Compared to gasoline operation, cars running on methyl fuel exhibit comparable nitrogen oxide emissions but one twentieth as much hydrocarbons and one tenth as much carbon monoxide. In one test, a 1972 automobile modified to run on methyl fuel almost met the 1976 air pollution emission standards. A possible limitation on methyl fuel, however, is engine corrosion. Because methyl fuel has a marked tendency to dissolve H_2O from the atmosphere, engines may be prone to rusting.

In power plants, methyl fuel would probably be used in the conventional fashion to generate steam, and the plant would have the low air pollution emissions discussed above.

The methanol sold commercially is made from natural gas, the production volume being about 1% that of gasoline and the cost being about 40% greater than that of gasoline. Other sources would be needed if methanol were to come into wider use. Municipal waste and farm waste can be converted into methyl fuel by a very simple partial combustion process using oxygen. If all United States municipal wastes were treated in this fashion, the methanol obtained would provide nearly 8% of the fuel used annually for transportation. A further advantage would be that the volume of solid wastes to be disposed of would be cut in half. Low-grade coal left over from mining operations is another potential source. These mine tailings would not ordinarily be used for fuel, so that generation of methanol from them does not seriously affect fossil fuel supplies. Wood is another source of methanol, and commercial forests might be operated as "energy plantations," serving in effect as solar sources of methanol.

Limitations to the present use of methyl fuel are price and supply. The increasing cost of fossil fuels will probably make the methanol option more attractive in the future, and its applications could be immediate.

Fuel from Wastes

The traditional United States approach to municipal wastes (garbage, trash, sewage) has been to discard it without further utilization. The volume of waste grows rapidly each year, creating problems of disposal. As noted in the previous section, useful fuels can be obtained from these. Several options exist. The decay of organic matter by bacteria in the absence of oxygen (**anaerobic decay**) results in methane gas, whereas partial combustion of the wastes with oxygen will produce methanol. If the wastes are incinerated with an abundance of air, the heat evolved can be used to generate steam that can be utilized for heating or for the generation of electricity. In Chapter 7 we shall discuss even more options.

Wastes are not a massive source of energy, but are of sufficient potential that they should be utilized. Municipal trash in the United States could produce an amount of energy equivalent to about twice that obtained from natural gas imports in 1970.

Clean Fuels from Coal

Our most abundant fossil fuel, coal, is also the dirtiest, being cursed with high sulfur and particulate emissions. As we saw in Chapter 4, removing these pollutants after combustion is expensive, often adding 20% to the cost of electricity. In view of the abundant energy available from coal, it is tempting to undertake the conversion of coal into a liquid or a gas that could be a cleaner fuel. Several fuels can be made from coal liquefaction and coal gasification, although at present such fuels are expensive and are at least a decade away from suitability for large-scale production. Coal conversion is also somewhat expensive from an energy standpoint: 25% of the coal's energy is lost in the conversion.

Conversion of coal into a liquid similar to fuel oil is possible. It requires hydrogen gas and yields about 2 or 3 barrels of oil per ton of coal; the oil has a greatly reduced sulfur content relative to coal. Widespread coal liquefaction is probably 20 years away because the processes are still in the experimental stage. However, if successful, the abundance of coal is such that converting recoverable coal reserves would yield ten times more "oil" than known worldwide fossil oil reserves.

Two types of gas, characterized by their different heating values, can be obtained from coal. A gas often called **high Btu gas** or **substitute natural gas (SNG)**, which has a high heating value almost identical to that of natural gas, can be obtained. At least six chemical processes are known which make SNG from coal, but they are complicated and the gas is expensive. Which process may be the most suitable for commercial application will probably be known by the late 1970s.

A more accessible derivative from coal gasification is **low Btu gas,** which has about one fifth the heating value of natural gas. Low Btu gas, sometimes known as power gas, is much more easily obtained from coal than is SNG. Indeed, before natural gas became abundant (about 1950), coal was converted into low Btu gas to supply those uses that demanded a clean fuel. A variety of reactions involving air, oxygen, and water, some known since the 1800s, may be used to gasify coal. Because of the low heat value, this gas cannot be transported economically, but it can be used as fuel for nearby power plants, and it has about one tenth the sulfur content of coal. Gasification is, in fact, one of the cheapest ways of removing sulfur from coal.

An inherent limitation of coal gasification is that the coal must be obtained in prodigious amounts. Electrical power plants now being planned will require 10 million tons of coal per day, more than the output of any single coal mine in the United States. The environmental problems with mining present barriers to expanded coal production. An intriguing alternative is to gasify the coal underground.

In Hanna, Wyoming, the U.S. Bureau of Mines has been conducting tests on underground gasification since the early 1970s. Holes are drilled in the coal seam, and the deposit is broken up with high-pressure water. The coal is then ignited, and combustion is maintained by piping air into the seam. Gases from the incomplete combustion are captured and piped to the surface. Typically the gas is about 5% methane and 15% hydrogen; these provide for the combustion properties of the gas. In the low Btu gas, carbon monoxide, carbon dioxide, and nitrogen are also present, providing the further possibility of combining carbon monoxide and hydrogen to obtain methanol or a variety of hydrocarbons. Thus coal may ultimately provide a source of petrochemicals.

In the western United States the deep, thick coal seams make conventional mining so difficult that often only 25% of the coal may be recovered. Underground coal gasification may be the best approach to utilization of these deposits. If so, the technology is expected to develop rapidly.

Oil from Shale and Tar Sands

In addition to the conventional pools which are tapped with drilled wells, oil can also be obtained from certain shale deposits and certain tarry sand deposits. Both types of deposits have been utilized on a small scale, and larger operations are planned for the future. Obtaining oil from either form is a gargantuan mining operation, however, and the possibility of environmental damage is great.

Producing Oil from Shale

Oil shale is composed of minerals bound together with an organic material called kerogen. Kerogen, the source of crude oil, must be separated from the rock to utilize the shale as an oil source. Oil shale is found in many areas of the United States, but the deposits with the most potential are those along the Green River Basin in Colorado, Utah, and Wyoming. The richest shale is in western Colorado in the Piceance Creek basin. Estimates are that this area contains an estimated 117 billion barrels (bbl) of oil, about twice the known recoverable resources of oil in conventional deposits.

Recovering the oil from the shale is a relatively simple process *after* the shale has been mined, and small plants have utilized this source of oil for over a century. Crushing the stone and heating it to 500°C (930°F) releases the crude oil. The Colorado deposits, about 70 ft thick and 1000 ft below the surface, are being utilized by several approaches. Underground mining followed by processing above ground has been accomplished successfully. Another method consists of breaking up a deposit with partial mining followed by blasting. The deposit is then ignited and the crude oil is recovered by means of special trenches dug under the deposit.

Oil from these sources has less energy content per ton than practically any other substance ever used for commercial fuel. Even for high-grade oil shale deposits, about 1.5 tons of shale will be required to produce each barrel of oil. In contrast, about half a ton of coal will produce 1 bbl of oil through liquefaction.

The upper limit of oil from shale in the Green River region appears to be 1 million bbl of oil per day (the 1974 United States oil usage was 18 million bbl per day). To extract this amount would mean mining, crushing, transporting, and processing 1.5 million tons of shale, then disposing of 1.3 million tons of residue each day, meaning that nearly 1 billion tons of material per year must be handled. (This is twice the United States coal tonnage of 1973—and more material than was moved to construct the Panama Canal.) Even a 100,000 bbl per day plant would require a production mining rate twice that of any existing underground coal mine in the United States.

A principal limitation appears to be the availability of water. Three barrels of water are required to produce 1 bbl of oil. Water is not abundant in that area of the United States and the shale oil operations will have to compete with man's other uses of water including extensive coal mining operations.

The environmental damage might be severe. Present plans are to dump the residue in nearby canyons. Water pollution from the runoff may be a problem, especially the buildup of salts in the Colorado River, the principal drainage of that region. Reseeding of the mined areas has been attempted on a test basis, and getting native grasses and shrubs to grow requires extensive fertilizer and water applications the first 2 years. Air pollution from the processing plant is a problem, and the population increase due to worker influx into neighboring towns will have a further impact.

The Department of Interior, which controls most of the shale oil mining, has recently leased 20,000 acres for development, and estimates that 80,000 acres may be involved over a 30-year period.

Producing Oil from Tar Sands

The Sun Oil Company has pioneered the production of oil from the Athabasca tar sands in northeastern Alberta, Canada. The problems encountered in mining these sands are challenging. The sand is hard to dig, sticky, and filthy. It is abrasive and destroys machinery quickly. The temperature in the area varies from $-45.6°C$ to $32.2°C$ ($-50°F$ to $90°F$), and much of the area is covered by muskeg swamp, a semifloating mass of decaying vegetation.

Before mining can begin, the trees must be removed from the area, and this can be done only during the coldest portion of winter when the muskeg is sufficiently frozen to support heavy equipment. Draining networks are then blasted in the swamp preferably 2 years before mining is to begin to drain off as much water as possible during the summer. Surface mining is done; the tarry sand is excavated with bucketwheels and dumped on a conveyor belt for transportation to the processing plant. The plant moves about 270,000 tons of material per day to obtain 50,000 bbl of oil per day.

The total yield from that field may be about 300 billion bbl of oil, about six times the known recoverable oil reserves from well sources. Other tar sand deposits are found in South America and the Soviet Union.

Alternative Power Sources

Thermonuclear Fusion

In our discussion of nuclear reactions in Chapter 4, we focused on fission, the splitting of the nucleus. Another type of nuclear transformation is possible: nuclear fusion, in which two nuclei react to form a third. Energy is released in the process, and fusion reactions offer a larger source of energy than any other available on earth with the exception of solar energy.

Several fusion reactions are candidates for the utilization of fusion energy. All involve light elements, those with mass numbers of 3 or less. Two of the best known involve isotopes of hydrogen undergoing reactions presented in Figure 5.3. The deuterium-deuterium fusion to form tritium would utilize the ocean as a source of the deuterium. Even though deuterium is a rare isotope (Table 4.10), the energy available from fusion from the deuterium in just $1 km^3$ (less than 1 mile^3) of seawater exceeds the energy available from the world's crude oil reserves. Fusion reactions have been caused by man: the hydrogen

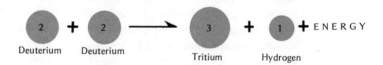

Figure 5.3. Fusion reactions. The reactions involve nuclei, not atoms. The mass number of each nucleus or particle is shown.

bomb is a fusion device. For power generation the reaction must be controlled, but making nuclei fuse under less than explosive conditions has not yet been achieved. To undergo fusion the nuclei must collide at very high speeds, which means very high temperatures, typically 100 million°C! Achieving fusion requires heating the nuclei to this temperature while maintaining sufficient density to cause frequent collisions. In the sun's fusion reaction, the nuclei are heated and forced together by the force of gravity of the sun's enormous mass. We could never achieve such strong gravitational forces on earth, so we are forced to find another means of containing the high temperature sample. All materials on earth melt at a few thousand degrees centigrade, so there is no hope of finding a mechanical container. Two approaches toward achieving controlled fusion are being pursued: magnetic confinement fusion and laser fusion.

Magnetic Confinement Fusion. Because charged particles, such as electrons or nuclei, move in response to applied magnetic fields, a gas of charged particles (a "plasma") can be confined by a magnetic field of the proper design to force the charged particles together. The stronger the magnetic field, the more tightly the particles are packed together. Such a device is then a "magnetic bottle."

One reactor design used to study magnetic bottle confinement of a plasma is the Tokamak. The name is derived from the Russian words for "toroidal magnetic chamber." In the late 1960s the Tokamak was promoted in the United States by Dr. Lev Artsimovich of the Soviet Union. It uses a torus (doughnut) shaped magnetic field that confines the plasma in the center of what would be the dough in a doughnut (Figure 5.4). The fuel, perhaps deuterium and tritium, is injected into the torus and heated using techniques similar to those utilized in microwave ovens. The gas is heated until it ionizes; that is, until the electrons are stripped from the nuclei. The resulting plasma consists of positively charged nuclei and negatively charged electrons. Continued heating and the application of a strong magnetic field force the nuclei together for the anticipated fusion reaction. For fusion to be a useful energy source, the plasma must be confined for a sufficient time that the energy released from fusion exceeds the enormous amount of energy required to heat and confine the gas. This will probably mean confinement for time periods of a few thousandths of a second.

Several Tokamak reactors are presently being used in the United States, the Soviet Union, and Europe, but at present, no controlled fusion reactions have been achieved. Most of the research is focused on the challenging problem of plasma heating and confinement.

The technical problems that must be solved before controlled fusion is a reality are enormous. The gigantic bursts of energy that will be created present problems in the design of the metal structure of the reactor. Designing appa-

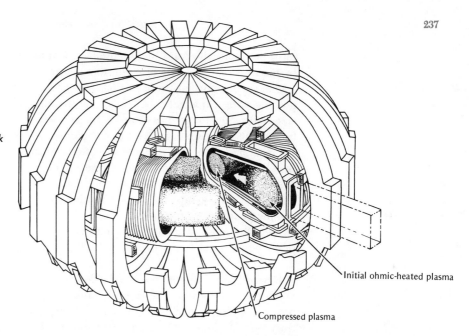

Figure 5.4. *The Tokamak fusion reactor (Adiabatic Toroidal Compressor, or ATC) at Princeton University. The plasma is heated in the region labeled, initial plasma, then compressed into the inner region by the magnetic field. [Courtesy of the Princeton University Plasma Physics Laboratory.]*

Initial ohmic-heated plasma

Compressed plasma

ratus that will withstand the intense heat and neutron bombardment created by fusion is certain to require considerable research.

One very important problem that must be solved in the event that fusion is possible with a magnetic bottle is how to convert the released energy into electricity. Several proposals are being studied. Conventional steam generation might be possible if the energy could be used to heat a fluid surrounding the reactor. Another approach is direct conversion, obtaining electricity from the charged particles emitted in the reaction.

After these and a host of other problems are solved, the reactor must be developed into a safe design for routine operation if it is to be used for commercial power generation.

Even the most optimistic magnetic confinement fusion researchers do not predict commercial applications before the year 2000.

Laser-Induced Thermonuclear Fusion. A newer and perhaps more promising approach to fusion avoids the magnetic confinement problems. This scheme, again pioneered by the Soviet Union, utilizes pellets of frozen fuel, such as deuterium (D_2). The pellet is dropped into the reactor and is struck simultaneously with intense light beams from several giant lasers. A laser provides a tiny beam of electromagnetic radiation of a very precise wavelength. As a consequence of the electromagnetic "purity" of the wave, the energy density of the beam is very high. Even small lasers can burn holes in stacks of razor blades, and large lasers can weld steel several inches thick or bounce light beams off the moon. In laser fusion, the pellet is heated by the laser pulses, and fusion occurs before the pellet can melt and become dispersed as a gas.

Should fusion be possible, problems of power generation, materials design, and safety similar to those mentioned for the Tokamak remain to be solved before laser fusion will be a commercial energy source.

Progress in laser fusion since 1970 has been sufficiently rapid that the AEC recently announced that laser fusion as a commercial source might be available in the mid-1990s, well ahead of projections for magnetic confinement.

Since the precise form in which controlled fusion, if successful, might be used is unknown, as assessment of environmental impact is difficult. If one of the present approaches is successful, fusion should not create the radioactive waste disposal problem that plagues fission. Radiation may be a problem, however. Tritium is a radioisotope and will be created in the reaction or used as a fuel. Its containment presents problems somewhat similar to those of the radioactive gases released by nuclear fission plants. Some radioisotopes may be created in the metal surrounding the reactor by the intense bursts of high energy neutrons released during the reaction, but the size of the problem should be smaller than that faced by fission plants.

Solar Energy: The Largest Resource

Solar radiation is the most abundant energy source on earth. The energy falling annually on 0.5% of the land area in the United States exceeds the total energy needs of the United States projected to the year 2000. Obtaining energy directly from the sun has fascinated scientists for centuries, but applications have been hampered by the fundamental limitations that on the surface of the earth sunlight is diffuse and intermittent. These limitations have dictated that tapping solar power necessitates a large land area and a provision for energy storage. However, space age technology may provide solar energy under conditions that eliminate, or at least minimize, these liabilities. Use of solar energy offers the advantages of a virtually infinite (and free) source of "fuel" and the potential of greatly reduced environmental side effects relative to present sources of power.

Over the years interest in solar energy has undergone periodic resurgence, but until this decade relatively little research has been done. In contrast to nuclear power, for example, which was backed by a tightly knit segment of scientists and had research and development funds borne by the government, solar energy has been the major interest of a handful of scientists and the research hobby of a somewhat larger number. Small solar ovens and cigarette lighters existed, solar water heaters were used in some countries, and a few innovative souls built homes in which half or more of the energy was derived from solar devices. As a consequence of the fuel shortages and prices in recent years, the status of solar energy is changing, and increased research funds are available. Several estimates give solar energy the potential of providing 25% of United States heating and cooling needs and 20% of the United States electrical needs by the end of this century.

Commercial use of solar power may follow various routes. Space heating and space cooling in homes and buildings are feasible now, and the first major utilization will probably be in this area. Generation of electricity with solar radiation is possible using satellites or a variety of land installations. Combustion fuels and petrochemicals can be produced by using biological processes dependent upon sunlight. In what follows we discuss some of the major proposals for using solar resources.

Solar Collectors. The utilization of solar energy begins with a device to collect the energy. Usually the energy is collected in the form of heat by a **solar collector,** which can be an arrangement as simple as a black metal surface covered with glass plates and exposed to the sun. Black absorbs radiation better than other colors, and the glass restricts the infrared emission of the black surface, so the device operates similarly to the greenhouse effect (discussed in Chapter 3, Air Pollution) to produce high temperatures at the plate. The temperature attained by the plate is very important because thermodynamics

Figure 5.5. One type of concentration solar collector. The pipes running the length of each collector are located at the focus of the reflector, so that a fluid in the pipes receives a concentration of solar rays and is heated to a high temperature. An automatic drive system rotates the reflectors to follow the sun's path. [Courtesy of Industrial Research, March 1973.]

guarantees that the higher the temperature of the collector, the more efficient the conversion of the solar radiation into whatever other form of energy is desired. Technological advances since 1965 have produced **selective surface** coatings that have greatly reduced infrared emission and consequently get much hotter in the sun than previously used surface coatings. Several types of selective surfaces exist; all are composed of thin layers of metals and metal oxides or perhaps plastic.

Two designs are used for the collectors. The **planar collector** is a flat plated surface. It performs well either in direct sunlight or in diffuse light such as that which exists on a cloudy day. The **concentration collector** (Figure 5.5) uses a lens or a magnifying glass to focus as much of the sunlight as possible upon the surface. It must be designed so that the lens will follow the sun across the sky and keep the image focused on the surface. In direct sunlight, the concentration collector gives higher temperatures than a flat plate collector. However, the concentration collector is efficient in direct sunlight only; its performance is greatly reduced by the presence of even a few clouds in an otherwise bright sunny sky.

Space Heating and Cooling. The advantages of sunlight for heating purposes have been recognized for many years. Roof-mounted solar water heaters have been used in Japan for years, and solar swimming pool heaters are available in the United States.

Using solar energy to heat buildings has become economically competitive with electric heating in recent years and, as fossil fuel costs increase, solar power will become the cheapest form. Typically a roof-mounted collector is used to create a reservoir of hot water that is stored in an insulated tank. The

Figure 5.6a.   *A solar-heated home in the Connecticut shoreline community of Westport. The solar system, designed in competition with more conventional practices, is expected to provide nearly two thirds of the total heating needs; the remainder will come from an auxiliary oil system.* [Courtesy of Donald Watson, AIA, Guilford, Conn.]

reservoir is then used to warm air that is passed through the building. Cooling with solar energy is accomplished with an absorption refrigerator, which operates on a principle similar to that utilized in gas refrigerators. Heat is used to separate and liquify ammonia from an ammonia-water liquid mixture. When cooling is needed, the ammonia is allowed to vaporize and expand into a set of cooling coils, a process identical to that used in conventional home air-conditioning systems.

Most of the solar heating applications to date have been used as supplements to more conventional heating in single family dwellings. The solar auxiliary sytems supply up to half of the energy needs for space heating. Figure 5.6a shows a Connecticut home whose solar heating system was designed in competition with more conventional practices. In the design, three rows of solar collectors on the roof face south. On the north sides of the raised roof collector enclosures are clerestory windows which provide natural ventilation and summer cooling. The solar system is expected to provide 45% of the space heating and almost all of the hot water, which together could total 65% of the home heating requirements. The additional requirements will be met by a conventional oil heating system. Over a 6- to 10-year period, fuel savings are expected to pay for the solar installation.

Applications to larger buildings are now appearing (Figure 5.6c). The General Services Administration is presently constructing two office buildings in

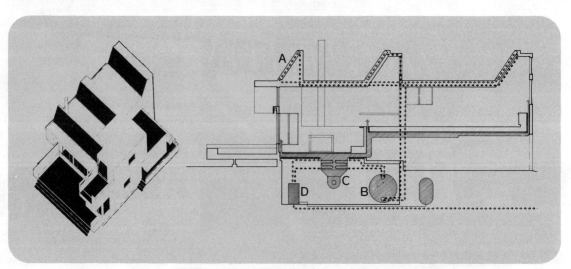

Figure 5.6b. *The essential features of the solar home are shown in the schematic at the right: collectors (A), a 2000-gal water storage tank (B), and a blower hot air circulation system (C). The auxiliary system is an oil-fired domestic water heater (D). The water flow is shown by the arrows, with hot water shown as the black arrows, cool water as the open arrows. [Courtesy of Sunworks, Inc. and Donald Watson.]*

Figure 5.6c. *Timonium Elementary School in Baltimore County, Maryland, has installed a solar heating system, a project sponsored in part by the National Science Foundation. Water flowing through the banks of solar collector panels on the roof of the school is heated by the sun, collected in a 15,000-gal tank, and piped to the classrooms. One of the school's three wings will be heated by the collectors, and the data collected will be evaluated for future systems. The project has also captured the imagination of the schoolchildren, who have decorated their hallways and classrooms with models and pictures describing the uses of solar energy. [Courtesy AAI Corporation, Cockeysville, Maryland.]*

the northern United States that will utilize solar heating and cooling with more conventional systems providing backup capacity. Many industrial firms are also adding solar auxiliary systems.

Proponents believe that solar systems could ultimately provide as much as half of the United States heating and cooling needs. Although initial development may be as an auxiliary system, even this function is important because it provides a reduction in the peak load requirement of more conventional fuel sources.

Solar Generation of Electricity. Sunlight can be converted to electricity using direct conversion (solar cells) or by using dynamic conversion, that is, by using solar radiation to generate steam which runs a steam turbine (see Figures 4.9 and 4.10).

Solar cells (photovoltaic cells), which produce electricity in direct sunlight, are used for space flights and for special applications such as light meters used in photography. At present, these are too expensive for commercial electrical use. To make a power plant with such devices would cost about 20 times more

Figure 5.7. *A satellite solar power system. The electrical energy created by the solar cells is converted to microwaves and beamed 22,000 miles to earth where it is collected by a large antenna. Each satellite has collector panels measuring about 5 × 5 miles, and a similar area is occupied by the land-based receiving antenna. The received energy would be converted at efficiencies of 90% and injected into conventional power transmission lines. [Courtesy of Peter E. Glaser and Arthur D. Little, Inc.]*

than a nuclear plant, although research and mass production might lower the cost and make such a plant more competitive.

The low efficiency (less than 20%) with which these devices convert sunlight to electricity, together with their reduced power production on cloudy days and at night, means that a large land area would be needed. These limitations could be reduced by placing the power station in orbit above the earth's atmosphere where the sun's energy would be stronger, and power generation could occur continuously. Compared to a land-based system, the satellite would give 10 to 15 times more electrical power per day from a given cell. Peter Glaser of Arthur B. Little, Inc., has proposed a satellite solar cell array measuring several miles on a side which would generate 15,000 Mw, enough energy to satisfy the needs of a city the size of New York in the year 2000 (Figure 5.7). The electricity from the cells would be converted into microwaves and beamed to the earth, where the signal would be received by an antenna and reconverted to electricity. Because the conversion of microwaves to electricity is very efficient, little waste heat would be produced on the ground. Glaser's proposal is designed to fit the space shuttle program of the National Aeronautics and Space Administration (NASA).

One interesting prototype house using solar collectors and solar cells has been constructed by the University of Delaware's Institute of Energy Conversion. "Solar One," designed to obtain 80% of its entire energy needs (including electricity) from the sun (Figure 5.8), has rooftop solar collectors and solar cells mounted on the roof and on the side of the house. Electrical energy from the cells is used directly in appliances or stored in batteries for future use.

Figure 5.8. *"Solar One," a prototype house designed by the University of Delaware's Institute of Energy Conversion. The house obtains 80% of its energy from the sun.* [*Courtesy of the Institute of Energy Conversion, University of Delaware.*]

Solar Thermal Power Plants. Technological developments are making it more feasible to generate electricity from solar energy using dynamic conversion. One of the more exciting proposals is that of Drs. Aden and Marjorie Meinel of the University of Arizona. The Meinels propose the development of land-based solar power "farms," large solar collector steam power plants (Figure 5.9). Although the proposal of such sites is not novel, the advantage of the Meinel approach over older proposals is that by using the selective surface coatings to obtain higher temperatures, the efficiency is improved and less land area is needed. They propose an array of long pipes covered with the selective coatings. Placing the pipes in evacuated glass tubes reduces heat losses while still permitting exposure to the sun. A gas is pumped through the pipes to transfer the heat to a large reservoir of a low melting point metal mixture which serves as a thermal storage system. Steam is generated by piping water through the hot metal liquid. The collector pipes would reach temperatures of several hundred degrees centigrade, and the resulting high temperature metal mass would provide power generation overnight and during several days of inclement weather. Careful design of the system ensures that it can be used with conventional turbines and transmission schemes. The solar station has no air pollution. In fact, air pollution must be discouraged in the area because particulates would reduce the intensity of the sunlight.

1. Sea water
2. Fresh water
3. Desalting plant
4. Cooling tower
5. Thermal storage A
6. Thermal storage B
7. Oil reserve
8. Maintenance
9. 250 MW(E) Turbine
10. Boiler

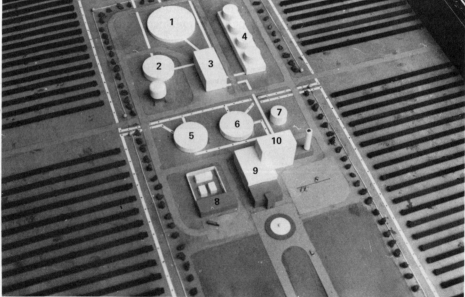

Figure 5.9. *The solar farm power generation proposal of Aden and Marjorie Meinel. The black lines radiating outward indicate the collector pipes. [Courtesy of Aden and Marjorie Meinel.]*

Crucial to the proposal is the land area required. The Meinels estimate for illustration that for a 1 million Mw plant, which would represent well over half the United States electrical demand by 1990, about 10,000 square miles of land area would be required. About half of this would actually be covered with collectors. Although this may sound large, the land area for strip mining for the Four Corners coal plant alone is almost one third of this, and the contrast between the strip mining and solar farm use of the land is striking. The solar farm would be for the most part a nonpolluting, unimposing operation. Farming and grazing could be done under the elevated arrays assuming the geographical location were suitable.

Smaller land areas would conceivably be needed if concentration collectors were used. However, experiments in 1973 indicated that even slight cloudiness reduced the efficiency of the focus collectors to the point that their larger costs probably cannot be justified.

Technological questions about the proposal focus on the lifetime of the coatings under high temperature conditions and on concerns about the handling of the several million tons of storage solution.

One major environmental concern is thermal balance because the plant results in increased absorption of the sun's energy by the earth. Since some (about 20%) of the energy normally reflected by the earth is absorbed by the collector, this might be expected to result in environmental change. However, the increased absorption can be almost exactly balanced out by altering the color of the paint on the array support or on panels on the desert floor near the arrays such that the reflection is increased over what would occur naturally. Thus in the local area of the plant the net energy absorbed is unchanged by the presence of the plant.

The conversion to electricity will of course result in the discharge of waste heat into the desert area. It would be possible to use this to desalinate seawater should a large source of water be available.

Economic estimates indicate that the cost of electricity from such a plant would be two to five times 1972 rates of more conventional systems. Construction costs on the plant are much higher than for the present plants, but operating costs would be much lower. Some federal-industrial-utility economic cooperation may be necessary to provide the developmental costs as was done with nuclear power reactors.

The Meinels propose not a single massive site, but rather a distribution of smaller solar farms over an 8- or 10-state area with each farm supplying electricity for 35,000 people. Adequate research funding might make the plants operable in the 1990s.

Solar Power from the Sea. Air- or land-based collector systems are not the only options for solar power. It is possible to generate electricity by utilizing the differences in temperature that exist between the top and bottom of the ocean. This is an indirect form of solar energy.

Utilization of these ocean thermal gradients was originally proposed nearly a century ago by the French physicist Jacques D'Arsonval and has recently been carefully studied by several scientists and engineers. Professor Clarence Zener of the University of Pittsburgh has updated a proposal of Hilbert and James Anderson for a thermal gradient plant in which a fluid is vaporized near the surface of the ocean (23°C or 74°F), passed across a turbine, and condensed at the temperature of the bottom of the ocean (7°C or 45°F). The operation is in principle similar to an ordinary steam plant except that the temperatures are lower and a fluid such as ammonia is used instead of water. Volumes of flow typical of a hydroelectrical plant of equivalent capacity would be required. Figure 5.10 shows a picture of the plant proposed by Zener. It appears that about 75 times the United States electrical demand expected in 1980 could be generated from plants located in the Gulf Stream off the coast of Florida.

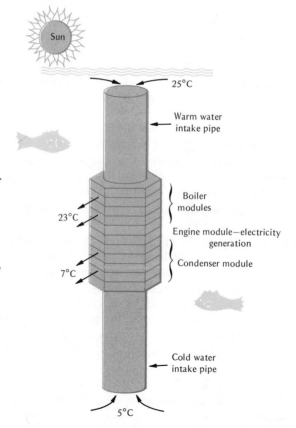

Figure 5.10. A proposed electrical power generator using thermal sea gradients. As proposed by the Andersons and by Zener, the plant uses warm surface water to vaporize a fluid such as ammonia, which drives a turbine and is condensed at the cool temperatures near the bottom of the ocean. The entire plant is neutrally buoyant at a depth of about 200 ft. [After C. Zener, Physics Today, January 1973.]

Electricity generation by such a plant may not be the most beneficial use of the facility, however, because electrical transmission from an ocean site is difficult. The electrolysis of water to generate hydrogen as discussed earlier in this chapter may be preferable. The hydrogen would be generated at high pressure near the bottom of the ocean and piped to shore to be used as a fuel in a land-based plant.

A solar sea site may have considerably more potential than that resulting merely from the generation of electricity. Upwellings of cold water from the ocean's bottom are very high in nutrients, and utilization of this water to accelerate the growth of aquatic plants and animals may be possible. Furthermore desalination of seawater could be done simultaneously.

Estimates indicate that the cost of a solar sea plant of the type proposed by Zener is approximately the same as the cost of a nuclear plant. If solar plants are sufficiently economical to become widespread, it would be necessary to learn much more about the flow of oceans and the impact of ocean temperatures on climate before embarking on a large-scale program.

Geothermal Energy: Heat from the Earth

The earth's geothermal reservoirs are a valuable source of energy. Radioactive decay of materials within the earth's center is the primary source of the molten rock which, under pressure, forces its way upward. When the heated rock encounters groundwater, natural deposits of steam and heated water result. Occasionally this naturally heated water breaks the surface as geysers or warm springs.

The first electric power plant utilizing natural steam was constructed in Italy at the beginning of this century. Now seven countries have operating geothermal power stations, eight other nations are installing plants, and interest in this source of energy is accelerating. One of the most extensive programs is in New Zealand, where 10% of the electrical power production is from geothermal plants.

Three approaches to geothermal utilization are being investigated: dry steam sources, wet steam sources, and hot rock sources.

The most easily tapped source is a **dry steam** deposit, from which only steam is emitted. These are the sources used in most of the operating plants, such as the only United States geothermal plant, the Geysers Field, north of San Francisco. No boiler is needed for such sites: the steam is fed directly to the turbines. Pacific Gas and Electric Company opened the first unit at the Geysers Field in 1960 (Figure 5.11). By the late 1970s it will produce almost 1000 Mw, or about as much as the largest nuclear plant under design. Estimates are that the ultimate capacity of the Geysers plant may exceed 4000 Mw. Construction costs and electrical rates for this plant are lower than those for any other source.

Figure 5.11.   *The Geysers Field geothermal steam power plant located north of San Francisco. [Courtesy of Pacific Gas and Electric Co.]*

Dry steam beds offer the fewest barriers to utilization and consequently have been developed first. A much more abundant (by ten times or so) geologic structure is a **wet steam** field in which steam and hot water are both emitted. To utilize such a source requires separation of the steam and water. In Mexico, just south of the United States border near Cerro Prieto, a wet stream field 75-Mw power plant has recently been opened which cost about half the cost of a more conventional plant of the same capacity.

Finally, there are massive regions of hot geothermal sites a mile or two below the surface of the earth. No groundwater exists near the deposits, but they could be utilized by pumping surface water into them to be heated and then pumping the water out for use in steam generations. Much of the western United States has underlying hot rock forrmations. A program is under way at the Los Alamos Scientific Laboratory under the direction of Dr. Morton Smith to investigate the feasibility of utilizing the hot rock sources to generate electricity. A well is drilled into the thermal region, and water at high pressure is pumped into the well to open cracks in the rock. The cracks increase the volume of the site and allow for more efficient heating of the water. Water is pumped into the site in one pipe and removed after heating by another pipe.

Geothermal energy can, of course, be used for other purposes than the generation of electricity. Space heating using dry steam sources is used on a small scale in several United States cities. Air conditioning can be accomplished at a cost several times less than that of conventional electric compressors by using an absorption refrigerator system. Desalination of seawater is a further use. The Cerro Prieto plant purifies 140 tons of water per day, and the Mexican government has plans for an integrated industrial complex at the power site.

As with any power generation method, geothermal plants have limitations and environmental liabilities. The lifetimes of the sources are usually estimated at 30 to 50 years, although the Italian plant has operated for 70 years. It is generally agreed that running out of water is the limitation of the dry and wet steam sites. Replenishing the steam beds by pumping surface water into the formation may provide an extension of the lifetime.

Removal of groundwater may result in land sinking, a phenomenon called land subsidence. This can be halted by recharging the formation with surface water. A potential water pollution problem exists with many geothermal plants. Several tons per day of minerals and salts are accumulated and some, such as boron, should not be dumped into the environment. These wastes are often discharged into the bottom of the geologic formation. Finally, hydrogen sulfide emissions from some geothermal plants may equal those from coal plants.

The ultimate potential power generation available from geothermal energy in the United States and the rest of the world is quite uncertain, in part because extensive mapping of the hot rock areas has not been done. Several estimates of possible United States development by 1985 are in the several hundred gigawatt range. Aerial surveys indicate that deposits in Central and South America may have more power potential than that region could use, and Ethiopia may have sufficient geothermal potential to provide electricity for all of Africa.

The United Nations has actively pursued development of these sources because of their special advantages for underdeveloped nations.

Electricity from the Wind

One relatively pollution-free and virtually limitless source of electrical energy is wind power. Electricity has been generated with wind-driven windmill turbines since the 1890s. In the 1920s and 1930s, thousands of windmills in the Great Plains provided most of the electricity on farms, and they are still used in isolated areas.

The energy per square mile of the wind at the earth's surface in the United States is about equivalent to that obtained from the sun on a 24-hr basis. Electricity is obtained at the windmill by using the rotary motion to run a generator, a very efficient device as we noted earlier. A fundamental problem is how to store the energy in order to provide electricity in the absence of wind. Although charging batteries during windy periods is an obvious approach, it is rather expensive. Present proposals emphasize using the windmill's electrical power to electrolyze hydrogen from water. The hydrogen can then be used as a relatively clean fuel source. In a slightly different twist, the North American Rockwell Company has proposed burning the hydrogen in a small version of the Saturn rocket engine, which would run a turbine. The advantage of the rocket engine is that the high combustion temperature ensures a higher efficiency of conversion to electricity than can be obtained in a conventional steam plant.

Another problem in building wind machines has been structural design, making the large machines sufficiently sturdy to handle the enormous stresses created. In the early 1940s, a 1 Mw aerogenerator was operated on a hill called Grandpa's Knob in Vermont. The turbine had a two-bladed propeller nearly 175 ft long mounted on a tower. The blades were made of stainless steel and weighed 8 tons each. Structural problems ended the Grandpa's Knob project after 4 years of operation, but it provides the only data on large systems, although wind generators of several hundred kilowatt capacities have been operated in Holland and in the Soviet Union.

By applying computerized design and the aerodynamic developments of space research, wind generation of electricity might provide a major source of electricity. Professor William Heronemus of the University of Massachusetts has recently proposed the installation of thousands of windmills in the United States to provide electrical power. These could have a variety of sizes and could be in backyards, in fields, along roads, or offshore. Such a system could be switched into present power grids to provide additional electrical capacity and thus reduce fossil fuel demand. For large-scale generation of electricity Heronemus proposes using windmills to obtain hydrogen by electrolysis. He calculates that 300,000 wind turbines in the Great Plains could generate as much electricity as 200 large nuclear plants. He suggests putting 20 turbines each on 850 foot high towers spaced a mile apart on a route from Texas to the Dakotas.

A serious problem is the lack of experience in the structural design and equipment maintenance required by such large towers. The array of towers might not be an appealing sight, and the utilization of the generated hydrogen remains to be implemented. However, his Great Plains proposal is only intended to be illustrative of the potential of this source. The cost of electricity from wind generation depends rather critically on the number of turbines installed per year, since the unit cost could be greatly reduced by mass production. Recent estimates (1972) indicate that if 20,000 of the machines were produced each year, the cost of electricity from the wind turbine-hydrogen proposal would be about 40% greater than the cost from nuclear plants.

Timetables for Some Options

The future use of the options discussed in this chapter will depend upon many factors: energy use, fuel reserves, economics, new technologies, and policies. Figure 5.12 presents one estimate of when some of the more immediately available energy options could make significant contributions to United States energy production. Tar sands, shale oil, and fuels from wastes could have an impact now, with the other options becoming important in the 1980s.

One of the many possible predictions of the impact of these options is shown in Figure 5.13. Many assumptions have been made to draw this chart. A principal assumption is that the energy needed in 2000 will be 25% less than that predicted from current energy growth rates. New fuel options are not included, and only small contributions from solar, geothermal, and shale oil are assumed. A tripling of coal production, a 50% increase in hydroelectric power, and a nuclear capacity of 1 million Mw are all assumed for the year 2000. From Figure 5.13, it is also obvious that even with these rather optimistic conditions, continued imports of energy will be a requirement throughout this century unless energy demand is drastically reduced.

Developmental time frames for energy technologies

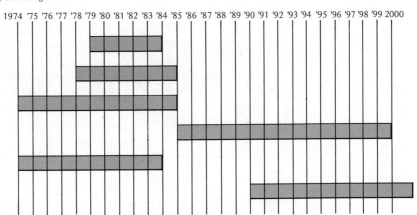

Figure 5.12. *The bars indicate the time frames in which some of the options in this chapter are expected to make contributions to U.S. energy use. [After "Building a Fire Under Energy Technology," Du Pont Context, **3**, 2 (1974), 15. Published by the Du Pont Company. Reprinted with permission.]*

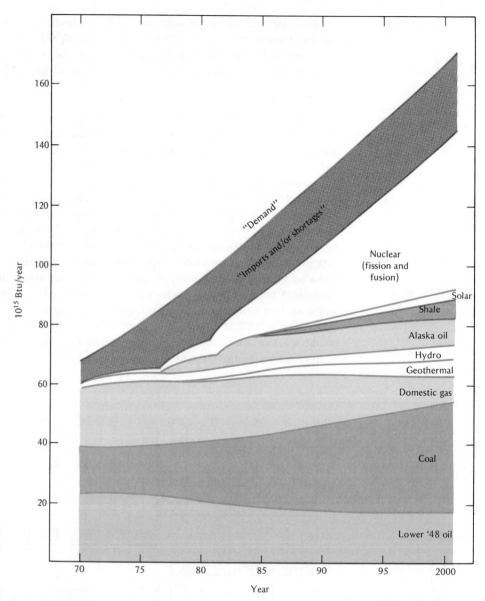

Figure 5.13. One of the many possible estimates of the use of the solar, geothermal, and shale oil options. Many assumptions have been used (see text). [From the Joint Committee on Atomic Energy, 1973.]

Energy Conservation   In all our previous discussions of energy, we have approached the problem from the point of view of attempting to meet anticipated energy demands. This is the approach that had been traditionally pursued in the United States until the mid-1970s. In fact, increases of energy consumption have in the past been encouraged by utilities and by oil companies. The policy of encouraging energy consumption is dangerously wrong in view of the limitations of energy reserves and production discussed in Chapter 4. A more farsighted policy is to investigate the reasons for the rapid growth in energy use and attempt to reduce the growth. A reduction in energy growth can be achieved by several means, but one of the most important is energy conservation: using our fuels more efficiently with less waste.

It is unlikely that even extreme conservation measures can entirely halt the need for more energy, but it is undeniably a poor idea to perpetuate wasteful and inefficient uses of energy. Five sixths of the energy used in transportation, two thirds of the fuel used to generate electricity, and nearly one third of the remaining energy consumed in the United States is discarded to the environment as waste heat. Although the second law of thermodynamics predicts that waste heat must be created, the efficiency of our energy use is much less than that which the second law predicts is possible. In all, as we saw in Chapter 4, well over half of the energy consumed in the United States is discarded as waste heat, yet little research has been devoted to energy utilization. Buildings are designed with inadequate insulation, large windows, and excess ventilation, all of which increase the need for energy in heating and cooling. Automobile engines are often more powerful than necessary, and the automobiles themselves are often heavier than necessary. Both more efficient technologies and more rational utilization of energy could save significant amounts of energy. The potential for reducing energy requirements by means of more efficient energy utilization appears to be enormous, amounting ultimately to more than 25% of what would otherwise be consumed. Saving even 1% of the energy consumed in the United States this year is equivalent to more than 100 million bbl of petroleum, more than the total expected production from shale oil and tar sands by the year 2000, more than our projected solar heating use by the year 2000, more than the output of 30 of the largest nuclear plants in 1 year. Conservation is clearly a major energy option.

Energy Use and Conservation in Transportation

Transportation constitutes the largest single end use of energy, accounting for one fourth of the total annual energy budget, and energy consumption for transportation grows at 4% per year.

Energy use in transportation is discussed in terms of **energy intensiveness,** the energy required to move a vehicle plus its load of passengers or freight 1

TABLE 5.1.   Energy intensiveness (EI) for various transportation modes in the United States

Mode	EI (Btu/passenger-mile)	Efficiency[a] (passenger miles/gal)	Load Factor[b] (%)
Urban			
Bicycle	200[c]	630	100
Walking	300[c]	420	100
Mass transit	3800	33	20
Automobile	8100	15	28
Intercity			
Bus	1600	79	45
Railroad	2900	43	35
Automobile	3400	37	48
Airplane	8400	15	50

[a] Calculated assuming all the energy comes from gasoline.

[b] The load factor (percentage of transport capacity utilized) listed for each mode is the average for that mode in the United States in 1970.

[c] Calculated in terms of the energy required to produce food.

Source: Eric Hirst, "Energy Intensiveness of Passenger and Freight Transport Modes," Oak Ridge National Laboratory, 1973.

mile. Energy intensiveness is the inverse of energy efficiency: the *lower* the energy intensiveness, the *greater* the efficiency because less energy is being used to accomplish the same task. Typical units for energy intensiveness (EI) are Btu per passenger-mile and Btu per ton-mile. Table 5.1 lists the average EI values in Btu per passenger-mile for various modes of transportation in the United States. Load factors, the percentage of transport capacity utilized, are important in EI. Doubling the occupancy of a vehicle, for example, reduces the EI by about one half.

In urban use, the average occupancy of automobiles in the United States is about 1.4 persons, which corresponds to a load factor of 28%. As shown in Table 5.1, the most efficient means of transporting people is by bicycle or walking, either mode being at least 25 times more efficient than the automobile. For motor sources in urban use, mass transit, even under present conditions of low ridership, is more than twice as efficient as the automobile. The EI values for intercity use are less than those for urban use because the load factors are higher and because engine operation is more efficient at steady speeds. Buses offer the most efficient transportation, railroads are second best, autos are third, and airplanes are the worst.

For the transportation of freight, railroads are 60 times more efficient than airplanes and four times as efficient as trucks (Table 5.2).

TABLE 5.2.   Energy intensiveness (EI) of intercity freight transport

Mode	EI (Btu/ton-mile)	Efficiency (ton miles/gal)
Pipeline	450	280
Railroad	670	188
Waterway	680	185
Truck	2,800	45
Airplane	42,000	3

For efficient energy use in transportation, the EI values for passengers and freight should be as *small* as possible. The trend in the United States since 1950, however, has been toward increasing EI values. The average EI in 1973 was 15% larger than in 1950. One reason for this increase is that there has been a modal shift of passengers from modes with low EI (railroads and buses) to modes with higher EI (autos and airplanes). Over the last 20 years total passenger miles have doubled and the automobile's share of urban transportation traffic has increased by 17%, while the mass transit share has dropped fivefold. For intercity transportation, the least efficient mode, air travel, has increased fivefold, whereas rail travel, which is 100 times more efficient, has dropped sevenfold since 1950.

Another important aspect of increasing EI trends is that the EI of individual modes has increased in recent years. The automobile EI, for example, has increased 15% since 1950. Part of this is due to a decrease in average passenger occupancy, but the major contributor is the nearly 14% *decrease* in automobile efficiency as measured in miles per gallon (mpg). In order to determine the major contributors to the decline in auto efficiency, the EPA has assembled data on more than 2000 cars from model years 1957 to 1973 tested with the Federal Driving Cycle, a standardized urban cycle. Their analysis concludes that increasing automobile weight is the biggest cause of declining fuel economy. The increased use of air conditioning, the installation of air pollution emission controls, and the increased use of automatic transmissions, in about that order, are further contributors to decreased economy.

Doubling the weight of an automobile cuts the fuel economy (mpg) about in half, and automobile weights have increased markedly in recent years. The Chevrolet Impala, for example, underwent a 38% weight increase and a 30% mpg decrease from 1957 to 1973. The average weight of the best-selling standard-sized cars increased about 800 lb from 1962 to 1973. Some examples of miles per gallon penalties for average auto use in the United States are given in Table 5.3. Laboratory tests indicate that air conditioning can reduce fuel

TABLE 5.3.   Variables in auto economy measured in miles per gallon (mpg)

Weight (25% increase)	20–30% decrease
Air conditioning (70°F day)	9% decrease
Air conditioning (90°F urban)	20% decrease
Air pollution controls (since 1968)	10% decrease
Change to automatic transmission	5% decrease
Steel belted radial tires	up to 5% increase

economy by 9% to 20% and options such as automatic transmissions and power steering also result in fuel penalties. The impact of emission control devices began with their initial installation in 1968 models, and a comparison of vehicles of equal weight by the EPA indicates that the impact has stayed relatively constant during the 1968–1973 years. That is, the 1969 and 1973 models each have a fuel economy penalty due specifically to emission controls of from 7% to 10% relative to pre-1968 (uncontrolled) models.

Freight transportation over the past 25 years has also been characterized by a trend toward higher EI values due in large part to increasing use of truck and air modes at the expense of the more efficient rail, pipeline, and waterway modes.

One obvious suggestion can be made to conserve energy in transportation: increase the efficiency of energy use by decreasing the average EI. This could be started immediately by reversing the historical trend toward increased use of high EI modes. (If present trends continue, by 1985 nearly three fourths of the energy used in transportation will be consumed by the least efficient modes: automobiles and air freight.)

Another obvious suggestion for energy conservation is to decrease the EI of the major modes by either usage changes or by technological changes. The EI of the automobile can be made quite low by doubling or tripling the average occupancy. By decreasing the weight of the car, and by using more efficient engines (see Chapter 3) the EI can be reduced further.

Some estimates of energy that could be saved by various proposed energy conservation schemes have been made by A. C. Malliaris and R. L. Strombotne of the U.S. Department of Transportation (Table 5.4). It is evident from this table that the greatest potential for energy saving in transportation lies in the changeover to smaller cars or in changes in automobile design that improve the fuel economy. The 1972 United States auto population could be approximated by 90% large cars (standard size or larger) operating at 13.1 mpg, and 10% small cars operating at 22 mpg. Conversion to a 50/50 mixture would yield the 9% saving in gasoline economy indicated in Table 5.4, assuming no

TABLE 5.4. **Energy savings as a percentage change in total transportation energy for 1970**

Proposal	Transportation Energy Conservation
Increase fuel economy of half the highway vehicles by 30%	11.5%
Convert 50% of all passenger cars to small cars (22 mpg)	9.0%
Persuade 50% of urban commuters to car pool	3.1%
50% of cars never exceed 50 miles per hour	3.0%
Shift 50% of commuters to buses	1.9%
Shift 50% of intercity auto passengers to intercity bus and rail, evenly	3.0%
Shift 50% of intercity trucking to rail	3.4%
Persuade 50% of the people to walk or bike all trips of 5 miles round trip or less	1.6%

change in passenger miles or work-trip distances. The savings for commuter car pooling may appear to be smaller than expected on the basis of the EI change that occurs in the operation of the automobile; this small percentage energy saving arises because only one third of all passenger miles are due to commuting.

Energy savings in transportation can be massive. The following illustrative scenario, based on calculations of Eric Hirst, an engineer at the Oak Ridge National Laboratory, results in a 50% decrease in transportation energy consumption with no reduction in total travel:

1. Shift half of intercity freight presently carried by truck and air to rail.
2. Shift half the intercity passenger traffic carried by air and one third of that carried by car to buses and trains. Alter the design of autos, planes, and trains to increase vehicle economy by 33%.
3. Shift half the urban auto passenger traffic to mass transit. Increase the mass transit and urban auto load factors by 10 percentage points each. Alter design of urban auto to increase mpg by 33%.

The purpose of Hirst's scenario is to demonstrate that large decreases in energy consumption in transportation can be achieved. The technological changes suggested for the various transportation modes are possible, but the time scale for these changes and for the modal shifts would probably be a decade or so. Whether or not we improve the system efficiency depends more on our collective wills and judgments than on scientific and technological breakthroughs.

Reducing Electrical Demand

Electrical utilities consume 25% of the energy used in the United States and electricity is the fastest growing component of energy consumption, having exhibited a 10-year doubling time since 1950. Historically electrical energy has been cheap, and its consumption has been aggressively promoted by the utility companies. Consequently there has been little tendency to restrict its growth or to encourage conservation. In fact, 1971 marked the first year since 1946 that the average price of electrical power, measured in constant dollars, increased. Escalating fuel costs since then have made the nation more aware of the desirability of increasing the efficiency of energy use.

In discussing electricity it is important to distinguish between **fuel conservation** and **power conservation.** A city with a fuel shortage can reduce electricity consumption by urging its citizens to turn off air conditioners, turn off lights, and alter thermostat settings. A reduction in electrical demand will follow, and less fuel will be consumed. However, the possibility remains that in the future, the citizens will slowly return to their profligate ways, and the electricity consumption will return to its previous value. Fuel conservation would have resulted during the reduced load, but such results are often temporary. Power conservation, however, means altering future demands so that the city's *maximum power* needs regardless of living standards will be less than they would have been without alteration. Although both types of conservation are important, the most far-reaching results can be achieved with power conservation: reducing future demands for energy either by improving the efficiency of energy use or by altering the way in which energy is used.

Projected energy demands, made regularly by the Federal Power Commission and by numerous utility and consumer groups, are to a certain extent self-fulfilling. A projected demand is accompanied by efforts to build an electrical generating capacity sufficient to handle that estimate. Creation of the capacity then results in a sales effort by the utility to sell the available power. The exact relationships between projected demands and the actual energy needs are difficult to establish. However, how the actual future energy demands might be reduced is fundamental to any energy conservation scheme.

A major factor in the growth of electrical energy is its price. Utilities have at least three categories of customers depending upon consumption: residential, commercial, and industrial. Industrial customers, the largest consumers, pay the cheapest rates per kilowatt-hour, often only one fifth as much as those paid by the residential group. In general, the more electricity used, the cheaper the cost per kilowatt-hour. Such a pricing scheme encourages increased consumption.

Duane Chapman and Timothy Mount of Cornell University and Timothy Tyrrell of the Oak Ridge National Laboratory, under a National Science Foundation grant, have evaluated the growth of electricity demand and its dependence on the price of electricity, on population growth, on income, and on the price of natural gas. Their work included the impact of these on each class of utility customers. They concluded that the most important factor in electrical growth seems to be the price of electricity, followed by population growth, income, and gas prices in about that order. Their estimates indicated that substantial electrical price increases and a reduction in population growth in the United States will noticeably lower electrical demands by the 1980s. Chapman, Mount, and Tyrrell made projected energy demands for four possible conditions and found that variations in price can cause as much as a fivefold reduction in future electrical needs.

In Figure 5.14 the upper curve is the electrical demand forecast by the Federal Power Commission in 1970; it predicts a sixfold increase in electrical needs by the year 2000. The lower curves were predicted by Chapman, Mount, and Tyrrell for various population growth and electrical energy price changes. Population growth rates in the United States were found to have almost no impact on future power demands; the difference between continued population growth at the present annual 1.4% rate and decreasing growth to

Figure 5.14. *Energy demand projections of Chapman, Mount, and Tyrrell. The pricing refers to real dollars, that is, to changes in the purchasing power of the dollar in any given year. The width of each of the lower curves reflects differences in demand for different population growth rates.* [After D. Chapman, T. Tyrrell, and T. Mount, Science, **178** (1972), 703.]

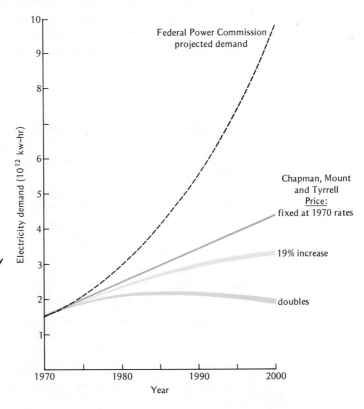

achieve zero population growth (ZPG) by 2040 is indicated merely by the width of the lower curves plotted in Figure 5.14! Thus in this model, price is the most important factor in future growth. The difference between these estimates and the Federal Power Commission projections is staggering. If the price of electricity (in real dollars) doubles by the year 2000, Chapman and coworkers predict that the demand will be only 33% larger than that of 1970. They also predicted that the Federal Power Commission projection could result only if the price of electricity were cut in half.

The importance of pricing in the growth of electrical demand has led to many proposals to use pricing to reduce future demands. One proposal is to do away with rate structures that provide incentives for high volume customers and instead use a "flat rate" where all consumers pay the same rate. Because business and industrial consumers are affected most by this change and because they have the most sensitive cost accounting, such a proposal might have a large impact in total consumption. However, increased production costs of products would probably be passed on to consumers. Other proposals are more drastic. Walter Heller of the University of Minnesota has proposed that large depletion allowances, capital gains shelters, and special tax deductions should not be allowed for energy-producing industries. The cost of energy would then more closely reflect the environmental and social costs of its production. The price would probably rise, and improved efficiency and increased conservation would result.

Possible Savings in Heating and Cooling

Over half of the energy consumed in residences is for heating and cooling; the amount of energy required can be reduced by increasing the insulation. On an economic basis, one pays increased costs for extra insulation, but recovers these over the lifetime of the house in terms of reduced utility bills. Of course, a balance must be struck between how much can be spent for insulation and how much of a reduction in utility bills will occur. There are no national requirements for home insulation, the nearest approach being a guideline written by the Federal Housing Administration (FHA) in 1965 and revised in 1971. Almost no buildings in use today meet the new standard, which applies only to new construction built with FHA loans. Many older buildings, especially in the South, have little or no insulation.

John C. Moyers, an engineer at the Oak Ridge National Laboratory, has calculated the impact of insulation on energy usage and utility bills. Using computer modeling, he determined the economic optimum insulation for various regions of the United States, that is, the insulation to install which affords the minimum total cost of insulation plus utility bills for the lifetime of the house. His results showed that additional insulation above the 1971 FHA revised standards (plus, in certain sections of the United States, the addition of storm windows) could be economically justified. The optimum insulation is 3.5 in. in the walls and 6 in. in the ceiling, about twice the 1971 FHA revised standards and well over twice the amount recommended before that revision. For a New York residence, for example, the optimum insulation reduces annual energy consumption by almost 50% compared to what would occur with the pre-1971 FHA standard. Furthermore the optimum insulation results in monetary savings each year for the life of the house. After the insulation costs are paid, the monetary *savings* are $49 per year for a home heated with gas and $155 for a home heated with electricity if the insulation is optimum rather than that recommended under the pre-1971 standard.

Commercial heating is another area in which savings could be obtained. Estimates of potential savings of 40% are often made.

It is often difficult to insulate older buildings, but efforts in that direction combined with optimum insulation in new buildings would slowly result in savings of as much as 7% of the nation's energy budget.

As we noted in Chapter 4, electrical resistance heating is half as efficient in converting fossil fuel energy into space heating as is direct combustion of the same fuels in a central furnace. However, electrically heated homes have been widely promoted in recent years.

Air-conditioning energy requirements are also reduced by improved insulation, but perhaps an even more important factor for homes can be the efficiency with which air conditioners convert electricity into cooling capacity. The efficiency of commercial air conditioners, in Btu of cooling per kilowatt-hour of electrical consumption, varies almost threefold. Making such information readily available to consumers can provide savings of both dollars and energy.

An example of how energy conservation and solar systems can be incorporated into private residences is shown by a private residence in Connecticut completed in 1974 (Figure 5.15), a three-bedroom structure built as a functional home. Solar collectors, which provide a heated water storage system, are expected to provide 60% of the energy required to heat the house (a conventional oil burning system provides a backup). The collectors and most of the windows are oriented to a southern exposure to maximize the rays of the sun during winter and to minimize the rays during the summer. Less glass is used than in conventional homes, and that which is used is double paned, insulating glass (even this type of glass has up to ten times the heat loss of a well-insulated, solid wall). Instead of drapes, insulated shutters are used. Summer cooling is afforded by a special belvedere ventilation system that encourages natural (convective) flow. Heat normally wasted in conventional homes is utilized in this residence. The fireplace and the chimney are used to heat water in the solar storage system, and the condenser coils on the refrigerator are used to preheat water going to the domestic water heater. Electric dryers are used for dishes and clothes, but the air supplied to these is preheated by the

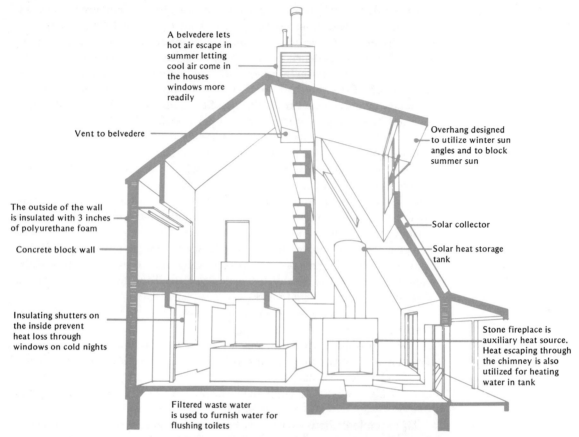

A belvedere lets hot air escape in summer letting cool air come in the houses windows more readily

Vent to belvedere

Overhang designed to utilize winter sun angles and to block summer sun

The outside of the wall is insulated with 3 inches of polyurethane foam

Concrete block wall

Solar collector

Solar heat storage tank

Insulating shutters on the inside prevent heat loss through windows on cold nights

Stone fireplace is auxiliary heat source. Heat escaping through the chimney is also utilized for heating water in tank

Filtered waste water is used to furnish water for flushing toilets

Figure 5.15. *Some of the design features of a solar-heated, semiself-sufficient home constructed for Everett Barber in Guilford, Connecticut. [Courtesy of Sunworks, Inc.]*

solar system to lessen the amount of electrical heating needed. Two 10-ft diameter windmills are used without storage to augment the electricity obtained from the local utility; about three fourths of the electrical demand is expected to be handled by the windmills. Even water is partially conserved; filtered waste water is used for the toilets.

Energy Savings in High-Rise Office Buildings

A 100-story office building in New York requires more electrical power than the entire city of Schenectady, New York (population 100,000). Common architectural practices in such buildings lead to extensive energy waste. The illumination levels in buildings have tripled in the last 20 years, and this has been combined with a decreased emphasis on natural lighting from windows. (There is little biological evidence that the higher lighting levels are necessary; they may in fact be harmful.) Because lighting uses about one quarter of all electricity consumed, savings in this area can be significant.

A reduction in lighting levels gives dual energy savings: from decreased demands by the lights themselves and from the decreased cooling required in the building. Reductions in lighting levels to the adequate 1960 standards, increased use of natural lighting, and large reductions of lighting levels in areas such as hallways and cafeterias could reduce the national energy demand by 4% according to New York architect Richard Stein. Stein has studied energy use and conservation in office buildings and has pointed out many changes which could be made to reduce energy consumption. Use of direct fossil fuel heating instead of electrical heating and use of air conditioners run by steam rather than by electricity are areas of major savings. In all, Stein believes that careful design could cut in half the energy requirements of high-rise office buildings. If all the office buildings planned for New York in the next decade followed Stein's suggestions, the reduced electrical demand would be equivalent to the output of four large nuclear power plants.

Other Potential Energy Savings

Energy savings in industry can occur through improved efficiencies of processes, through the use of more energy conscious design, and through improved maintenance on boilers and steam systems. Energy conservation programs in large plants, such as those devoted to automobile assembly, often result in annual savings of several million dollars.

The use of recycling for materials such as aluminum, as discussed in Chapter 7, represents a considerable savings of energy over obtaining the material from the raw ore.

Many life-style changes can produce significant energy savings accompanied by a cash bonus. For example, merely setting the thermostats in all Americans homes 2° higher during the summer and 2° cooler in the winter would produce energy savings in 1980 equivalent to 200 million bbl of oil.

Time Scale for Effective Conservation

A broad application of the major energy conservation ideas in this section would make enormous changes in our energy demand in the 1980s and 1990s. In 1973 the U.S. Office of Emergency Preparedness evaluated the potential of conservation and analyzed the times required to achieve change. Their study concluded that the most efficient savings could be achieved by improved insulation, by the adoption of more efficient air conditioning systems, by shifting transportation to more efficient modes, and by improving the efficiency of industrial processes. All of these changes are possible with present technology, and they could result in energy use reductions of 10% to 20% by 1980 and of as much as 30% by 1990.

Energy and Economic Costs of Pollution Controls

An important question that must be resolved in any comprehensive energy policy is the influence of antipollution controls on energy consumption; the benefits resulting from the controls must be weighed against their energy costs. Closely related to this is the impact of antipollution controls on the economy, especially on the inflation rate. It is instructive to consider both of these quantitatively, particularly in view of the advertising emphasis that private industry and utilities have put on these two effects in recent years.

The energy implications of several environmental quality improvement strategies were evaluated in 1973 by Eric Hirst of the Oak Ridge National Laboratory. His results, although somewhat preliminary because of the rapid technological and legal changes that are occurring, indicate that the total energy consumption increase needed for all anticipated environmental controls in the 1970s would be small, amounting to less than 4% of the 1970 energy consumption, and that these increases could be almost completely offset by compensating energy savings tactics; in fact, all of these environmental energy costs are dwarfed by the amount of energy which could be saved by an effective energy conservation program embodying many of the proposals outlined in this chapter.

For transportation, Hirst estimates that even if a 20% fuel penalty (relative to pre-1968 cars) resulted because of emission controls on all cars manufactured after 1976, this would mean a 1.9% increase in total energy consumption in the United States (Table 5.5). This increase could be more than offset by several energy savings tactics in transportation. For example, a 2% decrease in total energy consumption could result from the following: a 25% shift to small cars with only one half the weight of the 1970 models, plus a 25% shift from internal combustion engines to the more efficient diesels, plus a 10% shift of urban passengers from autos to mass transit (Table 5.5).

The energy required to give all communities secondary sewage treatment (Chapter 6) would require less than a 0.5% increase in energy consumption, while expanding sanitary landfill operations (Chapter 7) to include all cities would cost less than 0.1%. Energy savings would be possible by incinerating solid wastes and by increasing by about 30% the amount of steel, aluminum, and paper recycled annually (Table 5.5).

Increasingly, strict air quality standards proposed for stationary power sources in the 1970s, including the use of electrostatic precipitators and scrubbers, would increase the total United States energy consumption about 3%, while equipping half of all power plants with wet cooling towers in 1970 would have increased energy consumption by only 0.24%.

TABLE 5.5.   Energy use for hypothetical 1970 environmental quality measures

	Increase (10^{12} Btu)	Savings (10^{12} Btu)	Percentage of Total Energy Use[a] Increase	Decrease
Urban passenger travel				
Mass transit		320		0.47
Automobile pollution control	1100		1.63	
Redesign of automobiles		1100		1.63
Secondary level sewage treatment	290		0.43	
Solid waste management				
Landfilling	75		0.11	
Incineration, electricity generation		290		0.43
Increased recycle		440		0.65
Air pollution control at stationary sources	840		1.25	
Waste heat dissipation with cooling towers at power plants	160		0.24	
Totals	2465	2150	3.66	3.18

[a] Total 1970 energy consumption was 67,400 trillion Btu.

Source: Eric Hirst, "The Energy Cost of Pollution Control," Environment, October 1973, 37–44.

The predicted gross energy consumption increase of 3.66% (Table 5.5) due to environmental controls, which is in approximate agreement with an independent estimate by the National Petroleum Council, could be almost completely offset by the 3.18% energy savings that could be achieved in the same areas. Hirst's study indicates that the energy required to achieve the environmental goals is relatively small. The 3.66% increase, the cumulative maximum amount predicted for all of the 1970s, amounts to about 1 year's normal growth in energy use. This suggests that present 'nergy growth rates cannot be attributed to the operation of environmental protection systems. Furthermore, energy conservation measures, as presented in this chapter, could reduce energy consumption by typically 20% by 1980, which represents about six times more energy than that required for environmental controls.

The dollar costs of pollution controls for the 1970 decade are in Table 5.6. Although the estimated $280 billion for pollution controls is a considerable sum, it pales beside the total gross national product (GNP) of $13,200 billion over the same period. Russell W. Peterson, Chairman of the Council on Environmental Quality, has pointed out that emission controls represented

TABLE 5.6.   Total pollution control expenditures

	Cumulative Costs, 1971–80 (billions of 1971 dollars)
Air pollution	106.5
Water pollution	87.3
Noise	(0.9–2.7)
Radiation	2.1
Solid waste	86.1
Land reclamation (surface mining)	5.1
Total	287.1

[a] Total projected cumulative Gross National Product is $13,200 billion.

Source: Third Annual Report, Council on Environmental Quality, August 1972.

around 0.7% of the 1974 GNP and added no more than 0.5% to the inflation rate, which was over 10% in that year. A study by Chase Econometric Associates similarly showed environmental spending to be contributing 0.3% to 0.5% to the inflation rate in 1974. The amount that would be spent on proposed environmental protection in the next decade is no more than what the nation will spend on alcohol, only a third of what it will spend on recreation, and less than it will spend for defense in the next 2 or 3 years. Environmental spending shows compensating gains. The EPA estimates that for less than $4 billion spent on air pollution abatement alone, the nation would save $13 billion in reduced medical costs and property damage. These and the other benefits indicate that the relatively small energy and economic costs of environmental controls can be easily justified.

Looking Forward:
Policy and Future
Energy Growth

Future energy use and conservation will be affected by governmental regulatory policies and by the collective actions of citizens. One central aspect in energy policy is a long-standing tenet of the energy industry that economic growth requires increasing energy consumption; the impact of this belief on energy policies is substantial. However, there is some evidence that this tenet might be incorrect. The Rand Corporation, in a study of electrical energy growth in California, recommended that the state attempt to reduce future demands. The Rand report suggested that price structure alterations, conservation measures, improved efficiencies, and development of solar and geothermal sources could cut the projected energy demand by one half by the year 2000. These savings would reduce the need for new power plants from an estimated 127 to 45 or less. Perhaps more important, the Rand report concludes that only relatively minor economic dislocations would occur and that the growth of the state's economy would not be affected.

The Ford Foundation financed an ambitious Energy Policy Project beginning in 1971. The final report issued in 1974 evaluated future energy demands according to three scenarios. The report states no preferences for the options available but suggests that energy growth can be divorced from economic growth.

The three scenarios are proposed to compare policy choices and consequences. The "historical growth" scenario assumes that the use of energy will continue to grow much as it has in the past. It assumes that the nation will not deliberately impose any policies that might affect our ingrained habits of energy use, but will make a strong effort to develop supplies at a rapid pace to match rising demand. The political, economic, and environmental problems of obtaining the amount of energy required would be formidable.

The "technical fix" scenario shares with "historical growth" a similar level and mix of goods and services (Figure 5.16). But it reflects a determined, conscious national effort to reduce demand for energy through the application of energy saving technologies and policies. The slower rate of energy growth in "technical fix," about half that of "historical growth," permits more flexibility of energy supply, but still provides a quality of life at home, a travel convenience, and an economic growth that differ little from that of the "historical growth" scenario. Only one of the major domestic sources of energy—Rocky Mountain coal or shale, or nuclear power, or oil and gas—would have to be pushed hard to meet the energy growth rates of this scenario. The resulting flexibility could be used to limit severely, for environmental or security reasons, the role of nuclear energy, or of coal, or of imported fuels in the national energy mix.

The "zero energy growth" scenario is different. It represents a real break with our accustomed ways of doing things, yet it does not represent austerity. It would give everyone in the United States more energy benefits in the year 2000 than he enjoys today, even enough to allow the less privileged to catch up to the comforts of the average American. It does not preclude economic growth. Zero energy growth would emphasize durability, not disposability of goods. It would substitute the ethic "enough is best" for the idea that "more is better." The use of energy would rise slightly, leveling off at about 100×10^{15} Btu per year in 2000. Cities and transportation systems would be redesigned, and economic growth would be concentrated in the provision of services rather than in manufacturing.

The Energy Policy Project report concludes that all three growth rates are technologically possible. Which occurs will be the result of the actions of the government and of the citizens. Reductions in energy growth of the size indicated by the Office of Emergency Preparedness conservation study, by the Rand report for California, and by the Ford Foundation Energy Policy Project will not occur without careful, farsighted policies.

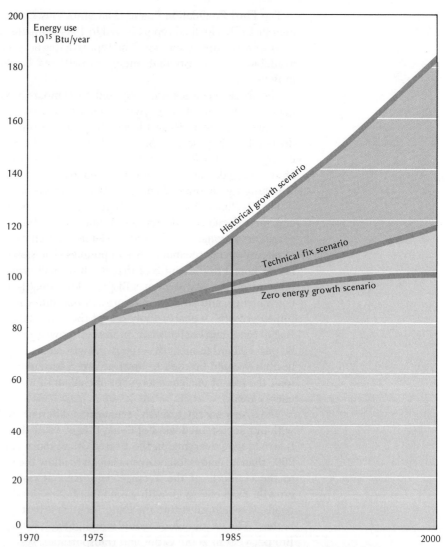

Figure 5.16. *Energy demands predicted by the Ford Foundation Energy Policy Project's three scenarios. [From Exploring Energy Choices, Energy Policy Project of the Ford Foundation, 1974. Copyright 1974 by the Ford Foundation, reprinted with permission.]*

For example, the energy savings available from improved home insulation are not achieved at present because there is no incentive to encourage insulation. Although monetary savings are possible, the builder, who usually applies the insulation, is usually interested in keeping construction costs low, and may elect to use less insulation. The home buyer, although he pays a lower purchase price, is stuck with the higher utility bills. A policy requiring optimum insulation in all new homes or at least a policy requiring an accurate evaluation of home heating and cooling costs before purchase would provide an incentive for energy savings.

Trends in transportation are often not directed toward a reduction in energy needs. If the government can regulate automobile exhaust emissions and safety equipment, it might also press for lighter weight and better fuel economy. A partial assist in this direction was the recently initiated requirement (by the EPA) that information about the predicted gas mileage of new cars must be available to prospective purchasers.

The technology exists to affect a reduction in energy use of as much as 25% with little change in creature comforts or life-styles. Achieving such a reduction is a matter of policy and incentives.

Summary   Our future energy sources may be quite different from those used presently. Hydrogen is a superb fuel that can be considered a renewable resource if obtained from water, since the combustion of hydrogen forms water. This fuel has no sulfur oxide, carbon monoxide, or particulate emissions, can be transported by conventional pipelines, and can improve the efficiency of space heating. Stored as a liquid in a Dewar vessel, it could be used to fuel jet airplanes. As a liquid, hydrogen possesses more energy per pound than any other combustion source. Hydrogen can also be converted into fertilizer and various petrochemical products. If the hydrogen economy were to become a reality, the sheer volume of hydrogen required means that electrolysis of water would be the source.

Methyl fuel, a mixture of alcohols including methanol, is a clean, versatile fuel that can be made from wood or from wastes. It burns with a clean flame, emits no sulfur oxides or particulates, and can be transported as a liquid. As a gasoline additive (15% by volume) methyl fuel increases fuel economy and acceleration while decreasing emissions. Automobile engines can be modified to run on methyl fuel alone.

A variety of clean fuels can be obtained from coal, our most abundant fossil fuel. Coal can be liquified, using hydrogen, into an oil similar to fuel oil. Synthetic natural gas, which has a heating value about identical to that of natural gas, can be obtained from coal in a variety of ways, but is relatively expensive. A cheap, lower Btu gas is more easily obtained, but it has only one fifth the heating value of natural gas. An intriguing possibility is underground gasification of coal, an approach that precludes mining.

Petroleum can be obtained from shale oil deposits and from tar sands. Both resources involve gargantuan mining operations; often several tons of shale or sand must be processed to obtain one barrel of oil. The availability of water may limit shale oil processing because the best deposits are in western Colorado and southwestern Wyoming. Tar sand deposits are presently being mined in Alberta, Canada. Both of these sources of petroleum contain in total several times more oil than conventional oil fields.

Thermonuclear fusion reactions involving the combination of deuterium nuclei or the reaction of deuterium with tritium provide a source of energy second only to solar energy, but controlled fusion has not been achieved. Magnetic confinement research attempts to heat the fuel to the necessary 100 million° C, then force the nuclei together by means of a "magnetic bottle." Laser fusion research involves the striking of solid fuel pellets with energy bursts from several lasers. Still to be solved in the event that controlled fusion becomes a reality is the means to produce electricity from such a source of energy. The ocean could provide the fuel.

Solar energy is our largest resource; the energy falling annually on just 0.5% of the United States land areas exceeds the total energy needs of this nation projected to the year 2000. Recent technological advances may overcome the two traditional barriers to solar energy use: the relatively large land areas required and the difficulty of energy storage. Solar heating and cooling are economically competitive with fossil fuel systms, hence this application will probably be developed first. Auxiliary solar systems can often provide 80% of the space heating and cooling needs in new buildings. Several means can be used to provide solar generation of electricity. Solar satellites containing solar cells could be used. High above the earth's atmosphere, a satellite in a synchronous orbit could generate electricity continuously and take advantage of the fact that sunlight is stronger there than on the surface of the earth. The electricity could be converted to microwaves, then beamed to an antenna on earth where the energy would be reconverted to electricity.

Aden and Marjorie Meinel have proposed a land-based solar power plant. Pipes coated with selective coatings to attain high temperatures would be placed in larger, evacuated glass pipes. An array of pipes, a "solar farm," in a desert area would be a heat source for a steam power plant.

Solar power could also be tapped by operating a turbine generator system in the ocean. A fluid such as ammonia could be vaporized at the temperature of the ocean's surface, passed through a turbine, and condensed at the temperature at the bottom of the ocean. The temperature difference of about 15°C (30°F) means that the volume of flow necessary is about the same as that in a hydroelectric plant of the same capacity. The transmission of electricity from a sea-based plant is difficult, so such a station might generate hydrogen by electrolysis; the hydrogen would then be piped ashore.

Geothermal energy is used in many nations. Both wet and dry steam sources have been tapped, but many more remain. These, combined with the possibility of pumping surface water to underground hot rock deposits to create steam, provide a large future source of energy.

Aerogenerators afford an almost pollution-free energy source. Structural problems stand as a barrier to large aerogenerators, and the storage of energy for use during calm periods is a barrier for all aerogenerators. The total amount of energy available, however, rivals that of solar energy.

Energy conservation offers an enormous potential for reductions in future energy demands. By altering designs and placing more emphasis on energy requirements, substantial reductions can be achieved in all areas of energy use. The largest energy savings in transportation could be achieved by increasing the efficiency of automobiles (more miles per gallon) and by shifting people and freight from autos, trucks, and planes to buses, railroads, and waterways.

The price of electricity is the most important factor in the growth of electrical power. Studies by Chapman, Mount, and Tyrrell indicate that Federal Power Commission projections of the power needed in the year 2000 might be reduced two- to fivefold by a combination of pricing and conservation techniques.

Increasing the amount of insulation in homes can reduce fuel consumption by 30% to 50% and provide annual monetary savings, even after the insulation costs are deducted.

By reducing lighting levels, altering interior designs, and installing the most efficient heating and cooling units, energy use in high-rise office buildings can be reduced by 50%.

The benefits arising from environmental controls justify their implementation, since the implementation of all controls projected for the 1970s would increase total United States energy consumption by less than 4%, an increase which could be more than offset by energy savings tactics.

Several studies indicate that overall energy savings of 30% could be made by 1990 if realistic programs focusing on energy conservation were initiated. The Rand Corporation recommended that the State of California initiate a conservation program and also alter electricity rates to achieve an overall 50% reduction in their projected power needs for the year 2000. The study could find little evidence that the economy of the state would suffer if such a program were instituted. The Ford Foundation's Energy Policy Project indicated that power demands could easily be lowered by 30% over the next 25 years and that it was technologically possible to achieve zero energy growth by the end of this century if a determined program were undertaken.

For energy conservation to become a reality, a broad, well-conceived federal policy will probably be necessary.

Review and Study Questions

1. Define or identify: Dewar flask, electrolysis, SNG, kerogen, Tokamak, wet steam, energy intensiveness (EI), selective surface coatings.

2. State the importance of each as related to the topics of this chapter: Richard Stein, Eric Hirst, Duane Chapman, Grandpa's Knob, Lev Artsimovich, Aden and Marjorie Meinel, H_2indenburg Society.
3. Describe why hydrogen can be considered a renewable fuel.
4. In what transportation system(s) is hydrogen not likely to be used? Why?
5. What is a major disadvantage of obtaining clean fuels from coal?
6. Describe the conditions that limit oil production from tar sands.
7. When might electricity from a fusion reactor be available?
8. What advantages and disadvantages (relative to a planar collector) are possessed by a solar concentration collector?
9. Describe the solar farm concept of the Meinels.
10. List two ways energy storage can be provided for windmill-powered generators.
11. What is the most energy efficient means of transporting people?
12. What two major effects have contributed to the increases in the average EI values for passenger transportation since 1950?
13. What is the most important contributor to the decline in automobile efficiency (miles per gallon) since 1960?
14. Describe the findings of the Chapman-Mount-Tyrrell study of electrical power growth.
15. What changes have occurred in suggested lighting levels in buildings since 1950? What evidence suggests the necessity for this change?
16. Describe the three scenarios presented in the Ford Energy Policy Project.

Questions Requiring Outside Reading

17. How has United States water usage in the southwestern United States affected the importance of geothermal sources of electric power? (See John Lear, Suggested Outside Reading)
18. How many kilowatt-hours of electricity were generated by the Grandpa's Knob, Vermont, aerogenerator? Why was it stopped? (See Julian McCaull, Suggested Outside Reading.)
19. Where can one buy a windmill-driven electrical system? (See Allen Hammond, *Science* (Apr. 19, 1974), "Energy" issue; Wm. and Ellen Hartley, Suggested Outside Reading.)
20. What are the relative costs to the household of electricity from nuclear power (all-electric rate), from coal converted to hydrogen, and from nuclear energy converted to hydrogen? (*Chemical and Engineering News*, October 2, 1972, 33–34.)
21. An energy storage system suitable for use with steam power plants, with windmill generators, with trolley buses, and with electric automobiles is the superflywheel. What is it? What are its advantages? (David Lampe, *Popular Mechanics*, November 1974, 124–127, 169; Richard F. Post and

Stephen F. Post, "Flywheels," *Scientific American* (December 1973), 17–23, and "Letters," March 1974, 8–10.)

22. Electricity can be generated from fuel cells. In one widely used cell, hydrogen can combine with oxygen to form water. What are the prospects for large-scale energy generation with fuel cells? ("Fuel Cell Research Finally Paying Off," *Chemical and Engineering News,* January 7, 1974, 31–32; Thomas H. Maugh, "Fuel Cells," *Science* **178** (December 22, 1972), 1273–1274B.)

23. What architectural design changes in homes result in energy conservation and facilitate the use of solar energy? (H. R. Hay, "Energy, Technology, and Solarchitecture," *Mechanical Engineering,* November 1973, 18–22; Richard G. Stein, Suggested Outside Reading.)

24. How efficiently would satellite solar power stations convert solar energy into electricity? What are the projected costs? (*Chemical and Engineering News,* January 1, 1973, 17.)

25. What are the relative capital costs of nuclear, coal, oil, and orbiting solar electric power plants? (*Chemical and Engineering News,* December 20, 1971, 39.)

26. The Environmental Defense Fund (see Chapter 8) and Capitol Community Citizens intervened in a routine application procedure for electricity rate increases in Wisconsin. The groups requested a "flat" rate structure (all users paying the same amount per kilowatt-hour) and other pricing policy changes in order to promote conservation. What was the result? (*Science* **185** [September 20, 1974], 1031.)

27. How would energy consumption be affected if the money in the highway trust fund were reinvested in railroads and mass transit, educational facilities construction, and a variety of social programs? (See Bezdek and Hannon, Suggested Outside Reading.)

28. Building houses below ground level offers the advantage of decreased energy consumption. How much reduction can be accomplished? What special problems are encountered? (V. Elaine Smay, "Underground Living," *Popular Science,* June 1974, 88–89, 132.)

29. Large industrial plants can often save millions of dollars by emphasizing energy conservation. How are effective programs operated? (Urban C. Lehner, "Manufacturers Save Millions by Increasing Efficiency in Energy," *The Wall Street Journal,* March 11, 1974, 1, 23; Ward Worthy, "Dow Reaps Benefits from Energy Conservation," *Chemical and Engineering News,* January 14, 1974, 25–26.)

30. The Ford Foundation Energy Policy Project's Advisory Board included environmentalists, scientists, and industrial executives. Many on the board disagreed with the final report of the Project, especially its emphasis on conservation. Who disagreed and why? (A. L. Hammond, "Energy: Ford Foundation Study Urges Action on Conservation," *Science* **186** (Novem-

ber 1, 1974), 426–428; "Energy Report Likely to Stir Controversy," *Chemical and Engineering News,* October 28, 1974, 27–28.)

31. One proposed use of solar power is to grow at maximum efficiency a crop, such as sugar cane, which can be used as a source of energy as well as a source of some products ordinarily obtained from petrochemicals. What is the potential of such a project? (Melvin Calvin, "Solar Energy by Photosynthesis," *Science* (April 19, 1974), 375–381.)

32. What energy problems exist in Europe and what energy choices must be faced there during the remainder of this century? (Wolf Häfele, "Energy Choices that Europe Faces," *Science* (April 19, 1974), 360–370.)

33. Rationing of energy or an allocation system giving priorities to certain uses of energy are methods often proposed to handle energy shortages. What are public attitudes to such approaches? Which uses of energy have the highest priorities in the view of the public? (James R. Murray, et al., "Evolution of Public Response to the Energy Crisis," *Science* (April 19, 1974), 257–263.)

Suggested Outside Reading

BARNEA, JOSEPH. "Geothermal Power," *Scientific American* (January 1972), 70–78. A very general discussion of the sources of geothermal energy, the costs of electricity from geothermal plants, and the future possibilities for this energy source.

BERG, CHARLES A. "Energy Conservation Through Effective Utilization," *Science* **181** (July 13, 1973), 128–138. An analysis of the potential for energy conservation. Included are estimates of the contributions which solar energy, fuel from wastes, and materials recycling might make.

BEZDEK, ROGER, and BRUCE HANNON. "Energy, Manpower, and the Highway Trust Fund," *Science* (August 23, 1974), 669–675. Analyzes the energy consumption changes if the $5 billion annually put into the highway trust fund were reallocated into such things as railroads and mass transit, law enforcement, national health insurance. In some cases energy consumption might increase, in others it might be reduced significantly.

CHAPMAN, DUANE, TIMOTHY TYRRELL, and TIMOTHY MOUNT. "Electricity Demand Growth and the Energy Crisis," *Science* **178** (November 17, 1972), 703–708; "More Resistance to Electricity," *Environment* **15** (October 1973), 18–20, 32–36. A detailed analysis of the major factors in the growth of electricity consumption. Cost is the major factor. Several projections of future power needs are made.

DIALS, GEORGE E. and ELIZABETH C. MOORE. "The Cost of Coal," *Environment* **16** (September 1974), 18–24, 30–37. Human and environmental costs of coal mining are from 9% to 18% of coal's price. Cleaning up coal by gasification or liquefaction would still provide a bargain fuel.

EMMETT, JOHN L., JOHN N. NUCKOLLS, and LOWELL WOOD. "Fusion Power by Laser Implosion," *Scientific American* (June 1974), 24–37. A general discussion of this approach to fusion.

Exploring Energy Choices. Ford Foundation Energy Policy Project, 1776 Massachusetts Avenue, N.W., Washington, D.C. 20036. A preliminary report of the project.

GREGORY, DEREK P. "The Hydrogen Economy," *Scientific American* (January 1973), 13–21. A comprehensive article on hydrogen and its many uses.

GRIMMER, D. P., and K. LUSZCZYNSKI. "Lost Power," *Environment* 14 (April 1972), 14–22, 56. An analysis of the efficiency of energy use in transportation and of the changes necessary to improve the efficiency. Included is a comparison of electric cars vs. gas engine cars. The total energy requirements of transportation, including lubricating oil, are considered.

HARTLEY, WM., and ELLEN HARTLEY. "The Wind Shifts to Windmills," *Popular Mechanics,* November 1974, 80–84. A description of aerogenerators, their use and their construction.

HIRST, ERIC, and JOHN C. MOYERS. "Efficiency of Energy Use in the U.S.," *Science* 179, March 30, 1973, 1299–1304. A review of the efficiencies of energy use in transportation, home heating, and lighting. some policy proposals are made which would encourage an increase in the efficiencies.

LEAR, JOHN. "Clean Power from Inside the Earth," *Saturday Review,* Dec. 5, 1970, 53–61. A close look at geothermal power, especially the potential in southern California and Cerro Prieto, Mexico.

LOVINS, AMORY B. "World Energy Strategies," *Bulletin of the Atomic Scientists,* May 1974, 14–32. A general review of world energy resources, policies, and prospects.

MALLOY, MICHAEL T. "Is Your 1973 Car Guzzling Gasoline? It May Have A Serious Weight Problem," *The National Observer,* June 30, 1973, 9. An analysis of the reasons for the decline in automobile miles per gallon. Various auto industry executives are interviewed.

McCAULL, JULIAN. "Windmills," *Environment* 15 (January/February 1973), 6–25. An extensive history of aerogenerators and a look at present proposals.

———. "Wringing Out the West," *Environment* 16 (September 1974), 10–17. An article stressing the limitations that will be imposed on coal mining and shale oil mining due to water shortages in the areas where mining is being considered.

MEINEL, ADEN B., and MARJORIE MEINEL. "Physics Looks at Solar Energy," *Physics Today,* February 1972, 44–50; "Solar Energy—The Possible Dream," *Aware,* February 1972, 3–5. A general description of the "solar farm" electrical power station proposal of the Meinels.

METZ, WM. D. "Ocean Temperature Gradients: Solar Power from the Sea," *Science* 180 (June 22, 1973), 1266–1267. A variety of proposals for solar power from the sea are discussed, including the Zener-Anderson scheme presented in this chapter. Costs, problems, and prospects are the focus of the article.

————. "Oil Shale: A Huge Resource of Low-Grade Fuel," *Science* **184** (June 21, 1974), 1271–1275. A look at the problems of and prospects for shale oil mining in the western U.S.

PERRY, HARRY. "Coal Gasification," *Scientific American* (March 1974), 19–25. A review of the history of and a look at the future of obtaining clean gas fuels from coal.

"Potential Energy Sources Pose Mining Problems," *Chemical and Engineering News*, April 15, 1974, 16–17. A report on the environmental problems encountered in shale oil and tar sand mining operations.

"Potential for Energy Conservation, The." Office of Emergency Preparedness, Superintendent of Documents, U.S. Government Printing Office, Washington, D.C., October 1972. A staff report analyzing the practical possibilities of conserving energy through a variety of policies. The study considers long- and short-term measures in transportation, industry, residential/commercial, and electric utilities energy sectors. Energy conservation measures can reduce U.S. energy demand by 1980 by as much as 7.3 million barrels of oil per day, about two-thirds of the projected oil imports for that year.

REED, T. B., and R. M. LERNER. "Methanol: A Versatile Fuel for Immediate Use," *Science* **182** (December 28, 1973), 1299–1304. A comprehensive article on this versatile fuel.

"Report on Automotive Fuel Economy, A." U.S. Environmental Protection Agency, November 1972 and February 1974. Analyzes the reasons for the decrease in miles per gallon of automobiles since 1950. Data on over 4000 cars driven over a federal driving cycle are analyzed. Vehicle weight is the most important factor.

STEIN, RICHARD G. "A Matter of Design," *Environment* (October 1972), 17–29. An excellent article on the potential for energy conservation in homes and office buildings. Building codes, lighting standards, insulation standards, and heating/cooling efficiencies are emphasized.

Time to Choose: America's Energy Future, A. Ballinger, Cambridge, Mass., 1974. The final report of the Ford Foundation Energy Policy Project.

ZENER, CLARENCE. "Solar Sea Power," *Physics Today* (January 1973), 48–53; also "Letters," *Physics Today* (June 1973), 11–13; (September 1973), 9–13; (January 1974), 15–16. Zener's updating of the Andersons' proposed undersea ammonia turbine power plant.

Many of the Suggested Outside Readings given in Chapter 4—especially *Science* (April 19, 1974); *Scientific American* (September 1971); Hammond et al., *Energy and the Future;* and *Resources and Man*—discuss many of the topics in this present chapter.

6 Water Pollution

It is a paradox that man has tended from earliest times to dispose of his wastes in the water courses from which much of his drinking water is to come.

Barbara Ward and Rene Dubos
Only One Earth

Introduction   Water, the most abundant substance on earth, is also perhaps the most important and most versatile. It is an integral part of living tissues and has historically determined the placement of agriculture, industry, and civilizations.

As viewed by the astronauts, the most impressive feature of the earth is the abundance of water: 71% of the earth's surface is covered by oceans and 3.5% is covered by ice. Much of the land area is capped by clouds. The seeming global abundance of water is misleading, however, for the distribution is uneven, a fact that leads to disastrous flooding in some areas and desperate shortages in others. As human population and water usage increase, it becomes increasingly important to understand the environmental impact of man's water appropriations. In this chapter we shall consider water resources, the impact of water pollutants, and some alternatives to our present water practices.

Water pollution is a more complex topic than air pollution. In contrast to the easily located exhaust pipes and smokestacks of air pollution, sources of water pollution are often poorly defined. Our waterways serve as sinks for our sewage and industrial effluents. Water pollutants may come from submerged pipes, from drainage runoff, from underground leaks in sewer systems, or from air pollutants washed down in the rain. In addition once pollutants enter the water, they often undergo chemical reactions with water itself or with one of the other pollutants present in the water. The wide variety of water pollutants and water pollutant sources requires a similar variety in pollution control techniques. This is in contrast to air pollution where many of the five major pollutants were common to many sources.

Because water has so many uses, our definition of a water pollutant is linked to the intended application. We define water as being polluted when it is unfit for its intended use (Figure 6.1). This means that water that may be classified as polluted for drinking purposes might be usable for irrigation or for power plant cooling.

Figure 6.1.   *Water is polluted when it cannot be used for its intended purpose. [U.S. Environmental Protection Agency, photo by Bert Emanuele for The Rotarian.]*

In Chapter 1 we noted that most, though not all, of the pollutants created by man exist naturally on the planet and are integrated in biogeochemical cycles but that man's activities in localized areas tend to overload and unbalance these cycles. Such is the case for many water pollutants, although there are others—for example, chlorinated hydrocarbons (see Chapter 8)—that do not exist in nature and may tend to concentrate in food chains.

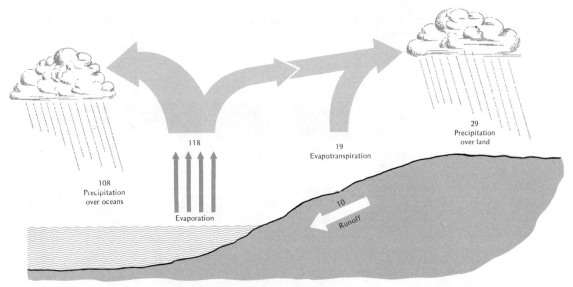

108
Precipitation
over oceans

118

Evaporation

19
Evapotranspiration

29
Precipitation
over land

10
Runoff

Figure 6.2. *Global hydrologic cycle as measured in quadrillion (10^{15}) gallons per day.
[From M. I. Lvovitch, World Water Balance, vol. 2, UNESCO, Geneva, 1972.]*

Hydrologic Cycle

The hydrologic cycle, the movement of water through the environment, is driven by energy from the sun. This cycle provides for continuous freshwater supplies and affords a natural purification scheme. Water is evaporated continuously from any wet surface and is also released into the atmosphere through transpiration of plants. The combination of evaporation and transpiration is called **evapotranspiration.** As water evaporates, most of the impurities are left behind, so relatively pure water is provided by rainfall. More water is evaporated from the oceans than is returned to them by rainfall (Figure 6.2). To maintain a balance, on land there is more precipitation than evaporation, and the excess is returned to the oceans as runoff. About two thirds of rainfall is evaporated; the rest, about 10×10^{15} gal per year, is available for use year after year. It is a renewable resource that can be utilized without affecting the hydrologic cycle. This average freshwater supply represents less than 1% of the total amount of water on the planet. Over 96% of the earth's water exists as ice and as salt water in oceans. The rest exists in the atmosphere, in the soil, in underground water supplies, or in rivers and lakes.

Water Use in the
United States

Of the 4.2×10^{15} gal of water falling each day on the United States, 3×10^{15} gal are lost to evapotranspiration and 1.2×10^{15} gal are available as runoff. Before discussing the use of this water, we should define terms of water

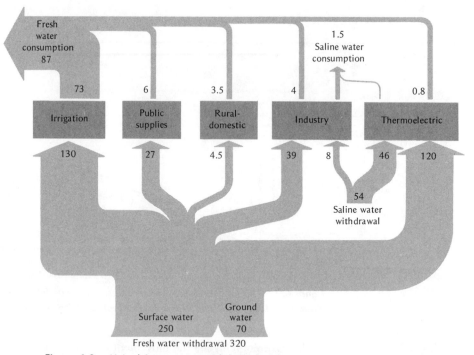

Figure 6.3. *United States water withdrawal and consumption in 1970. The data are given in billion gallons of water per day (bgd). [Data from C. R. Murray and E. B. Reeves, Estimated Use of Water in the U.S. in 1970, U.S. Geological Survey, Circular 676, 1972.]*

usage. **Withdrawal** denotes the use of water for any purpose. Some of the withdrawal will be returned to the waterway and can be reused subject perhaps to purification. The rest of the withdrawal will be **consumed**; that is, it will not be returned directly to the water source. Consumption of water may involve evapotranspiration or it may involve the incorporation of water in some entity, for example, in a living organism.

Figure 6.3 shows the consumption and withdrawal of water in the United States in 1970. About 315 billion gallons per day (bgd) represents an upper limit to what might be *consumed* in the United States on a sustainable basis. The rest of the fresh water is either required for natural processes or is not economically suited for development. Although the withdrawals in 1970 exceeded 315 bgd, the important category, consumption, is 87 bgd, well below the 315 bgd limit.

Figure 6.3 shows that the major user of water in the United States is agriculture, which accounted for 41% of the withdrawals and 84% of the consumption in 1970. Power plants are the second largest source of withdrawals of

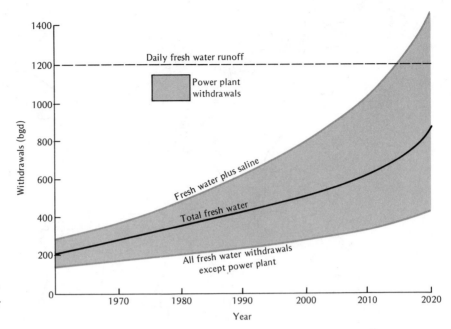

Figure 6.4. *Estimated water withdrawals in the United States. The power plant withdrawals are for cooling.*

fresh water, whereas municipal and industrial withdrawals are relatively small. The 320 bgd of total withdrawals represents about 1600 gal per person. Of this, about 130 gal is for municipal use: washing dishes takes 10 gal, flushing a toilet 4, a load of washing 20 to 30, a shower 10 to 30, and a bath 30 to 40 gal.

Estimates of future water usage indicate that by the year 2000 withdrawals in the United States will be equal to half the fresh water runoff, with power plant cooling accounting for half of these withdrawals (Figure 6.4). By early in the next century water *consumption* in the United States will be about half the maximum sustainable level of 315 bgd.

The above data indicate that on a national basis water resources in the United States should be adequate until well into the next century. However, the uneven distribution will result in severe problems in many regions, as shown in Figure 6.5. This figure, derived from the studies of the Commission on Population Growth and the American Future (Chapter 2), indicates the regions where fresh water deficiencies exist now and where they are expected in the future. The deficit begins in the southwestern portions of the nation and grows outward.

Groundwater Use   About one fifth of the water withdrawals in the United States are derived from underground sources commonly referred to as **groundwater** resources. These are fresh water sources that accumulate when rainfall percolates

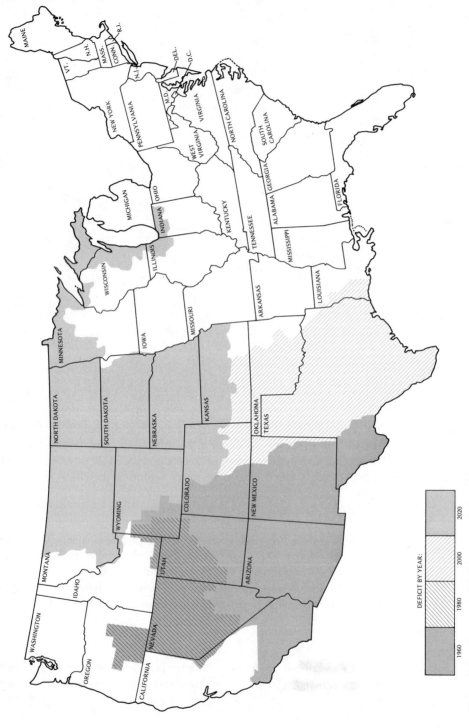

Figure 6.5. Present and future water-deficient regions in the United States as projected assuming *three* children per family. [From *Population and the American Future, The Report of the Commission on Population Growth and the American Future,* U.S. Government Printing Office, 1972.]

DEFICIT BY YEAR:

1960 1980 2000 2020

through the soil until it reaches impermeable rock where it builds up and slowly flows toward the oceans. The distance from the earth's surface to the top of the groundwater is called the water table.

Using such sources is not without problems. Excessive withdrawals may result in a settling, or subsidence, of the land. The U.S. Geologic Survey estimates that San Jose, California, sank more than 5 feet from 1948 to 1962. Portions of Houston are sinking at the rate of 1.5 in./year with damage to buildings, pavement, and flood control systems. Land subsidence can sometimes be halted by replenishing groundwater supplies through constructing percolation ponds. Near coastal areas, some groundwater reservoirs may become polluted with salt water if too much fresh water is consumed. Another problem with certain groundwater reservoirs is that the natural rate of replenishment is so slow that the reservoir can be pumped dry.

Groundwater sources usually provide very pure water because of the purity of rainfall and the action of bacterial processes that occur in the soil during percolation. However, pollution of groundwater is possible through the use of poor sewage treatment plants (discussed later in the chapter), urban development, industrial discharges, and faulty radioactive waste storage (Chapter 4). Such encroachments must be viewed with extreme caution, as pollution of this source of fresh water forces reliance on other sources, which are usually more expensive and often represent a greater environmental impact.

The Aquatic Ecosystem

In Chapters 1 and 2 we discussed ecosystems, focusing on those that exist on land. Aquatic ecosystems, the basis of life in the sea, are also based on plants. Both rooted plants and minute floating plants (usually algae) called **phytoplankton** are included among the producers. The consumers include minute animal plankton (**zooplankton**), insects, frogs, and fish. Also feeding on this chain are birds and land carnivores including man. Bacteria and fungi facilitate the decay of dead organisms in the sea, recycling them as nutrients back into the ecosystem (Figure 6.6). Aquatic food chains exhibit the energy relationships discussed in Chapter 2: each level has an energy efficiency of about 10% (Figure 6.7), so that an enormous production in all levels is necessary to support man and the large carnivores at the top.

Another important aspect of aquatic food chains is that biological magnification of certain pollutants can occur, causing their concentrations in the top levels of the chain to be thousands of times larger than in the water. We discussed the possibility of biological magnification of radioisotopes in Chapter 4. Other examples include metals and certain pesticides such as DDT (Chapter 8). Phytoplankton continuously ingest trace amounts of all substances, including pollutants, that are present in the water. Magnification of a pollutant can occur, for example, when the pollutant is more soluble in fatty tissue than in water; the pollutant, if not utilized in metabolism, may remain in

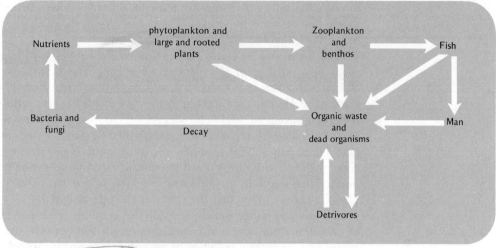

Figure 6.6. A typical aquatic ecosystem.

the organism, perhaps held in fatty tissues or substituting for some other element in a biological molecule. In the extreme case, all of the pollutant ingested by the plankton would be retained and all of this would be transferred to the next higher level when the phytoplankton are consumed. The consuming organism thus ingests the traces of pollutants collected by thousands of plankton, and if this organism retains most of these pollutants, it can transfer them to the next higher level. Repetition of this process gives magnification upward through each level in the chain. Table 6.1 lists some typical biological magnification factors (wide variations can occur).

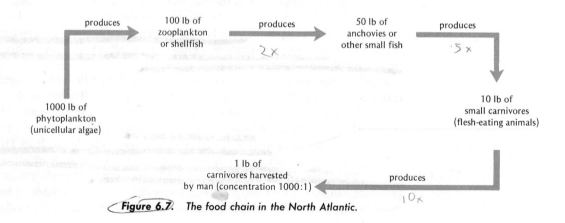

Figure 6.7. The food chain in the North Atlantic.

TABLE 6.1. Biological magnification of pollutant concentrations in food chains over their concentrations in the surrounding water

Pollutant	Concentration Increase
Aluminum	100,000
Iron	45,000
Copper	30,000
Radium	12,000

Dissolved Oxygen

The amount of oxygen in the gaseous form (O_2) present in water determines important properties of the ecosystem. A healthy body of water usually has dissolved oxygen concentrations of 5 to 7 mg/liter. A well-balanced warm water fish population, for example, requires a dissolved oxygen level of about 5 mg/liter, but areas highly polluted with sewage or other industrial wastes may have levels near zero. In 1971 more fish (71 million) were killed by water pollution than in any previous year on record, the major cause being a low dissolved oxygen content (Figure 6.8). In general, game fish such as trout and bass require larger amounts of oxygen than do rough fish such as carp and catfish.

The amount of oxygen in water is determined by the balance of several processes. Oxygen from the atmosphere dissolves in water (aeration), and oxygen is released in the water by plants during photosynthesis. Respiration of aquatic animals and the decay of organic wastes both remove dissolved oxygen from the water.   *Fish   suffocate*

Aerobic and Anaerobic Decay

If the decay of organic matter occurs in the presence of oxygen, aerobic bacteria are responsible and **aerobic decay** is said to occur. Aerobic bacteria use organic matter as food, breaking down the organic matter into simple end products that can be utilized by other organisms. Decay of this type removes oxygen from the water, and the more organic matter present, the greater the oxygen consumption. Decay also occurs in the absence of oxygen; biological degradation of this type, caused by anaerobic bacteria, is called **anaerobic decay**.

Aerobic decay, a relatively fast process, produces carbon dioxide, water, and new bacterial cells from organic waste. Anaerobic decay is much slower and more obnoxious because the end results typically include methane, hydrogen

Figure 6.8. The EPA estimates that the major cause of fish kills is a low dissolved oxygen content. These dead menhaden are in Chesapeake Bay. [EPA-DOCUMERICA-James H. Pickerell]

sulfide, and ammonia. In Chapter 5 we discussed the anaerobic decay of trash to give methane, but as indicated in Table 6.2, the process also gives odors and noxious gases such as phosphine (PH_3).

Both types of decay can occur in any organic system, but aerobic decay usually begins first. As the aerobic bacteria grow on organic waste, especially sewage, the oxygen is removed faster than it is replenished, creating an oxygen "sag curve" as shown in Figure 6.9. After the decay is completed and bacterial

TABLE 6.2. Comparison of aerobic and anaerobic decay

Compounds Containing	Will Decay Aerobically to Give	Will Decay Anaerobically to Give
Carbon (C)	CO_2	CH_4
Phosphorus (P)	H_3PO_4	PH_3
Nitrogen (N)	NH_3, NO_3^-	NH_3, amines
Sulfur (S)	sulfate	H_2S

growth ceases, the dissolved oxygen content slowly increases due to natural replenishment processes. The horizontal axis of Figure 6.9 is labeled as distance downstream from the point of waste discharge, but it might also be labeled as time from the dumping if the waste is discharged into a compounded water system. As the oxygen level decreases, the aquatic life changes. Those organisms requiring large amounts of dissolved oxygen will leave to be replaced by organisms that can live in the lower levels.

Temperature Effects

Oxygen solubility, the maximum amount of oxygen that can dissolve in a liter of water, decreases drastically as temperature increases. The solubility at 35°C (95°F) is about half that of 5°C (41°F). This is coupled with increased oxygen demands on the part of living organisms at higher temperatures because of increased metabolic rates. Thus, oxygen depletion at higher temperatures may have more serious consequences than depletion at lower temperatures.

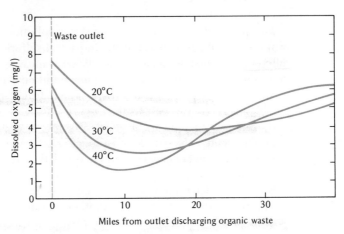

Figure 6.9. The oxygen sag curve caused by bacterial action on organic wastes. [From U.S. Dept. of Interior, Industrial Waste Guide on Thermal Pollution, September 1968, Federal Water Pollution Control Administration.]

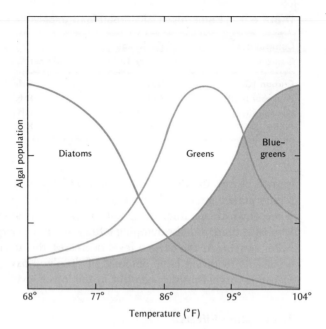

Figure 6.10. *Variation of algal species with water temperature. [After J. Cairns, "Effects of Increased Temperature on Aquatic Organisms," Industrial Waste, 1:150 (1956).]*

Another important change occurs in aquatic ecosystems as the temperature increases: the type of algae at the bottom of the food chain are different. Figure 6.10 shows that higher temperatures favor blue-green algae, in many respects the least desirable type. Blue-green algae are responsible for taste and odor problems in water, and they are even toxic to some aquatic organisms. Changes such as this at the base of the food chain can have far-reaching consequences at the top.

Measuring Water Pollution

The quantitative determination of water pollution mirrors the wide variety of possible pollutants. Because the impact of sulfuric acid, for example, is so different from that of sewage, quite different measurements are used to determine the potential harm of each. One traditional measurement of organic waste content, particularly of sewage, derives from the oxygen consumption of aerobic bacteria. The **biochemical oxygen demand (BOD)** measures the oxygen depletion that occurs in a sealed sample in a specified period, usually 5 days. The greater the amount of sewage or other organic matter present, the greater the amount of oxygen consumed and, thus, the larger the BOD value of the sample.

Bacterial action is not particularly effective for many pollutants, in particular for many industrial wastes, so another pollution test, the **chemical oxygen demand (COD)** is often performed. Measurements of COD use a chemical

COD is bigger than BOD

TABLE 6.3. Some typical BOD values in milligrams of O$_2$ consumed per liter

Drinking water	3
Municipal sewage before treatment	200–400
Milk processing and canning wastes	5000–6000
Wood pulping wastes	15,000

agent, an oxidizer, to convert organic wastes and many other pollutants to simpler end products. The amount of oxidizer required is then numerically converted into the amount of oxygen that would have been required to accomplish the task. Because the chemical agent degrades sewage and organic wastes as well as many other pollutants, COD values are always *larger* than the BOD values for a given sample.

Both BOD and COD are collective parameters in that they give no specific information on the pollutants present, reflecting merely the amount of pollutants subject to bacterial or oxidizer action. A BOD measurement, for example, gives no indication of what the bacteria are feeding upon; it may be sewage, dead animals, or paper mill wastes. Some typical BOD values are given in Table 6.3.

Dissolved oxygen is also a parameter used to specify water purity as are measured concentrations of many ions such as chloride, sulfate, phosphate, and nitrate. Color and turbidity (or clarity) are often specified. Both are measured using standardized testing devices. A measurement of total suspended solids, the tiny particles of all types including sand, mud, and sewage carried by the water, is an important parameter for drinking water standards and for discussing sewage treatment systems.

The presence of fecal matter in water is indicated by the presence of coliform bacteria. These are harmless bacteria that are present in both animal and human feces. There may be hundreds of millions of coliform bacteria per liter of sewer water. Because the harmful pathogenic bacteria are much less abundant and are considerably harder to test for, the easier test for the coliforms is usually done. The absence of coliforms is usually taken to mean the absence of pathogenic bacteria as well.

A sample water analysis of the type usually obtained when drinking water is tested by a state agency is listed in Table 6.4. The parameters we have just discussed are included as well as the results of pH and alkalinity tests which describe the acidity and basicity respectively of the water. **Hardness** is a measurement that reflects the concentration of calcium and magnesium ions. "Hard" water has high concentrations of these metal ions, and "soft" water has very low concentrations.

Any given "water analysis" can never be complete. Analyzing for every

TABLE 6.4.   Sample water analysis

Temperature	17.8°C
Dissolved oxygen	8.2 mg/liter
pH (hydrogen ion activity)	8.0
Biochemical oxygen demand	5.2 mg/liter
Chemical oxygen demand	19 mg/liter
Chloride	20 mg/liter
Alkalinity	71 mg/liter
Hardness	112 mg/liter
Color	7 (units)
Turbidity	35 (units)
Sulfate	70 mg/liter
Phosphate	0.1 mg/liter
Total dissolved solids	200 mg/liter
Coliform bacteria	100 per 100 ml

Source: "Water Pollution Surveillance System, Annual Compilation of Data, October 1, 1962–September 30, 1963." Public Health Service Publication No. 663, Vol. 2, Government Printing Office, Washington, D.C., 1963.

known substance is, in a practical sense, impossible. Furthermore, as emphasized in Chapter 1, the lower limit of present analytical techniques is usually a few parts per billion, so most pollutants cannot be detected if their concentration is less than this. Many substances not listed in Table 6.4 could be present in the sample: mercury, cadmium, radioisotopes, and pesticides are but a few examples. Water can be analyzed for these and many other substances, but this usually is not done unless their presence is suspected (Figure 6.11).

Sources of Water Pollution

The sources of water pollution before treatment are given in Table 6.5. It is very difficult to obtain data on the extent and adequacy of the treatment of many of the sources, especially industrial discharges. Industry and manufacturing account for about three fourths of the waste volume and of the total BOD. Municipal sources account for the remainder.

Sewage Treatment

Efforts to handle sewage by means of sanitary sewer systems are several thousand years old. The first known sewer system was constructed with clay pipe in ancient Babylonia over 5000 years ago. The Cloaca Maxima sewer line, constructed 2400 years ago in Rome, is still in service. The ancient systems were often quite advanced, based on principles still followed today. Much of this emphasis was lost during the Dark Ages, and several centuries were re-

Figure 6.11. *A variety of pollutants can render a water supply unfit for drinking as has happened to this spring that is the main water supply for Steele City, Nebraska. [U.S. Environmental Protection Agency, Office of Public Affairs, Washington, D.C.]*

quired following the Renaissance before sanitary treatment became widespread. The city of London had no sanitary facilities until after 1700, and these were very crude until late in the nineteenth century.

The development of sewage treatment plants was begun to prevent the spread of disease. Bacteria transmitted from individuals to waterways by means of fecal matter in sewage are the source of the great epidemic diseases,

TABLE 6.5. Estimated volumes of industrial and domestic wastes before treatment

Industry	Wastewater Volume (billion gallons)	BOD (million lb)	Suspended Solids (million lb)
Food and kindred products	960	4,300	6,600
Textile mill products	140	980	(no estimate)
Paper and allied products	1,900	5,900	3,000
Chemical and allied products	3,700	9,700	1,900
Petroleum and coal	1,300	500	460
Rubber and plastics	160	40	50
Primary metals	4,300	480	4,700
Machinery	150	60	50
Electrical machinery	91	70	20
Transportation equipment	240	120	(no estimate)
All other manufacturing	450	390	930
All manufacturing	13,100[a]	22,000	18,000
For comparison: Sewered population of U.S.	5,300[b]	7,300[c]	8,800[d]

[a] Columns may not add, due to rounding.
[b] 120,000,000 persons times 120 gal times 365 days.
[c] 120,000,000 persons times 1/6 lb times 365 days.
[d] 120,000,000 persons times 0.2 lb times 365 days.

Note: The values are for the year 1964 and are the most recent available. The next comprehensive National Waste Assessment is scheduled for 1977.

Source: 1st Annual Report of Council on Environmental Quality, 1970, p. 32.

cholera and typhoid, and of many lesser diseases. Fortunately bacteria usually survive only a few days in the waterway. Viruses in sewage may also transmit diseases, a principal example being infectious hepatitis.

As originally designed and for the most part as operated today, sewage plants reduce the BOD and total suspended solids content of raw sewage. Bacteria are killed by the addition of a disinfectant to the effluent before discharge.

Primary and Secondary Plants. Municipal sewage is more than 99% water, and the average flow is 130 gal/day per capita. In communities with sanitary sewers, the sewage is transported by gravity flow to the treatment plant, if one exists, or to a nearby body of water, usually a river or the ocean. Sewage treatment plants are classified as primary, secondary, and tertiary, the degree of treatment increasing with the numerical order indicated by the classification.

A **primary sewage plant** (Figure 6.12) is little more than a coarse screen and a settling tank. The screen removes larger objects such as branches and stones, while the settling tank serves to remove the larger solids, those that settle quickly. The solids are periodically removed as sludge. Chlorine may be added

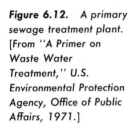
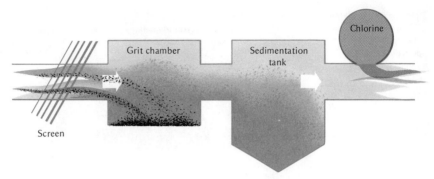

Figure 6.12. *A primary sewage treatment plant. [From "A Primer on Waste Water Treatment," U.S. Environmental Protection Agency, Office of Public Affairs, 1971.]*

to the effluent to kill some of the bacteria. There is no provision in primary plants for extensive bacterial action on the sewage.

Secondary plants have the equivalent of a primary plant (except for chlorination) at the input followed by some provision, either an **activated sludge** or a **trickling filter** section, for extensive bacterial action on the wastes. A trickling filter plant sprays the sewage over gravel which affords abundant air and extensive surface area for aerobic bacterial action (Figure 6.13).

In both types of plants, the settled solids are pumped back to the bacterial action section to improve biodegradation, then collected and removed as a solid sludge. The activated sludge system (Figure 6.14) provides for a retention time of the wastes of 5 to 24 hr during which air is bubbled through the wastes to provide for aerobic decay. Following treatment, the effluent is treated with chlorine and discharged to a waterway. The point of discharge is usually downstream from the city and is always downstream from the city's source of drinking water.

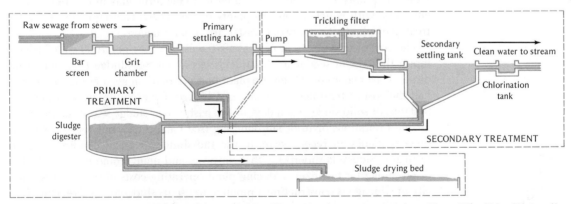

Figure 6.13. *A trickling filter secondary treatment plant. [From "The Living Waters," U.S. Public Health Service Publication No. 382.]*

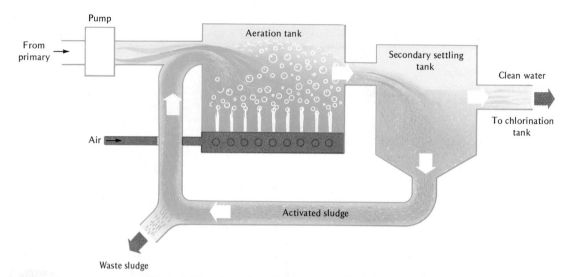

Figure 6.14. *An activated sludge secondary treatment plant uses an aeration section as the principal region of bacterial action. This section can be used in place of the trickling filter section. [From "A Primer on Waste Treatment," U.S. Environmental Protection Agency, Office of Public Affairs, 1971.]*

A comparison of the effectiveness with which primary and secondary plants reduce the pollution levels in sewage is shown in Figure 6.15. Primary plants reduce the BOD only about one third and the suspended solids by about two thirds. Secondary plants provide a 90% reduction in BOD and solids. Both types of plants are rather ineffective at removing other pollutants. For example, they remove less than half of the phosphates (present because of their use in detergents as discussed later) and only a small fraction of the pollutants from industrial plants. Industrial wastes, if lethal to bacteria, may in fact reduce the efficiency with which secondary plants reduce BOD. Consequently separate treatment plants for industrial wastes are usually necessary.

The cost of primary treatment in 1970 was about 4¢ per 1000 gal of sewer input; the cost of secondary treatment was about 8¢. Sludge treatment and disposal is a major cost of sewage plant operation, often accounting for nearly half the cost of treatment. The city of Chicago, for example, must dispose of 900 tons of solid sludge each day. In the past, sewer sludge was disposed of in landfill operations or in ocean dumping (see Chapter 7), but there has been some tendency in recent years to use the sludge as a soil conditioner and fertilizer. Sludge is superb for that purpose, and many communities sell the dried sludge for fertilizer, reducing plant operating costs in the process. The city of Chicago is constructing a pipeline to use its sludge in land reclamation projects in strip-mined areas in northern Illinois.

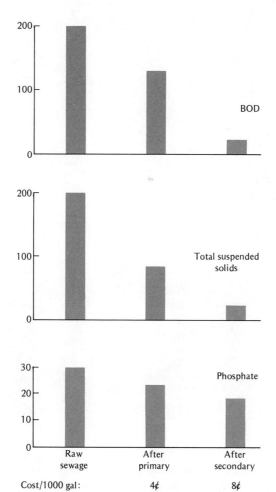

Figure 6.15.
*Effectiveness of primary
and secondary sewage
treatment. All units are
in milligrams per liter
(mg/l). The costs are
those for 1970.*

Most communities in the United States have secondary treatment but, as indicated in Figure 6.16, of the people served by sewers about 40 million rely on primary treatment and about 8 million have no treatment plant whatsoever. In 1970 over 50 million people in the United States had no sewer system; many of these use septic tanks as discussed later in this chapter.

Tertiary Plants. The design of primary and secondary plants is so conventional that the names alone are usually sufficient to indicate the level of sewage treatment. Tertiary treatment plants are not so conventional, however, because tertiary merely indicates some treatment in addition to secondary. The additional treatment may vary from simply filtering the effluent through sand

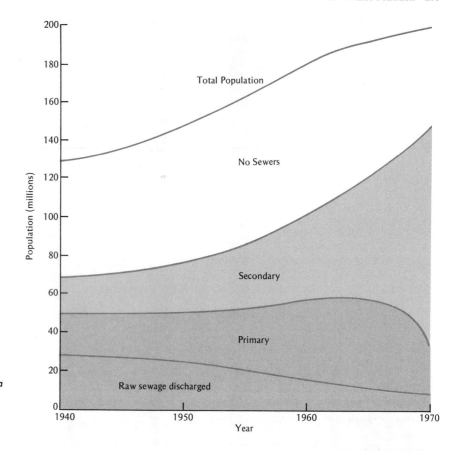

Figure 6.16. *Sewage disposal in the United States. The shaded area indicates the total population served by sewers.*

to elaborate processing stages for removing phosphates, nitrates, and industrial wastes. Only about ten sophisticated tertiary plants exist in the United States, although several hundred exist that are capable of improved phosphate removal relative to secondary treatment.

The South Lake Tahoe treatment plant is one of the most advanced in the

TABLE 6.6. Characteristics of the South Lake Tahoe sewage treatment plant (1970)

| | | Effluent | | |
Pollutant	Raw Waste Water	After Average Primary	After Average Secondary	After Lake Tahoe Tertiary
BOD (mg/l)	200–400	130–270	20–40	1 or less
COD (mg/l)	400–600	270–400	80–120	3–25
Phosphate (mg/l)	25–30	23–27	16–20	1
Cost per 1000 gal input (1970)		4¢	8¢	20¢

world. It uses advanced secondary treatment followed by additional stages to remove other organic wastes, phosphates, and nitrates. Table 6.6 shows the capability of this facility compared to that of average primary and secondary treatment. The cost is about 2½ times as much as that of an average secondary plant, but the pollutant levels in the effluent are about 1/20 of those from secondary plants.

Even tertiary treatment of this type does not produce water that meets drinking standards. To accomplish this would cost about twice that of the South Lake Tahoe treatment (Table 6.6). Even if the cost were not a factor, the use of treated effluent as a source of drinking water is usually discouraged unless absolutely necessary, principally because of the health hazard arising from the difficulty of killing viruses.

Septic Tanks. In areas without sewers, septic tanks are used to treat the sewage of individual homes. A tank with a capacity of about 1000 gal is buried

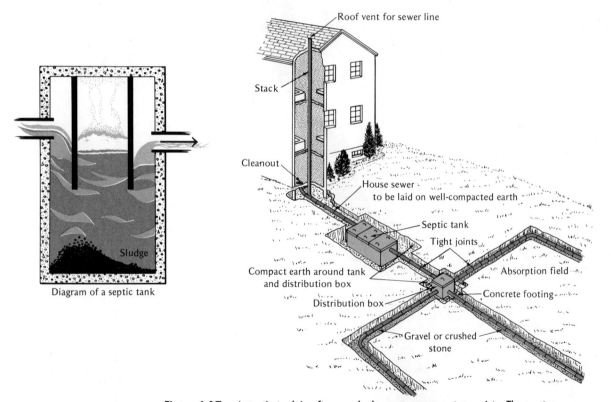

Figure 6.17. A septic tank is often used where no sewer system exists. The wastes enter from the left-hand side and exit at the right toward the tile drainage field. [From "A Primer on Waste Treatment," U.S. Environmental Protection Agency, 1971; "Manual of Septic Tank Practice," Public Health Service Publication No. 526, 1958.]

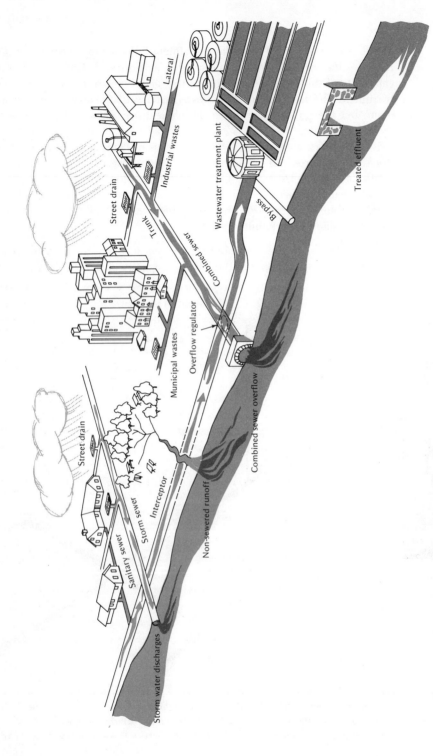

Figure 6.18. A typical storm and sanitary sewer system that might exist in a city. Some storm sewers drain directly into the waterway, and an overflow regulator allows both sanitary and storm sewers to drain to the waterway during high water periods. [From "A Primer on Waste Water Treatment," U.S. Environmental Protection Agency, Office of Public Affairs, 1971.]

in the ground to receive the wastes, as shown in Figure 6.17. Anaerobic decay occurs in the tank, and the sludge sinks to the bottom. The effluent water from the top of the tank passes through a drainage system to the soil. Septic tanks rely on bacterial action in the soil to degrade the sewage which passes out the drainage system. A major environmental problem with septic tanks is that the effluent may percolate down to groundwater supplies before the wastes are completely degraded. Such problems are particularly worrisome when the drainage is into an area with thin soil underlain with limestone rock, which is rather porous to water.

Problems with Sanitary and Storm Sewers. As noted earlier, sanitary sewer systems use gravity flow to transmit the sewage to sewage treatment plants. During periods of heavy rainfall, water pressure may build up in the sanitary sewer because of water backing up from the plant and because of water seeping into the sewer through leaks. To avoid the overflow of this water into houses, a provision is made to divert the sanitary flow directly to a waterway when water pressure builds up. Sometimes this provision is a bypass tube near the treatment plant, but it might also be an emergency overflow connection into another sewer system, the storm sewer (Figure 6.18).

The **storm sewer** system is a separate facility that drains runoff from lawns and streets directly to a nearby waterway. An emergency overflow connection means that during heavy rainfall, part of the domestic sewage is diverted without treatment to waterways. Sanitary engineers emphasize the value of the bypass and overflow systems by pointing out that the sewage would reach the storm sewers even in the absence of these, but the path would be comparable to toilet overflow in the home followed by drainage across the floor, out the door, and into the street. The overflow serves to eliminate homes from the path! On a yearly basis, about 3% of domestic sewage is estimated to enter storm sewers through crossover connections.

Rivers and creeks near cities may therefore receive sewage pollution from several sources. (Figure 6.18). Even good sewage treatment plants leave about 10% of the BOD in the effluent. The initial rain runoff in storm sewers contains fecal matter from lawns and streets, and later stages of rainfall may result in sewage diversion through the overflow or through the bypass. In addition, many sanitary sewer systems, especially older ones, have underground leaks that provide a source of water pollution. Finally, septic tanks in the area may be a source as noted previously.

Each of these sources requires a different solution, and sorting out which may be the largest source of pollution in any given area is no easy task. The overflow problem could be eliminated by providing treatment of the storm sewer system, but this would require an enormous storage capacity. In some cities, the storm sewer flow is stored in holding ponds, then fed slowly during dry periods into the sanitary sewer treatment plant. The cost of correcting sanitary and storm sewer effluent problems was estimated in 1970 to require about $50 billion. In the present political climate, this is an enormous amount to be diverted to environmental problems: it was in 1970 about two thirds of the annual budget of the Department of Defense, for example.

Advanced Treatment Methods. The secondary treatment plants we have just discussed use a very old concept. Many improvements could be made, and many are already available. The use of pure oxygen instead of air in activated sludge processes affords economic and technological advantages. The overall efficiency of the action of pure oxygen is such that it costs less to separate oxygen from the air and use it than to use the larger volume of air directly. Other advantages include the greater treating capacity for a plant of a given size, and the creation of smaller amounts of sludge. The use of oxygen in sewage treatment plants was pioneered in 1970 by Union Carbide, and four other companies now market such systems either as conversions of existing plants or as completely new systems.

A process called coagulation-sedimentation may be used to increase the removal of solids after secondary treatment. In this process, alum, lime, or iron salts added to the effluent cause the smaller particles to coagulate or "floc" into larger masses which either settle to the bottom of the tank or float on the surface. Flocculation has been used in various industrial operations for many years, but has only recently been applied to sewage treatment.

Organic materials may be removed by absorption by passing the effluent across activated charcoal (carbon). Activated carbon, often used to remove odors from the air, is a very porous substance with a large surface area that facilitates the absorption of organic molecules.

Rather complicated processes involving electricity and the flow of ions through membranes have been developed in the last fifteen years. Many of these "electrodialysis" techniques are applicable to wastes; they can remove the salt ions from water, for example.

Other improvements can be derived by changing our view of sewage effluents from one of disposal to one of utilization (Figure 6.19). One recent concept is that of "zero discharge": putting effluents back on land rather than discharging them into waterways. Many resort and recreation communities designed in the 1970s hold the sewage effluents in large ponds, then use the water for watering golf courses and lawns. The nutrients enter the soil, the water recharges underground sources, and the rivers do not suffer oxygen depletion. Of course, the effluent must be sufficiently well treated to reduce the spread of disease. Such techniques are not limited to municipal discharges. The Campbell Soup Company's food processing plant in Paris, Texas, for example, produces over 3 million gal/day of waste water containing organic pollutants. This is spread over 500 acres of cropland, and the hay and other crops harvested from it are sold.

Think of it, gentlemen—not only do we cease polluting our lakes and rivers, but we produce eighty million gallons a year of the best darn soft drink you've ever tasted.

Agricultural Wastes

In a natural cycle, animal wastes provide nutrients and humus for the soil. Modern farming operations tend to concentrate livestock in massive feedlot operations, and the natural spreading of the wastes does not occur. Water pollution problems are some of the many that may arise from this change in

nutrients

agriculture. Treatment of the wastes from large operations is now required in some regions of the United States.

Agricultural operations also result in a flow of nutrients to waters from the applications of fertilizer and from feedlot drainage. Nitrates present special problems because high concentrations in drinking water (usually the result of contamination by feedlot runoff) can cause methemoglobinemia in babies ("blue babies"). This disease occurs because in an infant's stomach the nitrate ion (NO_3^-) can be converted by bacteria to the nitrite ion (NO_2^-); the nitrite can replace the oxygen in hemoglobin, thus causing oxygen shortages in the cells. This same process can occur in cattle.

Mining Drainage

Another important water pollution problem is the acid drainage from underground mines. On exposed coal surfaces, minerals containing sulfur, especially pyrite (iron sulfide), come in contact with air and water and can be converted into sulfuric acid. This occurs in both abandoned and operating mines; acid mine drainage from coal mines east of the Mississippi is very common (Figure 6.20).

Eutrophication. The addition of excessive amounts of nutrients, such as phosphates and nitrates, to water creates a major water pollution problem. Lakes that have been overenriched are called eutrophic, meaning "well-nourished," and the process of overenrichment is called **eutrophication.**

The most visible result of eutrophication is an explosion of algal bloom which covers the surface of the water. Sunlight penetration into the water is reduced, and algae in the lower depths begin to decay. As the decay progresses, the dissolved oxygen content drops, so that anaerobic decay, with its malodorous and often toxic end products, occurs. Blue-green algae are often prevalent in the algal bloom and, as noted earlier, they are sometimes toxic to aquatic organisms. As a result of the toxic decay products and the reduced oxygen content, fish and other aquatic animals either die or leave the area. The turbidity of the water increases, and decaying vegetation builds up on the bottom of the lake. Over a period of years the lake gets shallower, becoming first an odorous marsh and ultimately a meadow.

Eutrophication occurs naturally on a geologic time scale, the nutrients being provided by sediment and runoff. A new lake, formed perhaps during glacial retreat, may suffer a shortage of nutrients and not be as supportive of aquatic life as it might be. Such a lake is termed **oligotrophic** by geologists. Over thousands of years, the nutrient level increases and a well-balanced aquatic ecosystem develops. The lake reaches its prime, supporting an abundance of life; such a healthy lake is termed **mesotrophic.** Finally, after many thousands of years the lake begins to age from eutrophication due to the excess of

Figure 6.20. Abandoned mines such as this one in Colorado are sources of water pollution, especially acidity. [EPA-DOCUMERICA-Bill Gillette.]

nutrients. In this stage, geologists call the lake **eutrophic.** This slow aging, which always occurs even in the absence of man, is called **natural eutrophication.**

Man's activities tend to accelerate the aging process by hundreds or thousands of times by adding to the waters such nutrients as nitrates and phosphates from fertilizers and phosphates from detergents. Sewage effluents also add nutrients and, in addition, deplete the oxygen content. Thermal pollution accelerates the algal growth rate and decreases the amount of dissolved oxygen. Such effects of man are termed **cultural eutrophication.**

Our knowledge of eutrophication is meager, but we do know that it is a complex process that depends on the water temperature, the amount of sunlight, and the presence of perhaps 20 or more nutrients including nitrogen, phosphorus, carbon, potassium, silicon, calcium, iron, manganese, boron, and vitamins. It is usually assumed that most or all of these nutrients must be present but the rate of eutrophication is limited by that nutrient least available, relative to the plant's needs. That is, reducing nitrates may have little effect if nitrates are so overly abundant that the algal growth rate is limited by the small amount of silicon present. The limiting nutrient may vary from lake to lake or with the time of the year.

The Great Lakes

The Great Lakes exhibit the three geological lake classifications. Lake Superior, the deepest and the possessor of the largest volume, is oligotrophic; Lake Huron and Lake Ontario are mesotrophic. The northern section of Lake Michigan is mesotrophic, but the southern end, near Chicago and Gary, Indiana, appears to be approaching the eutrophic classification.

Lake Erie is well along the road to the eutrophic classification. It has the least volume and the least depth (average 60 ft), so it affords the poorest dilution of wastes. Lake Erie also suffers the greatest impact by man. Located on its shores are such industrial centers as Detroit, Cleveland, Toledo, Erie, and Buffalo. Eleven million people live on its edge; it receives 1.5 million gal/day of sewage from treatment plants, and 39 overflow systems add an additional 50 million gal during rainstorms. A total of 400 industrial plants dump 10 million gal/day of wastes into Lake Erie. Many of the rivers that flow into the lake are officially declared to be fire hazards because of the high concentration of petroleum products and trash on the surfaces. The Cuyahoga River in Cleveland caught fire in 1969, and the Buffalo River has had at least three fires since then. The central and western portions of the lake are the most strongly affected. The fishing industry has changed considerably in Lake Erie in the last 50 years; the tonnage has remained relatively constant, but the Blue Pike and the Cisco have been replaced by perch, carp, alewife, and shad. Many beaches are grossly polluted, especially near Cleveland, and are posted as unsafe for swimming.

TABLE 6.7. Sources of phosphates and nitrates in the waters of Wisconsin

	Urban	Rural
Phosphate	2/3	1/3
Nitrate	1/3	2/3

Controlling Cultural Eutrophication

Efforts to reduce man's contribution to eutrophication have focused upon reducing his discharge of nutrients to the waters. There is little doubt that phosphorus and nitrogen contribute to eutrophication, although they may not always be limiting, and the efforts to control cultural eutrophication usually begin with these. Studies in Wisconsin indicate that in that state, two thirds of the phosphates come from urban sources, and two thirds of the nitrates come from rural sources, principally from fertilizers (Table 6.7). Controlling fertilizer runoff is very difficult, but considerable attention has been focused on the phosphates from urban sources, especially those present in detergents.

Detergents

A detergent is by definition anything that cleanses, but it is useful to consider soaps and synthetic detergents separately.

Soaps are derived from animal fats and oils. They are the sodium or potassium salts of long-chain organic acids (Chapter 1). Sodium stearate, Figure 6.21, is one example of a soap. Both soaps and detergents derive their cleaning ability from the fact that one end, the nonpolar or hydrophobic end, is soluble in oil or grease, whereas the other end, the polar or hydrophilic end, is soluble in water. This dual nature allows such molecules to act as an emulsifying agent, a chemical that allows two substances which usually will not mix, such as oil and water, to be suspended in each other. Figure 6.22 illustrates how the emulsifying agent forms a "bridge" that allows small microscopic oil droplets to "dissolve" in water: the long chain hydrophobic end of the agent is in the oil, and the polar end is in the water. The emulsified oil droplet is called a micelle.

The main drawback to soap is that in hard water it reacts with the calcium and magnesium ions to form a floating gummy scum (the "ring around the bathtub"). This removes the soap so that the remaining dirt and grease are not emulsified. The hard water complications of soap can be avoided by adding a "water softener" such as washing soda (sodium carbonate). The softener causes the formation of a solid, granular precipitate (calcium carbonate and magnesium carbonate) which sinks to the bottom, thus removing the metal ions before they can react with the soap to form scum.

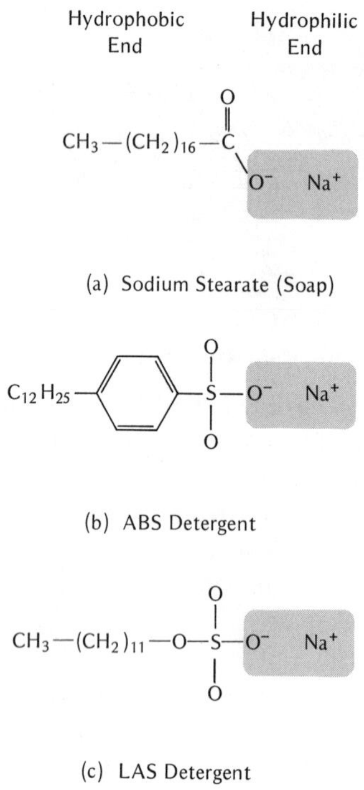

Hydrophobic End   Hydrophilic End

$$CH_3-(CH_2)_{16}-C \overset{O}{\underset{O^- \quad Na^+}{\parallel}}$$

(a) Sodium Stearate (Soap)

$$C_{12}H_{25}- \langle\text{ring}\rangle -S \overset{O}{\underset{O}{\parallel}} -O^- \quad Na^+$$

(b) ABS Detergent

$$CH_3-(CH_2)_{11}-O-S \overset{O}{\underset{O}{\parallel}} -O^- \quad Na^+$$

(c) LAS Detergent

Figure 6.21. *Soaps and detergents. The hydrophilic (shaded) end is the polar portion of the molecule, and the hydrophobic portion is the nonpolar or hydrocarbon chain portion of the molecule.*

Synthetic detergents, formed primarily from petroleum products, have a lesser tendency to form scum in hard water, especially when used in conjunction with a large percentage of phosphates. In water, the phosphates serve as **sequestering agents:** they react with the calcium and magnesium ions to form water soluble compounds called **complexes.** These complexes tie up the metal ions in solution and prevent their reaction with the detergent molecules, the end result being better cleaning power. Synthetic detergents were introduced in the United States in 1945 and now outsell soaps five to one.

The first problem encountered with detergents was not eutrophication but foaming of the waters into which the detergent flowed after use. This resulted from the fact that the detergent molecule ABS (Figure 6.21) was not biodegradable, principally because it contained an aromatic ring. Many fast-flowing streams in urban areas were covered with several feet of foam in the early 1960s (Figure 6.23). This problem was remedied by substituting LAS (Figure 6.21), which is biodegradable, for ABS. LAS degrades quickly in sewer systems and waterways, and consequently does not create a foaming problem. By a federal law, which became effective in 1965, all components of domestic detergents must now be biodegradable.

During the late 1960s eutrophication problems began to become apparent.

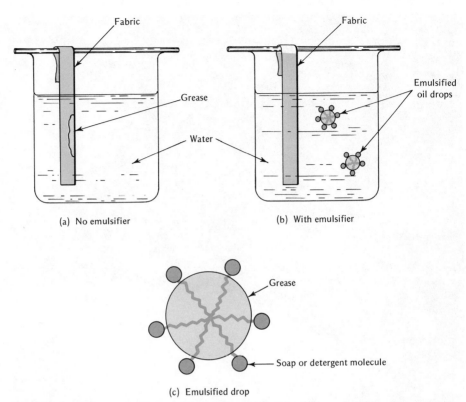

Figure 6.22. *Action of soap and detergent emulsifiers. In (a) the grease remains on the fabric because grease is not soluble in water. In (b) the emulsifier molecules made the grease soluble in water. The emulsified oil drops are greatly magnified; in reality they are too small to be visible. Part (c) shows an expanded view of the emulsified drop. The hydrophobic section of the soap or detergent molecule (*∿*) protrudes into the grease blob while the hydrophilic end (*-•*) of the molecule dissolves in the water.*

At that time many commercial detergents contained 75% phosphates, but the publicity about eutrophication resulted in large reductions of the phosphate levels by the early 1970s. Reducing the eutrophic effects of phosphates in detergents is a problem with many practical barriers. Phosphate detergents have been banned in some counties and cities (Miami and Chicago are notable examples) that had particularly bad problems with eutrophication. The state of Indiana has placed an upper limit of 3% phosphorus by weight in detergents sold there. In many communities where phosphate detergents were banned the eutrophication problems were reduced, but in Chicago the ban was overturned by a court ruling because in the eyes of the court no reduction in algal growth had occurred. On a nationwide basis, however, an outright ban on

Figure 6.23. Foaming on fast-flowing streams, a problem in the early 1960's, resulted from the fact that the ABS detergent molecules were not easily biodegradable. [Courtesy U.S. Department of Agriculture.]

phosphates in detergents may not be desirable for a number of reasons. The Soap and Detergent Association, which represents the industry, maintains that there are not sufficient animal fats available to permit a return to soaps in the amounts in which we now use detergents. The possibility of substituting some agent for phosphorus in detergents has often been proposed. For that, however, we need to find an agent that can substitute for the several billion pounds of phosphates used each year and will produce fewer environmental and public health problems.

Phosphate detergents have an appealing characteristic: they are nontoxic even in large dosages to infants. Some of the nonphosphate detergents presently sold and some of the proposed phosphate substitutes are not so appealing. NTA (nitriloacetic acid) has been proposed and occasionally used in the past as a substitute for phosphates in detergents. It is a good sequestering agent for calcium and magnesium, it is cheap, and it is biodegradable. However, NTA also forms complexes with mercury, lead, cadmium, and arsenic, leading to the possibility that these might be put into solution where they could enter the ecosystem. Furthermore, NTA complexes copper, so its use might lead to the deterioration of copper and brass bearings, piping, and fittings. Incomplete evidence suggesting that NTA may be carcinogenic and teratogenic led to its withdrawal for use in the early 1970s pending further tests.

In view of the difficulty of finding phosphate substitutes, many have suggested that the best approach would be to continue using phosphate detergents but develop tertiary sewage treatment plants designed to remove phosphates. As we have seen, such plants are relatively expensive and at present relatively rare.

One viable alternative is to put smaller amounts of phosphates in detergents themselves and to use smaller applications than those recommended by manufacturers. Many consumer groups maintain that using half the recommended amount of detergent produces cleaning results identical to those of the recommended amount. They also contend that since less phosphate is left as a residue in the clothing when smaller appictions are used, fewer health problems, such as rashes on infants allergic to phosphates, will be encountered.

Metals

In water, most metals exist as positively charged ions, having lost some of their electrons to other elements or to water (see Chapter 1). Some metals, such as sodium, potassium, and iron, are essential to life, but many others present health hazards. As we shall see, much of the impact of metals occurs because they bind to important biological molecules such as proteins. As noted earlier, metals can undergo biological magnification; consequently the release of metals as a water pollutant can result in their distribution throughout the food chain with especially large concentrations near the top of the chain.

Mercury

Mercury, familiar to nearly everyone because of its presence in thermometers, is the only metal that exists as a liquid at room temperature. Mercury forms many compounds, and their toxicities and environmental impacts vary widely. Some mercury compounds are deadly poisons; others somewhat less toxic have been used for medical purposes for nearly two millennia.

Pure liquid mercury (Hg°) is not toxic unless inhaled; however, its oxides are a health hazard if ingested, and since they are almost invariably present as a gray film on the metal's surface, the metal should for all practical purposes be considered toxic. Two of the many inorganic mercury compounds are very common. Calomel ($HgCl$) was in the recent past a commonly prescribed diuretic, but corrosive sublimate (Hg_2Cl_2) is a deadly poison. Mercury also combines with organic molecules forming such compounds as methyl mercury, dimethyl mercury, and phenyl mercury (Figure 6.24). These exist in the environment and can be ingested by man.

Although many inorganic forms of mercury are toxic, the organic forms, especially methyl mercury, have the worst impact on living organisms. There is often a latent period of weeks or months between exposure to methyl mercury and the development of positive symptoms of poisoning. Mercury compounds have a marked affinity for the sulfur atoms found in the sulfur-hydrogen (or "sulfhydryl") linkages in proteins present in cell membranes. Once attached, the mercury compounds interfere with normal cell operations such as the flow of nutrients and ions in and out of the cell. The central nervous system is affected when the mercury compound binds to the membrane sheath of the nerve. Muscle weakness, loss of coordination in gait, loss of vision, loss of hearing, and brain damage result from severe mercury exposure. The methyl mercury forms are especially dangerous because they are the least subject to degradation in the human system and because they form the strongest bond

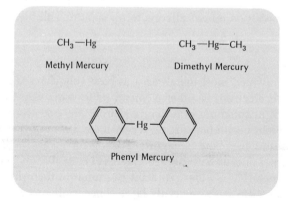

Figure 6.24. *Some important organic mercury compounds.*

CH$_3$—Hg

Methyl Mercury

CH$_3$—Hg—CH$_3$

Dimethyl Mercury

—Hg—

Phenyl Mercury

with the sulfur atoms. The half-life (Chapter 4) of the methyl mercury compounds in the human body is about 70 days. Mercury causes chromosomal damage and abnormal cell division in concentrations of 0.05 ppm, lower than that of any other substance. In addition, it can cross the placental wall and affect the fetus.

The modern history of mercury poisoning began in the fishing village of Minamata, Japan, in the early 1950s when nearly the entire village began to experience mercury poisoning. Residents of the village subsisted largely on fish caught in a bay into which were discharged mercury wastes from a plastics manufacturing plant. Biological magnification produced methyl mercury levels of 5 to 20 ppm in the fish. (In contrast, the maximum concentration permitted by the U.S. Food and Drug Administration in fish sold for food is 0.5 ppm.) By 1960, 116 people in Minamata had sustained such extensive brain damage that they would always be mentally deficient. Forty-three persons died from the poisoning.

In the United States in 1969 the Huckleby family in New Mexico was poisoned after eating pork from hogs that had been fed grain previously treated with a mercurial fungicide. Three children had irreversible nerve damage, and the Huckleby's unborn child was blinded. Seed that had been treated with mercury is dyed red as a warning to prevent its use as feed or as a starting product for bread, although the treated seed can be used for planting.

In 1972 the most catastrophic mercury poisoning occurrence ever recorded happened in Iraq because bread was made from wheat and barley that had been treated with mercury compounds. By August 1972 over 6500 cases of poisoning had been admitted to hospitals and nearly 500 deaths had resulted. The grain had been delivered to all provinces of Iraq in the fall of 1971 for use as seed grain. Because the dye washed off with water, some people assumed incorrectly that the mercury had washed off as well. The latent period between exposure and poisoning also contributed to the tragedy. Some farmers gave grain to their chickens for several days, noticed no effects, and assumed the grain was safe for human consumption. By the time the humans who had eaten the poisoned bread showed evidence of poisoning, toxic doses had already been consumed by thousands.

Mercury creates other problems in the environment. Concentrations of organic mercury compounds as low as 0.1 ppb are toxic to phytoplankton.

Environmental Sources. Mercury is obtained from the mineral called cinnabar (mercuric sulfide). It is not a particularly abundant element, but man's activities have increased its concentration in many areas of the environment 1000-fold. Mercury is used in paints, paper, and electrical equipment and is an essential component in many industrial operations, especially in the production of chlorine gas. About 3000 tons are used in industry each year in the United States, and prior to 1970 about three fourths of this wound up in the environment. It is present in trace amounts in coal and petroleum, and several thousand tons per year are put into the air during combustion. It has been estimated that since 1900 over 100,000 tons have been dispersed from this source, as much mercury as has been released by all agricultural and industrial operations during the same period.

Bacterial Methylation. For decades little concern was given to mercury discharges from industrial operations or to liquid mercury dumped down drains, because it was believed that the elemental mercury, which is chemically rather unreactive, would remain in the muds of the waterway. In the late 1960s, however, it was discovered that mercury is a surprisingly mobile element. Many forms of mercury, including the elemental liquid form, can be converted by bacterial actial into methyl and dimethyl mercury, precisely those forms most hazardous to living organisms. The methyl mercury compounds are water soluble and can be ingested by plankton. Thus mercury wastes, once considered harmless, are potentially very dangerous. Even though we have now reduced mercury discharges, the amount already present in the environment will be a threat far into the future.

Allowed Mercury Levels. Concern in North America about mercury occurred in the late 1960s when high mercury concentrations were discovered in fish. Swordfish and tuna often have the highest levels because of their location near the top of the food chain, and shipments of both have been confiscated in the past by the Food and Drug Administration because the levels exceeded the allowed 0.5 ppm maximum. Freshwater fish from Lake Erie, Lake Michigan, and other lakes with nearby industries sometimes show high levels.

The allowed levels of mercury in foods can be easily set once the **acceptable daily intake** (ADI) has been established. The ADI of a substance is the maximum weight of the substance which can be ingested each day over a period of years with no observable effects. Establishing the ADI is an inexact procedure at best, but from a summary of existing data, a Swedish panel recommended that the methyl mercury intake not exceed 0.03 mg/day averaged on a weekly basis, and this is now accepted as the mercury ADI. Eating ½ lb of fish containing 0.5 ppm of mercury results in the ingestion of 10 mg of mercury. Thus only two meals of this fish could be eaten per week to stay below the maximum average allowed consumption of 0.03 mg/day. The U.S. Food and Drug Administration estimated that the average American eats fish less than twice a week and hence concluded that an upper limit of 0.5 ppm would ensure safety for nearly all Americans. This seems particularly valid in view of the fact that most fish have mercury levels that are well below 0.5 ppm. No poisoning from mercury in fish has ever been reported in the United States. The average United States daily intake of mercury from all sources is less than

0.02 mg. In contrast, those hospitalized in the poisonings in Iraq had ingested 50 to 400 mg in a short period of time.

It is possible that high mercury levels in fish have existed for a century. Fishes caught in the 1800s and carefully preserved often have mercury concentrations approximately equal to those of fishes caught today.

Cadmium

The ingestion of cadmium can result in medical problems similar to many of those created by mercury. In addition, cadmium has an affinity for bones, gradually causing them to weaken and disintegrate. In regions of Japan where cadmium poisoning has resulted from drainage of a zinc and cadmium mine, the effects are called "itai-itai" ("ouch-ouch") disease because ribs and other bones are susceptible to multiple fractures after such slight traumas as coughing.

Cadmium occurs naturally with zinc, and the chemical similarities between the two elements cause cadmium to occur with zinc in galvanized coatings. Black plastic polyethylene pipe contains cadmium as an impurity, and it can be leached from the pipes by water standing overnight.

Lead

Although lead is not a common water pollutant, we discuss it here because of its chemical similarity to mercury and cadmium.

Lead poisoning is a source of brain damage, mental deficiency, and kidney problems, especially nephritis. Like most metals, lead attaches to proteins. In particular, it interferes with the enzymes that biosynthesize heme, an essential component of hemoglobin; the shortage of hemoglobin results in anemia and in oxygen deficiencies in cells.

Lead poisoning has been known for centuries but has been forgotten by some cultures, then rediscovered by others. The Greeks apparently knew about the possibility of lead poisoning from lead compounds used to glaze pottery. The Romans, however, were apparently unaware of this problem, and one theory of Rome's decline centers on medical, genetic, and mental problems induced by mass lead poisoning in the ruling upper classes. Supposedly the poisoning came because upper-class Romans consumed large amounts of wine from vessels glazed with lead compounds. The lower classes were spared the poisoning because they consumed very little wine. (Unfortunately they had little to say about running the affairs of the nation!) The Massachusetts Bay Colony forbade rum storage in lead-lined containers in order to prevent the occurrence of the "dry gripes," a disease with symptoms very similar to those of lead poisoning. In 1929 widespread lead poisoning occurred in Queensland,

Australia, because the drinking water was collected from roofs that had been painted with a lead-based paint.

In Chapter 3 we noted that lead in gasoline was a major source of lead in the environment. No known effects of lead poisoning in humans from that source are known, but the effects of mild lead poisoning are so subtle that they may escape diagnosis. As was mentioned earlier, there is evidence that animals in New York zoos suffer lead poisoning as a result of lead buildup from automotive sources.

Most of the cases of human lead poisoning result from the ingestion of lead from paints or from ceramics glazed with lead compounds. A less frequent source of lead poisoning is moonshine whiskey. The use of automobile radiators in illicit stills results in significant lead concentrations in the alcohol because a lead solder is used in the radiators.

The average lead dietary intake in the United States is about 0.3 mg/day, but one cola from a lead-glazed mug can contain 3 mg, and a thumbnail-sized chip of lead-based paint can contain 100 mg. Lead poisoning is especially common among young children in urban slums who eat flaking chips of old paint containing lead coloring agents. Over the past 40 years in both New York and Baltimore the reported cases of lead poisoning have displayed the seasonal variations shown in Figure 6.25. It is not known whether the variations are physiological, perhaps some effect of sunlight on lead metabolism, or whether they are social, perhaps arising from variations in the availability of medical examinations.

Lead poisoning in children may also result from the ingestion of lead contained in the paint on pencils. In 1971 the Department of Health, Education, and Welfare reported that the average pencil contained 0.2 mg of lead in the paint, a possible hazard to children who chew on pencils.

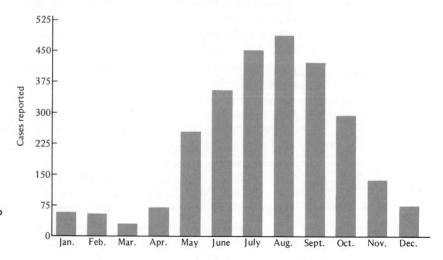

Figure 6.25. *Monthly variation in lead poisoning cases reported in Baltimore from 1931 to 1951. New York shows a similar variation.*

TABLE 6.8.   Oil production sources in the world in 1970

Source	Metric Tons	Percentage of Total Losses Due to Man
Manmade sources		
Used motor and industrial oils	3,300,000	67.2%
Tankers, normal operations	530,000	10.7%
Oil from bilges of other ships including pleasure craft	500,000	10.1%
Refineries, petrochemical plants	300,000	6.0%
Tanker and ship accidents	100,000	2.0%
Nonship accidents	100,000	2.0%
Offshore production, normal operations	100,000	2.0%
Natural sources		
Seepage	600,000	12.2%

Oil

Approximately two thirds of the world's oil production (over 400 million gallons per year) is transported by sea. In fact, oil accounts for about two thirds of all tonnage transported on the seas. About 0.1% of this is lost, some from normal transfer operations and some from accidents. An even larger source of oil pollution is the disposal of used motor and industrial lubricating oils, most of which ultimately wind up in water. Table 6.8 lists the sources of oil pollution in 1970. The man-made sources total nearly 5 billion metric tons (1.5 billion gal), and the largest individual source is disposal of motor and industrial oils.

Oil is a mixture of large and small hydrocarbons, and the impact of the various components is summarized in Table 6.9. The smaller alkanes and aromatics often evaporate soon after a spill, but both types cause considerable disruption in aquatic ecosystems, especially the aromatic hydrocarbons because of their high toxicity. Some of the larger alkane hydrocarbons in oil

TABLE 6.9.   The impact of oil hydrocarbons

Type	Molecular Size	Effect
Alkanes	small	narcosis and anaesthetic effects at low concentrations; cell damage and death at high concentrations
Alkanes	large	interfere with chemical communications systems
Aromatic	small	acute toxicity
Aromatic	large	long-term poisons and carcinogens

occur naturally in marine life. For example, they are similar to the compounds used for chemical communication by some fish for such activities as reproduction, feeding, and migration. A buildup of the long-chain alkanes in the ocean might lead to possible interference with these communications systems and hence have a drastic effect on the aquatic ecosystem. The larger aromatic molecules present long-term dangers due to the presence of carcinogens such as benzo(a)pyrene, a compound discussed in Chapter 3.

All oil spills, whether resulting from ship accidents, broken oil lines, or offshore well blowouts, share several common properties. There is an immediate kill of aquatic life and sea birds in the area of the spill due partly to suffocation and partly to the toxic properties of the hydrocarbons. As the spill spreads, the more volatile (smaller molecular size) hydrocarbons evaporate, but the remaining tarry residues continue to affect the environment. Crabs, oysters, lobsters, and fish that survive the spill usually ingest sufficient oil to have an oily taste. Consequently fishing in the area of the spill is usually suspended for months or years. Changes in the ecosystem occur: even several years after the spill, some aquatic life that was present before the spill may not have returned, and other species may be present that were not there before the accident.

Despite the frequency of oil spills, few have been carefully studied to determine the long-term effects. One of the most carefully studied spills was the wreck in 1969 of the barge *Florida* in West Falmouth Bay in Massachusetts. This relatively small spill (170,000 gal of fuel oil) occurred within a few miles of the Woods Hole Oceanographic Institute, an internationally renowned research laboratory. Researchers at Woods Hole knew a great deal about the ecology of the area and had the sophisticated research equipment with which to detail carefully the impact of the spill. They found, for example, that the kill of shallow water plants encouraged erosion, and that many plant and animal species had not repopulated the area 2 years later. Much of the oil sank to the bottom as small droplets and killed the plants there. This led to increased sediment drift in the bottom of the bay which provided a mechanism of movement for the oil particles throughout the area. Samples taken from the bottom of the bay 2 years after the spill showed that the oil had penetrated to a depth of 7.5 cm (about 5 in.) into the bottom muds.

Measurements of the oil concentrations in organisms, in water, and in sediments revealed that oil was more persistent than visual appearances might suggest. A surprising find was that the oil hydrocarbons ingested by shellfish, oysters, and other aquatic organisms were still present 2 years later in nearly the same concentrations and compositions, even though the oily taste had disappeared. Oysters transported to unpolluted water and analyzed after 6 months there showed very little loss of oil residue hydrocarbons from their tissues. It is possible that the hydrocarbons, which dissolve in the fats and lipids of the organisms, will remain in the organisms throughout their lifetimes and will be passed up the food chain. Sediment samples indicated that decomposition of the hydrocarbons by bacterial action began to become significant after 1 year but proceeded at a very slow rate.

Corrective Actions

The most obvious corrective action for oil spills is prevention. Better tanker and oil transfer equipment design, better navigational and safety devices, and better training of tanker pilots will reduce shipping losses. The installation of safety valves designed to shut off oil flow in offshore oil wells in case of accidents is required by federal law, but such valves, which cost over $1000 each, have in the past been omitted from many drilling rigs.

Used motor and industrial oils can be collected and re-refined. Most of the major oil companies have started such a system for their automobile service stations.

It is clearly of economic as well as of environmental concern to reduce oil spills.

A number of devices have been developed that facilitate cleanup after spills have occurred. The spill can be encircled by a special plastic fence which floats vertically in the water; pulling the fence closed can concentrate the oil sufficiently that it can be collected, or if necessary, burned. Special barges have been developed that pick up the oil on a carpet-covered conveyor belt arrangement and collect it. Another interesting development is the isolation of bacteria that metabolize the hydrocarbons rapidly but create no toxic products. Tests have shown that a type of bacteria called *Pseudomonas* is capable of breaking down 50% of oil into smaller compounds within 48 hours.

One possible action, which it is generally agreed should not be taken, is the dumping of detergents on spills. In 1967 the tanker *Torrey Canyon,* which carried crude oil, broke apart in the sea off the coast of England. The spill occurred during the holiday season in Europe, and as the oil slick drifted toward the vacation beaches of England and France intense publicity focused on the efforts to combat the spill. Napalm bombing failed to ignite the oil, so detergents were applied in an effort to disperse the slick. The detergents may have caused more problems than they solved, however. Much of the oil sank to the bottom, where it would last longer and perhaps have a greater long-range impact than if it had remained on the surface. The detergent killed limpets, which normally would have helped remove the oil by browsing on the rocks. An additional factor in the *Torrey Canyon* spill was that the detergents were dissolved in toxic aromatic solvents before use, so that more toxic chemicals were added to the area.

Oil cleanup on land appears to be done best with an old, well-tested method: straw placed on the sand absorbs the oil, then the straw is collected.

Efforts by volunteers are usually made to save birds affected by oil spills

Figure 6.26. Birds killed by an oil spill. [Courtesy U.S. Department of the Interior, Bureau of Sport Fisheries and Wildlife.]

(Figure 6.26). Feathers provide insulation for birds by creating many small air pockets. When the feathers are soaked in oil, they no longer provide insulation, and the birds become chilled. Oil is ingested during preening, and this results in nausea, illness, and a reduced appetite. Mortality is very high because of the combined effects of exposure, starvation, and poisoning. Rescue efforts require capturing the birds and removing the oil from the feathers by gentle washing. This removes the natural waterproofing oils also, so the bird must be confined and fed for approximately one molting period (weeks or months) until the natural oils are replaced. If the bird survives all this, at the end of the confinement its wing muscles may have atrophied; poor flying will occur for several weeks until the muscles regain their strength. Under optimum conditions only about 10% of the birds captured are saved. Certain species of birds that inhabit the North Atlantic shipping lanes, notably the auks and puffins, have experienced drastic population reductions in the last 30 years due in part to the effects of the oil present in the waters in that area. These birds are heavily feathered and hence suffer severe problems from oil.

Tanker Size

As the worldwide consumption of oil has increased, the size of the tankers has followed suit, giving rise to the possibility of even greater disruptions from future accidents. The *Torrey Canyon,* a large tanker for its time, was classified in the 117,000 ton class. Today's supertankers are in the 500,000 ton class, and many under design approach 1 million tons. Very few existing ports in the world will accommodate such large vessels. Offshore supertanker docking terminals are being constructed; the oil will be pumped to shore by means of pipelines several miles long.

Worldwide Monitoring

Very little data exist on oil and oil residue concentrations in the world's oceans. In an effort to provide such data, a worldwide oil monitoring network, sponsored jointly by the Intergovernmental Oceanographic Commission and the World Meteorological Organization, began operations in 1975. Oil slicks, tar balls, and dissolved hydrocarbons will be monitored at many stations around the globe, and the results will be tabulated.

An Example of Change: The Cleanup of the Willamette River

The Willamette River in central Oregon is the twelfth largest river in the United States and was once one of the most polluted (Figure 6.27). Fifty years ago men refused to work at riverside construction because of the water's intolerable stench. Now thousands regularly swim, fish, water ski, and boat on the Willamette on summer weekends, and for the first time Chinook salmon ascend the Willamette to spawn in the fall. Oregon's experience demonstrates how a major river can be restored if people are determined, the government committed, and the legal tools available. The river improvements were not, however, easily achieved.

Seventy percent of Oregon's population lives along the Willamette, and the river has always played a major role in the history of the state. Agriculture, lumbering, and food processing are the principal industries in the Willamette Valley.

The variation in river flow has been an important factor in the pollution of the Willamette. Summer rainfall is extremely low, leading to a pattern of very high flow in the winter and spring, followed by extremely low flow in July and

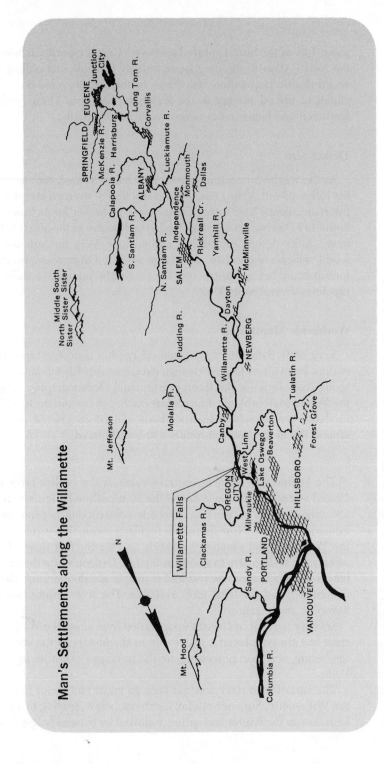

Man's Settlements along the Willamette

Figure 6.27. The Willamette River Basin. [From Pacific Northwest River Basins Commission.]

322

August, often one tenth of that occurring in early spring. Traditionally, therefore, the river had severe pollution problems during late summer, but few during the high flow periods.

In the 1920s five pulp and paper mills were in operation on the river, and their wastes plus the fact that the sewage of all cities in the valley was dumped untreated into the river created severe pollution problems (see Figure 6.28). The dissolved oxygen level measured in August 1929 was 8 mg/liter near the headwaters (Eugene) but was only 0.5 mg/liter where the Willamette dumped into the Columbia near Portland. A public survey taken that year showed that 50% of the people favored water pollution legislation.

In 1938 the voters approved a bill establishing a state Sanitary Authority that was authorized to establish and enforce water quality standards. The Sanitary Authority initially concentrated on requiring treatment of municipal wastes. Primary treatment was required for all cities. This was at first thought to be sufficient because a series of multipurpose storage reservoirs on the

Figure 6.28. The Willamette suffered from pulp mill effluents similar to these in Bellingham Bay, Puget Sound, Washington. [EPA-DOCUMERICA-Doug Wilson.]

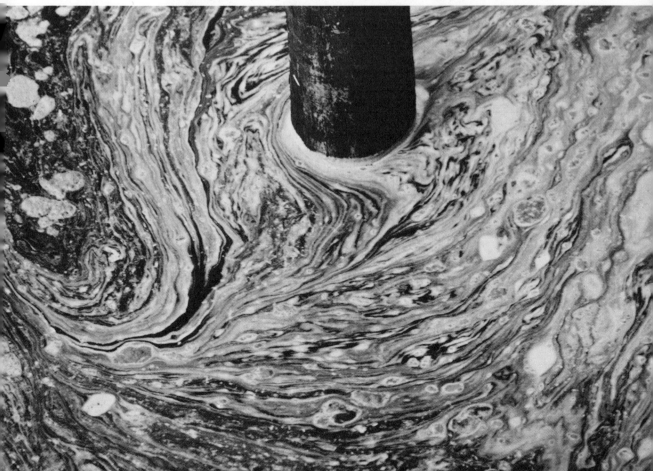

tributaries of the Willamette, being constructed by the U.S. Army Corps of Engineers, would provide for a doubling of the river flow during the summers and thus provide more dilution during that period. Efforts to institute industrial pollution controls proved more troublesome. The industries hinted that the technology to control the wastes did not exist and that stringent controls might force them to relocate in another state.

By the mid-1950s all cities had primary treatment, and in addition Portland's effluent discharge had been diverted into the Columbia, which had a much greater flow. By this time as well, the Sanitary Authority had succeeded in requiring the five mills to store effluent discharges in special holding ponds during the months of June through October, then slowly release the wastes during high water periods.

Even though these improvements had been instituted, the water quality in 1957 was still poor, in large part because a 73% increase in population and a doubling of the industrial waste load had occurred on the river over the preceding 20 years. New standards were imposed in 1958 which required secondary treatment of all municipal wastes.

Political leadership was important. From 1939 to the middle 1960s the legislature supported public demands for pollution control. In 1966 both candidates for governor listed the Willamette cleanup as the number one priority. The elected governor himself (Tom McCall) served for nearly a year as chairman of the Sanitary Authority, and over the next 8 years he made good on his campaign promises. McCall's determination resulted in new legislation in 1967 which required a waste discharge permit for all cities and industries using the Willamette and which established even more rigid controls. The mills were required to begin using secondary waste treatment. During this period, federal action on environmental problems offered financial assistance for waste treatment plants, and many communities received grants for treatment improvements. Federal laws passed near the end of the 1960s meant that other states also had water pollution standards; this lessened the possibility that the industries might leave Oregon for states with fewer restrictions. The waste discharge permits were issued under the conditions that the specified daily pollution loads could not be increased; if a municipality or an industry were to expand, treatment must be improved accordingly.

The Willamette is now a healthy river. The reduced pollution and the increased flows during the dry summers have permitted a new phenomenon: the fall spawning of the Chinook salmon in the river. Other changes were fostered by the "new" Willamette. Cities along the river had traditionally developed away from the foul-smelling, polluted river. Now the cities are regaining interest in the river; urban land along its banks is being developed for parks, museums, and recreation areas. Furthermore, a river-long series of recreational areas and state parks was planned and is partially completed. This "Greenway" plan, suggested by a dean at the University of Oregon during the

1966 gubernatorial campaign and endorsed by both candidates, provides a comprehensive land use policy along the river, ensuring that the Willamette will now be used for boating, camping, and hiking as well as for farming, industry, and transportation.

The changes in the Willamette are heartening. However, they required an active and enlightened public that would push for change, elect officials with similar goals, and pass the bonds required to improve the environment. There is no guarantee that the river will remain healthy, but the best insurance is continued public interest and vigilance.

Federal Water Pollution Requirements and Their Costs

The costs of water pollution controls increase rapidly as purity standards increase. A governmental study done on the Delaware Estuary indicated that it would cost one and a half times as much to assure a dissolved oxygen level of 5 mg/liter as it would to assure a 4 mg/liter level. A fourfold cost increase would be required to achieve 6 mg/liter instead of 4 mg/liter. Table 6.10 shows the costs of water pollution controls for this decade in the United States. The costs are those required to meet water quality standards adopted in the 1972 Amendments to the Federal Water Pollution Control Administration Act. The 1972 Amendments constitute one of the most comprehensive and expensive pieces of environmental legislation ever passed. President Nixon vetoed the bill, but Congress overrode his veto in October 1972, and the bill

TABLE 6.10.   Costs of water pollution control in the United States, 1971–1980 (in billions of 1971 dollars)

Sector	Capital Investment	Operating Cost	Cash Flow
Public			
Federal	1.2	2.7	3.9
State and local			
Treatment systems	18.9	23.6	42.5
Combined sewers	(17 to 56)	—	(17 to 56)
Private			
Manufacturing	11.9	14.2	26.1
Utilities	4.5	4.2	8.7
Feed lots	1.9	1.8	3.7
Construction sediment	0.9	0.1	1.0
Vessels	0.9	0.5	1.4
Total	40.2[a]	47.1	87.3[a]

[a] Does not include the estimates for combined sewers.

became law. It sets as a national *goal* the elimination of all pollutant discharges in United States waters by 1985, and sets as an *interim* goal the achievement of water quality safe for fish, wildlife, and recreation by 1983. The bill requires all municipal plants to provide secondary treatment by 1977. By July 1, 1977, all industries are required to use the "best practicable technology" to treat all discharges, and by July 1, 1983, the "best available technology economically achievable" is to be used.

About $42.5 billion is necessary for sewage treatment systems and as much as $50 billion might be necessary to correct the problems of combined storm and sanitary sewer systems. Costs to private industry will total about $41 billion. Total costs for the 10-year period will be in excess of $100 billion. The economic costs (and increased energy consumption) resulting from the implementation of these controls can be justified in terms of the health and social benefits that will result. (See Energy and Economic Costs of Pollution Controls in Chapter 5 for further discussion.)

Summary

Water is cycled in the environment by energy from the sun, and is said to be polluted when it cannot be used for its intended purpose. Water pollution is a more complex topic than air pollution because water pollution sources are more varied and because the pollutants often undergo reactions in the water.

About 315 bgd is an upper limit to the amount of water that might be consumed in the United States on a sustainable basis. Withdrawals exceed that now; consumption will reach about half this maximum sustainable level early in the next century.

The properties of an aquatic ecosystem depend to a great extent on the amount of dissolved oxygen present. The oxygen in water comes from the air or from plant respiration; it is removed by aquatic animals and by the aerobic decay of organic wastes. As temperature increases, the oxygen content of water decreases, but the oxygen demands of most organisms increase.

Pollutants can be ingested by the phytoplankton at the bottom of the food chain and subsequently undergo biological magnification in the higher levels of the chain.

Measurements of water purity usually rely on BOD and COD, which reflect, respectively, the bacterial and chemical actions necessary to degrade wastes. Coliform bacteria indicate the presence of fecal matter in water; the absence of these harmless bacteria is usually taken to mean the absence of pathogenic bacteria.

About three fourths of the polluted water in the United States, as measured before treatment, is created by industry, but the impact of the industrial pollutants may be quite different from those of the municipal wastes. The treatment of domestic sewage is usually accomplished in primary or secondary treatment plants. Secondary treatment removes about 90% of the BOD and suspended solids. Tertiary treatment plants, which are rare at present, provide advanced treatment and remove such pollutants as phosphates, nitrates, and industrial chemicals.

Eutrophication, the overenrichment of lakes, is accelerated by the nutrients released by man's activities. An explosion in algal bloom followed by decay, oxygen depletion, odors, and increased lake sedimentation are characteristics of eutrophication. Efforts to reduce eutrophication have concentrated on phosphates in detergents, but evidence exists that many other nutrients are essential to algal growth and that some of them may be the limiting factor in that growth.

Mercury and cadmium compounds are water pollutants that attach to the sulfur atoms in proteins and interfere with biological processes. Methyl mercury is an especially hazardous pollutant that can be formed from inert liquid mercury by bacterial action in the water. Mercury poisoning arises most frequently from fish or from the consumption of bread made from grain treated with mercurial pesticides. The ingestion of paints containing lead pigments is the principal source of lead poisoning in the United States; children of the urban poor show the highest incidence of the disease.

Oil is an increasingly important water pollutant. Some of the hydrocarbons in oil are toxic, some are carcinogenic, and some can possibly inhibit the chemical communications between fishes. Oil spills kill most of the aquatic life in the immediate area; repopulation may omit some species present before the spill and may include some species never before present in the area. Most of the oil ingested by aquatic organisms is retained, even though the oily taste may disappear.

Even heavily polluted rivers can be cleaned up, as the renewal of Oregon's Willamette River over the past 50 years has shown. Achieving such dramatic transformations requires a determined attitude on the part of the citizens, as applied through legislative acts.

The 1972 Amendments to the Federal Water Pollution Control Act present a series of deadlines for pollution abatement. The Amendments establish goals of water quality safe for wildlife and recreation by 1983 and of the elimination of pollutant discharges to U.S. waters by 1985. The cost of the Amendments from 1971 through 1980 is estimated to be $100 billion.

Review and Study Questions

1. How much of the earth's surface is covered with water and ice? What percentage of the water on the earth is freshwater?
2. What is evapotranspiration?
3. Distinguish between water withdrawal and water consumption.
4. What problems arise from the use of groundwater sources?
5. How is the oxygen balance in water maintained?

6. What are the characteristics of anaerobic decay?

7. What changes can occur in an aquatic ecosystem as the temperature is increased?

8. How are BOD and COD measured? What are the relative sizes of the two parameters?

9. What are the relative costs of sewage treatment with primary, secondary, and tertiary plants?

10. Why are emergency overflow systems present in sanitary sewer systems?

11. What is flocculation?

12. What causes methemoglobinemia?

13. What are the three geologic classes of lakes? Give examples of each from the Great Lakes.

14. Why are phosphates present in common household detergents?

15. Why don't we simply bar phosphate-containing detergents and return to soaps?

16. What is the chemical difference between LAS and ABS? What differences occur in the environment?

17. Which organic mercury compound is most harmful to humans?

18. What was the source of large-scale mercury poisonings in Japan? In Iraq?

19. How are the allowed mercury levels in foods established?

20. What variations in the reported cases of lead poisoning are found in Baltimore and in New York?

21. List four possible sources of lead poisoning.

22. How much of the oil consumed in the world is transported at sea?

23. How quickly does oil degrade after a spill?

24. What problems are encountered in attempting to save birds coated with oil after an oil spill?

25. What are the national goals established by the 1972 Amendments to the Federal Water Pollution Control Administration Act?

Questions Requiring Outside Reading

26. Forty one percent of United States drinking water supplies do not meet Public Health Service standards. Why? (Janice Crossland and Virginia Brodine, "Drinking Water," *Environment* **15** [April 1973], 11.)

27. What is the environmental impact of ship canals and river dredging on aquatic systems which exist in estuaries? (Luther J. Carter, "Galveston Bay: Test Case of an Estuary in Crisis," *Science* **167** [February 20, 1970], 1102.)

28. Six thousand years ago, the Mediterranean area had ten times more forest land than it now has. Deserts have spread in the area since then. How is man's water use involved? (R. L. Nace, "Man and Water: A Lesson in History," *Science and Public Affairs (Bulletin of the Atomic Scientists)* **28** [March 1972], 34.)

29. What evidence led to the withdrawal of NTA as a phosphate substitute in commercial detergent preparations? (S. Epstein, "NTA," *Environment* **12** [September 1970], 2.)

30. What was the impact of the opening of the St. Lawrence Seaway on aquatic ecosystems in the Great Lakes? (W. I. Aron and S. H. Smith, "Ship Canals and Aquatic Ecosystems," *Science* **174** [October 1, 1971], 13.)

31. What detective work was involved in establishing the theory that the fall of Rome resulted from lead poisoning? (See discussion of S. G. Gilfillan, *Saturday Review,* August 7, 1965, 36.)

32. What, if any, variations of bloodstream lead concentration are found in urban versus rural dwellers in the United States? (J. Julian Chisolm, Jr., Suggested Outside Reading.)

33. What sensitive analytical instruments were used by the Woods Hole Oceanographic Institute researchers to determine the oil hydrocarbon concentrations after the wreck of the barge *Florida?* (M. Blumer, H. L. Sanders, J. Fred Grassle, and G. R. Hampson, "A Small Oil Spill," *Environment* **13** [March 1971], 2.)

34. What is the environmental impact of oil spills in the Arctic which might arise from the Alaska pipeline? (R. O. Ramseier, "Oil on Ice," *Environment* **16** [May 1974], 6; Ron Moxness, "The Long Pipe," *Environment* **12** [September 1970], 12.)

35. What law allowed citizens who report polluters to share a reward if the polluters are fined? (*Chemical and Engineering News,* May 3, 1973, p. 21.)

36. What water pollution problems exist in the Soviet Union? (P. R. Pryde, "Victors Are Not Judged," *Environment* **12** [November 1970], 30.)

37. How were the inevitable conflicts over whether to use land for agriculture or for industry or for parks resolved by the State of Oregon's Greenway plan along the Willamette? (*Environmental Quality,* The Fourth Annual Report of the Council on Environmental Quality, U.S. Supt. of Documents, September 1973.)

38. Drinking water can be polluted by both man-made and natural sources. Is your city's water safe to drink? If not, what can you do about it? (R. H. Harris, E. M. Brecher, and the Editors of *Consumer Reports,* Suggested Outside Reading.)

Suggested Outside Reading

BAKIR, F., et al. "Methylmercury Poisoning in Iraq," *Science* **181** (July 20, 1973), 230. An extensive report of the tragedy which includes an analysis of the effects of methylmercury poisoning.

CHISOLM, J. JULIAN JR. "Lead Poisoning," *Scientific American* **224** (February 1971), 15. A review of lead's uses, environmental impact, and health effects.

Environment Staff, *Environment* (May 1971). This issue contains three articles on mercury and metals in the air and in the environment. Included are extensive discussions of the environmental sources and the environmental impacts of the metals.

Environmental Quality. Annual Reports of the Council of Environmental Quality, 1970–, U.S. Supt. of Documents. Useful discussions of environmental problems. The 1973 report includes a discussion of the cleanup of the Willamette River.

GOLDWATER, L. J., "Mercury in the Environment," *Scientific American* **224** (May 1971), 15. A general review article on mercury, its methylation, its toxicity, and its environmental impact.

HAMMOND, A. L. "Phosphorus Replacements: Problems with the Washday Miracle," *Science* **172** (April 23, 1971), 361. A discussion of the problems encountered in attempting to find a phosphate substitute for detergents.

HARRIS, ROBERT H., E. M. BRECHER, and the Editors of *Consumer Reports.* "Is the Water Safe to Drink?" *Consumer Reports,* June (436–43); July (538–542); and August (623–627), 1974. An excellent study of water purity problems, what improvements can be made, and what can be done to improve the local water supply.

HEYERDAHL, THOR. "Atlantic Oil Pollution and Biota Observed by the 'Ra' Expedition," *Biological Conservation* **3** (April 1971), 164. An account of the evidences of man's pollution observed in 1969 and 1970 by Heyerdahl and his colleagues during an Atlantic crossing in a papyrus sailing vessel.

MARX, W. *The Frail Ocean.* Ballantine Books, Inc., New York, 1969. A broad description of man's impact on the ocean, its coastal areas, and its food supplies.

MOSTERT, NOËL. *Supership.* Alfred A. Knopf, Inc., New York, 1974. A superb discussion of the supertankers developed for oil transport, the problems inherent in their use, and their impact on the environment. Included is a narrative of the voyage Mostert took round the Cape of Good Hope on a 214,000 ton supership. He issues an urgent call for strict safety standards and international control of the ships.

U.S. Department of the Interior, *Lake Erie Report: A Plan for Water Pollution Control,* U.S. Supt. of Documents, Washington, D.C., 1968. A well-documented breakdown of the sources of Lake Erie pollution and the steps necessary to reduce the pollution.

Readings on aquatic ecosystems can be found in Suggested Outside Reading for Chapter 1.

7 Solid Waste and Recycling

Keep Ohio Clean. Dump Your Trash in Michigan.
Sign along the Ohio Turnpike

Trash is our only growing resource.

Introduction   Solid waste is that material arising from animal and human life and activities that has no immediate use and is discarded as waste. Solid wastes form a trail of visible blight that leaves few corners of the country unblemished. Across the nation the same scenes repeat: litter on beaches and along roadsides, abandoned automobiles, rusting refrigerators and stoves in thousands of dumps scarring the landscapes. (Figure 7.1). Perhaps more critical, however, are the less visible aspects of solid wastes—wasted resources, needless energy consumption, water pollution, and air pollution. Proper management of solid wastes is an important factor in resource conservation and in upgrading environmental quality.

Solid wastes can be classified as urban, industrial, mining, and agricultural as in Table 7.1. Agricultural wastes created the largest tonnage in 1970, nearly 2.4 billion tons, 75% of which was manure. Mining wastes are the second largest source (1.2 billion tons in 1970). Urban waste (280 million tons in 1970) and industrial waste tonnages are much smaller, but these often receive more attention because of their proximity to urban areas.

Urban (or municipal) wastes include garbage, rubbish, demolition wastes, sewage sludge, and metal scrap. The composition of municipal waste varies rather strongly with locale in the United States, but typical figures are in Table 7.2.

All of the factors that increase energy consumption, air pollution, and water pollution—population growth, industrialization, increasing urbanization, and increasing affluence—contribute to an increase in solid wastes. In the last 50 years in the United States the per capita amount of solid waste has increased dramatically. The collected wastes alone have changed from 2.75 lb/day per person in 1920 to 4.5 lb in 1965 to 5.5 lb in 1970. (These figures do not reflect all wastes generated; uncollected wastes probably raise the total to 10 lb/day per person.) In contrast, the municipal waste in Sydney, Australia, is 1.75 lb/day per person, and in India it is about ½ lb.

Although some waste is inevitable, the rather cavalier approach the industrialized nations have taken toward solid waste has accelerated resource de-

Figure 7.1. *Littered countrysides are a common sight in the United States (photo courtesy of William D. McCormick).*

pletion and caused needless energy consumption. Our traditional approach to solid waste management has been one of disposal. Even the words we use to describe the materials denote this: waste, trash, rubbish, dump. Almost no emphasis has been placed on the potential value of wastes as sources of materials and resources.

A prime example is our approach to household wastes. We presently store these wastes in cans, collect them in specially designed trucks, and then dispose of them by either dumping, burning, or burying. In this scheme, collection is usually done best. Not coincidentally, it is also the most expensive step in the process, accounting for 80% of the cost of collection and disposal. The expertise with which collection is performed (and its relative cost) reflects citizen interest; sloppy garbage collection prompts protests to city councilmen and, ultimately, better collection. Relatively few citizens view the disposal process, which is usually performed in some remote, rural area. Our collective lack of

TABLE 7.1. Solid waste production in the United States in 1970

Category	Description	Estimated Production 1970 (million tons)
Urban		280
Garbage	animal and vegetable kitchen waste	
Rubbish	dry household, commercial and industrial wastes	(230)
Ashes	residue after burning	
Demolition waste	construction and demolition debris	
Sewage residue	sludge and other solids from sewage treatment plants	(50)
Metal scrap	automobiles, industrial machinery	
Industrial	industrial processing and by-products	120
Mining wastes	mill tailings, strip-mining wastes, washing-plant rejects	1200
Agricultural wastes	animal manure and carcasses, crop residues, logging debris	2400
	Total	4000

Source: Bureau of Solid Waste Management, U.S. Environmental Protection Agency.

TABLE 7.2. Composition of municipal solid wastes in the United States

Material	Weight Percentage
Paper products	50
Food wastes, yard wastes	20
Metals	10
Glass products	10
Plastics, leather, rubber, cloth	10
	100%

concern with the utilization of the resources in domestic trash and garbage has prompted little change from our traditional policy of "get rid of it where no one will notice."

In the next section we discuss the traditional, most widely used solid waste disposal methods, most of which place little emphasis on reclamation or resource utilization. The remainder of the chapter is devoted to methods emphasizing utilization of solid wastes.

Disposal Methods   ## Ocean Dumping

Historically, extensive use has been made of the ocean for the disposal of harbor dredgings, sewer sludge, industrial wastes, radioactive wastes, and military wastes such as nerve gas (Table 7.3). The EPA has approved 126 ocean disposal sites near the United States: 51 in the Atlantic, 47 in the Pacific, and 33 in the Gulf of Mexico. Disposal sites for nontoxic wastes are near shore in water less than 100 ft deep; the sites for toxic wastes such as cyanides are at least 300 miles offshore. In the Gulf of Mexico the industrial disposal sites are in water at least 2400 ft deep located more than 125 miles offshore. More than 20 major seaboard cities dispose of large amounts of solids in the oceans, and many inland cities barge their wastes to the coasts for ocean deposition. Special barges may be used for dumping, especially for such wastes as sewer sludge, harbor dredgings, and bulk industrial wastes. Other chemical wastes may be transported in drums that are sunk on the site. Military wastes, including chemical and biological warfare agents, have occasionally been transported in surplus naval craft which are then sunk on the site.

The characteristics of the ocean in the dumping area and the nature of the wastes, especially particle size, determine how rapidly the wastes are dispersed and in which direction. The sites are chosen in a "tradeoff," minimizing harm to aquatic and shore life while keeping transportation costs as low as possible. Ocean dumping is the cheapest means of solid waste disposal; the costs in 1970 varied from 40¢ per ton for dredging spoils to $15 per ton for explosives and containerized materials.

The entire matter of ocean dumping requires more study. The long-term impact of dumping is unknown. Metals and chlorinated hydrocarbons (see Chapter 8) can accumulate in the food chain, and their disposal in the oceans should be restricted. Nutrients, such as sewer sludge, may need fewer restrictions. In fact, because most of the ocean is starved for nutrients, the dumping of sludge (assuming it is free of metals and pesticides) in remote areas may be

TABLE 7.3. **Extent of United States ocean dumping in 1968**

Source	Tonnage
Dredging spoils	80%
Bulk industrial wastes	10%
Sewer sludge	9%
Construction debris	1%
Containerized industrials	1%
Refuse and garbage	less than 1%
Explosives	less than 1%
Miscellaneous	less than 1%

Figure 7.2. New York City garbage being towed to sea from the East River. At the disposal site, slicks a mile in diameter occur following individual dumping. [EPA-DOCUMERICA-Gary E. Miller.]

beneficial, unless the bacterial action and dissolved oxygen levels in the immediate area are severely depleted.

The city of New York disposes of nearly 5 million tons of sewer sludge yearly in fairly shallow water located about 10 miles into the Atlantic off Sandy Hook, New Jersey (Figure 7.2). The site has become marked by an easily visible discolored region several miles in diameter. In this case, the sludge disposal has overloaded the capacity of the ocean to decompose the wastes. Aquatic life in the immediate area of the site has died, and the metals concentration at the site is much higher than in the surrounding ocean. Fishing on the periphery of the dumping area is good, but some charge that the fish caught in that area have a bland and unusual taste. The general direction of water currents in the area is from the dump site toward the Long Island beaches. It has been suggested that the dumping site should be moved farther offshore, perhaps 100 miles, so that the sludge can be dumped in much deeper water. Thus, far, however, the site has not been changed, partly because of the increased transportation costs which would result and partly because the impact of the dumping can be more readily assessed at the present site than at a much deeper location.

A principal worry about sewer sludge disposal is of course the possible spread of disease.

Open Dumps

An open dump is a disposal site in which the waste is not covered daily with compacted earth. For most of our history, the disposal of urban wastes was accomplished by dumping the refuse at an isolated site, setting it on fire to reduce the volume, then bulldozing the accumulated ashes and noncombustibles into a ravine or into a clump (Figure 7.3). Open dumps provide a cheap means of disposal; in 1970 the cost averaged about 50¢ per ton. They do encourage reuse of items through salvage operations, but their environmental limitations are obvious: the possible spread of disease, air pollution from burning, water pollution during rains, and litter proliferation. Efforts have been made for years, most recently by the EPA, to close most of the open dumps in favor of sanitary landfills (next section). These efforts have not been very successful, however, because open dumps still account for the majority of waste disposal on hand.

Sanitary Landfills

This disposal technique compacts wastes, covers them with earth, and seals them underground. Ideally the cells are sufficiently air and water tight that slow anaerobic decay occurs with little loss of material to the air or water.

A sanitary landfill site, typically a large ravine, should be chosen so as to minimize water pollution problems, hazards to human and animal life, and transportation costs. Finding a suitable site is a problem in urban areas, where land is expensive. The rule of thumb for municipal waste is 1 acre of 10-ft-deep landfill per year for each 10,000 persons. The soil of the site should preferably be clay because its superior compaction properties make it almost impervious to air and water. Waste is dumped at the site and compacted with a bulldozer, the depth after compaction being 8 ft or less. At frequent intervals, certainly at the end of each day, the compacted waste is covered with 8 to 10 in. of compacted earth (Figure 7.4). Thus the entire site is filled with cells of compacted trash.

Figure 7.3 An open dump solid waste disposal site. [EPA-DOCUMERICA-Bruce McAllister.]

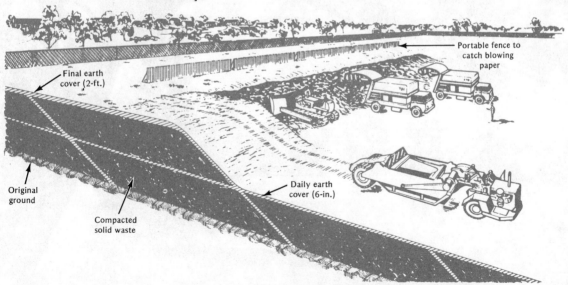

Portable fence to catch blowing paper

Final earth cover (2-ft.)

Original ground

Compacted solid waste

Daily earth cover (6-in.)

Figure 7.4. A sanitary landfill. [Courtesy Office of Solid Waste Management Program, Environmental Protection Agency.]

When the site is completely filled, it is covered with dirt and seeded. If the landfill is properly maintained, the trash slowly decays anaerobically with no appreciable air or water pollution. Because of the methane gas created during anaerobic decay, the construction of buildings on the reclaimed site is usually discouraged because of the danger of explosions should this gas seep into the buildings. The land is suitable for agricultural or recreational purposes such as parks and golf courses (Figure 7.5).

The cost of sanitary landfill waste disposal averaged $1 to $3 per ton in the United States early in this decade, but large individual variations occurred. The cost of collection and disposal by landfills for the city of New York was $35 per ton, and it had 30,000 tons of garbage alone to dispose of daily. In contrast, Austin, Texas, a city with a population of 250,000, used sanitary landfill disposal at a total cost (collection plus disposal) of $13 per ton during the same period.

Although they have advantages over open dumps, sanitary landfills also

Figure 7.5. *This golf course and park in Jackson, Mississippi was built on a filled sanitary landfill.* [EPA-DOCUMERICA-Bill Shrout.]

possess several liabilities. The reclamation of land may be beneficial or harm-ful. For example, filling a ravine to obtain agricultural lands or a park may be a benefit in an urban area, but filling a marsh, the center of aquatic life in a coastal area, may be an ecological disaster. A practical limitation of most landfills is that they do not operate under the ideal conditions implicitly assumed in our description. Compaction and earth covering are often not done daily, and litter, odors, and pest infestations may result. Water often enters a landfill by seepage, by erosion, or by means of underground sources. During wet periods such landfills can be a serious source of toxic, foul-smelling water pollutants. Another complication can arise if the landfill is located on lime-stone. Carbonic and sulfuric acids, formed if the decay products react with moisture, can decompose the limestone (as discussed in Chapter 3) and allow the wastes to flow downward to groundwater sources. In 1973 the Bureau of Solid Waste Management identified over 10,000 sanitary landfill disposal sites in the United States and found that over 90% of these failed to meet minimum standards because of litter, burning, surface or groundwater pollution, or pest infestations.

Finally, sanitary landfill operations do not encourage resource utilization; the emphasis is on disposal.

A Summary of Disposal Methods

The preceding methods account for 90% of the solid waste treatment in the United States. Table 7.4 shows the distribution of solid waste disposal methods.

Methods Emphasizing Resource Utilization

None of the traditionally used disposal methods discussed above encourage resource utilization. It is appropriate to ask if the use of wastes as landfill is the best use to which they can be put. The answer is a resounding "NO!" A better approach would be to utilize the solid wastes in some fashion. Utilization may take many forms: reclaiming and recycling resources, burning the wastes to generate electricity, converting the wastes into other, more useful forms, or perhaps combining several of these approaches. Recycling is a term usually

TABLE 7.4.   Solid waste disposal in the United States in 1970

Method	Urban Wastes (%)
Open dumps, including landfills covered less frequently than once a day	77
Sanitary landfills	13
Incineration[a]	8
Miscellaneous (hog feeding, composting, recycling, etc.)	2

[a] Incineration is discussed in the next section.

applied to the extraction, processing, and then reuse of a material in a form almost identical to that in which it was obtained. Although metals, paper, and glass can be recycled, other substances, including plastics, rubber, and food cannot be easily reconstituted into the form in which they are obtained; they must be converted into some other form. Several methods of solid waste utilization are described in the following sections.

Incineration

About 10% of the garbage collected in the United States is burned under controlled conditions in slightly more than 300 incinerators. Incineration affords a 90% reduction in the volume and a 75% reduction in the weight of trash. This approach requires less land than a sanitary landfill operation. An incinerator facility capable of handling the 400 tons of trash per day generated by a city of 100,000 persons would occupy perhaps 5 acres.

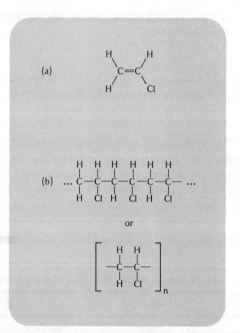

Figure 7.6. The plastic (polymer) polyvinyl chloride (PVC) is formed by causing many vinyl chloride molecules (a) to react together ("polymerize") to form a very long molecule (b). The double bond of vinyl chloride opens, allowing each molecule to bond to two others, and these to bond further, thus creating a long chain. The bottom structure in (b) indicates that n (often 100 to 1000) vinyl chloride molecules are combined to make the long polymer.

It is possible to utilize the combustion energy to generate steam or electricity that can be sold to reduce disposal costs. As a fuel, municipal trash has an energy content of nearly 5000 Btu/lb, about one third that of coal (see Chapter 5). Another advantage of incineration is that reclamation of metals and glass is easier after paper, garbage, yard trimmings, and plastics are burned.

Incineration has several disadvantages, however. Air pollution is a special problem: over 90% of the incinerators in operation do not meet present air emission quality standards. This failure to meet standards is partly due to the age of some of the incinerators and partly to the heterogeneity of the solid wastes, an aspect that puts enormous demands on the capabilities of air pollution control equipment. An increasing problem in recent years has been the ever-growing use of plastics, such as polyvinyl chloride or PVC (Figure 7.6). When burned, PVC, which contains chlorine (Cl), produces a strong acid, hydrochloric (HCl). Its corrosive action creates health problems and causes rapid deterioration of the incinerator. Hydrochloric acid is also formed when grass is burned.

Figure 7.7. Residue from a municipal incinerator. (Courtesy U.S. Bureau of Mines.)

Compared to landfill and dumping operations incineration is expensive. The cost is often $20 or more per ton. (The sale of electricity can reduce the cost by $2 to $4 per ton.) Furthermore skilled personnel are required for the operation and maintenance of an incinerator. An improperly operating incinerator will leave a large amount of unburned refuse to be reburned or disposed of in some other fashion. Damage from acids, molten metals, and exploding aerosol cans is the most frequent source of maintenance troubles.

The increasing demands on land areas probably mean an increase in incineration of municipal trash. A modern incinerator conveys the wastes from a temporary storage area into the combustion chamber. The gases are vented through a stack equipped with air pollution controls, which usually include a scrubber to remove the acids and an electrostatic precipitator to remove particulates (see Chapter 3). The residue or ash is removed for further utilization (Figure 7.7). Montreal and Chicago have recently completed construction of large municipal incinerators capable of handling 1500 tons/day of waste while generating useful energy in the process. The Chicago plant shreds the waste, removes the iron with a magnet, and burns the remainder. The waste heat is transferred to a boiler to generate steam that is sold to local industry. To forestall steam shortages if sufficient wastes are not available, the plant is also equipped to burn natural gas.

Pyrolysis

This approach to solid waste treatment uses a very old technique formerly known as "destructive distillation": heating in the absence of air. As applied to solid wastes, pyrolysis offers the principal advantage of no air pollution; the wastes are sealed in an oven and heated to a very high temperature, a process that converts them into a mixture of oils and solids.

Pyrolysis may be especially applicable to the treatment of tires and plastics. In the Firestone Company's pyrolysis plant in Akron, Ohio, tires are cooked at temperatures from 500° to 900°C (932° to 1652°F) and are converted to solids (45% of the original weight), liquids, and gases. The liquids, poor quality oils composed of more than 50 different aliphatic hydrocarbons, can be used in the manufacture of new rubber or in the manufacture of petrochemicals. Some of the solids can be used as a particulate in concrete, and others—those that are nearly pure carbon—can be used in the manufacture of new tires. The process is presently a borderline economic operation, its principal purpose being the diversion of tires from sanitary landfills.

Although steam or electricity might be generated by burning the oils obtained from pyrolysis, there would be little advantage in doing so because the net energy gained (after subtracting the energy required for pyrolysis) is very small.

Composting

Nearly all organic matter—including garbage, sewage, paper, grass, and leaves—can be converted by bacterial action into humus, the organic portion of the soil. The decay of vegetable matter on and in the soil is the usual source of humus, and composting allows solid wastes to be used as a soil conditioner and fertilizer. In European cities, up to 25% of the garbage is composted, but the process has not become widespread in the United States, partly because of the greater availability of relatively cheap land for landfills and partly because of the greater affluence in this country.

Although many pitfalls can arise, it is possible to achieve composting with few odors or other problems if aerobic decay is ensured. Organic matter is shredded, put in small piles, perhaps dampened with water, and stirred once or twice a week. The rate of bacterial action depends on particle size, moisture content, oxygen content, and the ratio of carbon to nitrogen in the refuse. Decay may take from a few days to several months, but the end result (compost humus) is a dark, odor-free, inert material with a consistency similar to peat moss or top soil.

The carbon to nitrogen (C/N) ratio (by weight) in the refuse is crucial to successful operation; a knowledge of the importance of this ratio is fundamental to understanding the decay process as well as the problems that have arisen with large-scale commercial composting ventures.

In **aerobic** decay the organic matter, principally carbon, nitrogen, hydrogen, and phosphorus, is consumed by the bacteria and converted into energy, cell protoplasm, and inert organic material. Much more carbon than nitrogen is needed, since carbon is used for energy as well as for protoplasm production (Figure 7.8). On the average, about two thirds of the carbon is utilized as an energy source for the bacteria, then respired as CO_2. Nitrogen, however, winds up primarily in the protoplasm; little is lost or diverted to other purposes. Consequently in aerobic decay the nitrogen is said to be "tightly cycled." The C/N ratio is a major factor in the rate of the bacterial action, the fastest compost decay occurring when the C/N ratio is approximately 28. If the C/N ratio is either much greater than or much less than 28, the rate of biological action is greatly diminished. (Under either of these less than optimum conditions, the system has a natural mechanism to alter the C/N ratio to a value nearer 28; it does this by undergoing biological or chemical reactions to lose carbon or nitrogen as necessary to bring the C/N ratio back to the optimum value.) As noted earlier in the text (Chapter 6), aerobic decay produces few odors.

In contrast to its uses in aerobic decay, carbon is converted in **anaerobic** decay into CO_2 (that is, used for energy), into protoplasm, *and* into methane (CH_4). Nitrogen is changed into ammonia (NH_3) as well as protoplasm; thus it is not tightly cycled in anaerobic decay (Figure 7.8). Another important dif-

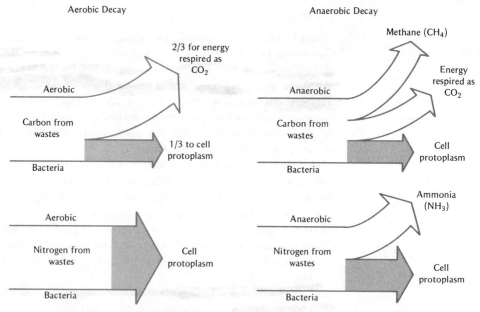

Figure 7.8. *A greatly simplified illustration of the uses of carbon and nitrogen in aerobic and anaerobic decay. Under the action of aerobic bacteria, the carbon is used for energy and protoplasm, while the nitrogen is tightly cycled, going almost entirely into cell protoplasm. In contrast, anaerobic bacteria convert carbon into methane in addition to using it for energy and protoplasm. Nitrogen in anaerobic decay is not tightly cycled; some of it is converted into ammonia.*

ference between the two types of decay is that converting carbon into CO_2 releases 20 times as much heat as converting the same amount of carbon into methane. Thus organic matter undergoing aerobic decay will get much hotter than it will if undergoing anaerobic decay. The temperature a few inches inside a properly operating compost pile is often 70°C (158°F), sufficiently hot so that the hand cannot be kept in the pile comfortably for more than a few seconds. Achieving this high temperature is very important; it kills pathogens in a few hours and weed seeds in a few days. Under these conditions the use of humus as a soil conditioner presents no health hazards. An expert composter can control the C/N ratio by a proper blending of materials. As shown in Table 7.5, paper and municipal refuse have large ratios, whereas manure and grass clippings have low ratios.

Flies are a potential problem when composting. They can be controlled by utilizing the fact that the reproductive cycle of flies is 7 to 14 days. If all the material is in the pile rather than scattered about, and if stirring is done twice a week, the fly eggs, usually laid on the cooler outside portion of the pile, are circulated to the higher temperature interior and killed. (Flies in garbage can

TABLE 7.5.   Carbon to nitrogen ratios in common materials

Compound	C/N Ratio
Urine	0.8
Blood	3.0
Manure	14.0
Grass clippings	19.0
Municipal refuse	35.0
Straw	128.0
Paper	very large (almost no N)

areas can also be controlled by keeping all the garbage in the cans, then having the garbage picked up twice a week rather than weekly.)

Achieving a successful, odor-free decay does not, however, mean the end of potential troubles with composting. It happens that the C/N ratio for optimum composting (about 28) is greater than the optimum C/N ratio of about 20 preferred by soil bacteria. The addition of compost humus to soil may therefore result in nitrogen "robbing" the first year. Soil bacteria exposed to the high C/N ratio undergo rapid growth, and they use nitrogen from the soil to accomplish it; in the process they "rob" the plants of needed nitrates in the soil. Crops grown during the first year humus is used are often poor, but by the second year production is vastly improved. The addition of humus to the soil increases the porosity and improves soil aeration; consequently crops often do better two or more years after humus is added than crops fertilized with chemical fertilizers. One means of reducing nitrogen robbing the first year is to plant a legume, such as alfalfa; legumes are unusual in that they have the ability to put nitrogen directly into the soil and compensate for "robbing."

In commercial applications, the municipal refuse is shredded, the iron is removed magnetically, and some sewage sludge may be added to hasten decay (Figure 7.9). The material may be arranged in long "windrows," piles 15 ft high and up to 100 yd long. The refuse is turned frequently to ensure aerobic decay. For better aeration and faster decay, it may be put in large revolving bins. When decay is complete, the plastics, metals, and glass that remain are removed, and the humus is sold.

Scarsdale, New York, began composting its leaves in 1966 under the direction of an experienced operator, Stan Bulpitt. Now each year it disposes of 200,000 yards of leaves at a cost of $15,000, which represents a savings of $40,000 per year over the cost of sanitary landfill disposal. The compost is sold for $4 per cubic yard (three fourths of a ton) to gardeners and farmers.

Other large-scale composting operations in the United States have not fared so well. Of the 20 plants started in the United States in the last 20 years, 17

Figure 7.9. *A windrow being turned in a commercial composting operation. [U.S. Environmental Protection Agency-Don Moran.]*

have gone out of business. Frequently the shutdown is because of complaints about odors from citizens living nearby or from customers. Commercial operations often contract for a daily delivery of wastes. If the decay slows down for some reason, the operator, faced with a space problem, may release partially decayed compost which later develops odors when anaerobic decay begins. Such large-scale composting operations require the services of an expert "compostmeister," one who can tell by sight and feel whether a compost is operating properly and knows what to add to correct any deficiencies. Such expertise is rare in the United States.

Composting is now done by many householders in this country. A common recommendation for home composting is to avoid adding meat scraps, as they often cannot be shredded sufficiently to avoid anaerobic decay in the interior. Another recommendation is to avoid adding feces to the compost pit because improper decay could lead to the spread of disease.

The World Health Organization has for years promoted composting of human and animal wastes in underdeveloped nations as a means of reducing health problems while simultaneously improving agricultural productivity.

The recycling and reutilization of resources offer many advantages. Resource lifetimes are extended, pollution resulting from mining and manufacturing is often lessened, and the volume and cost of solid waste disposal are decreased. An increasingly important aspect of recycling is that in most cases the energy required to manufacture a product from recycled materials as opposed to raw materials is often substantially less. For example, making 1000 tons of steel reinforcing bars from scrap instead of from virgin ore takes 74% less energy and 51% less water, creates 86% fewer air pollution emissions, and generates 97% fewer mining wastes.

Whether extensive recycling of any particular resource will be realized, however, depends on economics, including the availability of markets. The use of virgin raw materials is favored by economic policies dating to the industrial revolution in the United States. Transportation rates for raw ore are often only one half to one third those charged for finished products, a category that unfortunately includes material being transported for recycling. Furthermore the use of raw materials affords an economic benefit in the form of depletion allowances similar to the oil depletion allowance discussed in Chapter 4. It has been estimated that in the steel industry, for example, prices would have to be 13% higher to produce the same profits now obtained with the help of depletion allowances. Should recycling of a particular resource still appear potentially profitable, despite the economic favoritism shown raw materials, little recycling will be done unless the material can be easily separated and sold to a nearby market. Expensive separation techniques or high transportation charges can easily make recycling unprofitable. Market availability depends partly on confidence on the part of the commercial buyer that recycled materials are a steady, long-term source of goods. For many resources, recycling is not sufficiently common to provide that confidence. *glass isnt economical to recycle.*

Cellulose Fiber Recovery

Paper, cardboard, wood, and similar organic substances can be recycled by "pulping" the material and recovering the fibers, a process similar to that used to make paper from wood. In 1946 about one third of the paper used was recycled. By 1970 only 18% of the 59 million tons of paper and cardboard consumed in the United States was recycled. Of the remainder, about 40 million tons was of recyclable quality, but was reused directly, buried, or, more probably, simply disposed of. The reclaimed cellulose fibers can be made into paper of almost any desirable type, from the highest quality bond stationery to wrapping paper (Figure 7.10).

The limitations on paper recycling are largely economic. Over the past 30 years, the paper industry has located mills closer to the forests and has improved the technology of harvesting wood: chain saws, bulldozers, large

Figure 7.10. Environment Magazine is printed almost entirely on recycled paper. The subscription envelope emphasizes no-waste mailing. The envelope itself serves as the subscription form. When the envelope is returned, it is recycled.

trucks, and snow cats for year-round work have assumed dominant roles in the paper industry since the 1940s. These advances have served to minimize the costs of paper made from virgin materials relative to paper made from recycling processes.

Furthermore, over the same time period, several other technological developments have tended to *increase* paper recycling costs. To make white recycled paper, the reclaimed fibers must be deinked. The development of improved, more permanent inks has make deinking more difficult and more expensive. In recent years the increase in plastic lamination of paper and the development of stronger glues have rendered many paper products unsuitable for recycling. Casual contaminants, such as food, oil, and paint, hamper the

TABLE 7.6. Recycled cellulose fiber

Country	Consumption (%)
Japan	38
West Germany	32
France	29
United Kingdom	26
United States	18
Canada	11

recycling of paper from municipal trash. Consequently the majority of the paper presently recycled comes from relatively clean sources such as binderies, publishing houses, and newspapers. As a result of all of these factors, recycled paper is often more expensive than paper made from virgin sources, so there is little economic initiative for recycling.

Some paper products can be recycled more easily than others. Magazines are difficult to recycle because of the presence of clay, the ingredient used to produce the glossy finish of most magazines. Newsprint can be easily recycled, but economic competition with virgin sources is a major factor that determines the extent to which recycling is done. Depending on locale and conditions, used newsprint can be sold for $10 to $30 per ton, and recycling operations usually show a profit only if volunteer labor such as the Boy Scouts is used.

Because virgin materials are more expensive in foreign countries, paper recycling is more extensive there as shown in Table 7.6.

The paper industry estimates that 1 ton of paper requires the harvesting of 17 trees. Thus the 11 million tons recycled in 1970 represented a savings of 200 million trees. If the percentage of recycling could be increased to 50%, the savings would be equivalent to the sustained yield of combined commercial forests in Texas, Louisiana, Arkansas, and Alabama.

Glass

Glass can be easily remelted and recycled. It is formed by melting a mixture of sand (70%) with smaller amounts of naturally abundant materials, soda ash and limestone. There is a small market for recycled glass in the glass container manufacturing industry. Mixing about 5% crushed glass (or "cullet") with the sand and other raw materials in the glass furnace improves the quality of the final product. Only a small fraction of all the glass consumed in this country is recycled, however, because of the low cost of virgin raw materials. Sand suitable for glass manufacturing costs $15 per ton delivered to the door of the plant. A recycling operation must collect the glass (2000 to 4000 bottles), sort it by color, crush it, and transport it to the plant. Partly as a public service, glass companies often pay $20 per ton for such deliveries. Even so, if the plant is 50 or more miles away from the collection site, the recycling operation can usually succeed only with extensive volunteer labor. Sorting expenses alone, if done by paid manual labor in a city, might be $35 per ton. Mechanical sorters exist but are sufficiently expensive that they can be justified only with large-scale integrated recycling operations that utilize all the resources in trash (such systems are discussed later in this chapter).

Other uses can be found for crushed glass. It can be used to replace gravel in road surfaces; the paving material that results is called "glasphalt." Although glasphalt surfaces are equivalent to those of the more common gravel asphalt, glass is again competing with an abundant and cheap raw material, and this use can be economically successful only in urban areas. Waste glass can also be used to make building blocks and bricks for the construction industry.

On an energy basis, the recycling of glass beverage containers cannot be justified, as it requires more energy than making the glass from raw materials (see Energy and Recycling later in this chapter).

Tires and Plastics

As noted earlier, rubber and plastics cannot simply be remelted then re-formed. Both are polymers, formed by the binding together of thousands of small molecules into long-stranded structures (Figure 7.6). Heating these to a high temperature disrupts the bonding and degrades the material to a variety of oils and solids. As discussed earlier in the chapter, pyrolysis provides one means of utilizing tires and plastics, but alternate possibilities exist.

Over 200 million tires are discarded in the United States each year. About 10% of the rubber discarded is reclaimed for use, some of which after processing can be used in the manufacture of new tires. Only a few reclamation plants are in operation, so transportation costs become an important factor. For example, it costs $15 per ton to ship old tires from Madison, Wisconsin, to the Akron, Ohio, reclamation plant where they are worth $6 per ton.

The Goodyear Company points out that tires have 50% more Btu per pound than coal and has constructed a tire destruction furnace in Mississippi that will consume 1 million tires annually while generating useful energy. The furnace is not a company money saver, however, because the boiler and the air pollution controls are so specialized that the plant construction and operating costs are considerably greater than those of conventional steam plants.

Two University of Wisconsin engineers have designed a very low temperature (or "cryogenic") scheme that may simplify resource recovery of tires. This particular process freezes tires in liquid nitrogen, which has a boiling point of −196°C (−321°F). At this low temperature the rubber is very brittle; it will shatter when struck with a hammer, for example. In this scheme the frozen tires are ground into bits with a hammer mill; the polyester and steel cords are easily separated from the rubber and can be recycled. The reclaimed rubber is in sufficiently pure form that it might find use in artificial playing field surfaces or might be mixed with asphalt to form road surfaces. Liquid nitrogen is easily handled using vacuum insulation; it is sufficiently cheap that the costs of decomposition are estimated to be only 3¢ per tire.

Other uses exist for tires. They can be recapped and reused, but less than one fourth of the tires in the United States follow this route compared to the 60% to 80% which are recapped in Europe. The lack of biodegradability of tires makes them useful components of underwater reefs, a use that consumes over 300,000 tires per year.

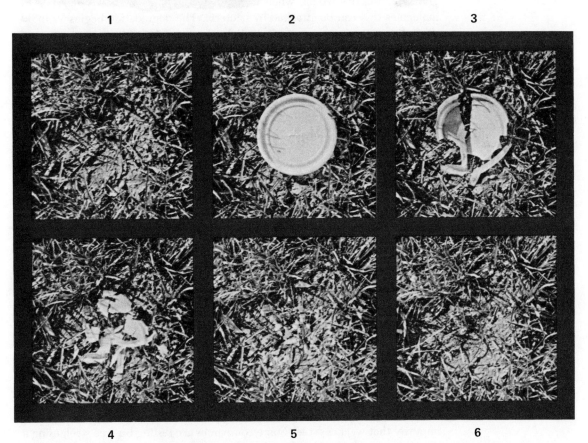

Figure 7.11. *A special additive causes this plastic (polystyrene) cold drink cup lid to undergo photodegradation in sunlight. It begins to disintegrate in 30 to 90 days (3) and is completely disintegrated in 150 days (5). Several common soil and airborne microorganisms digest the residue after degradation. [Courtesy of Bio-Degradable Plastics, Inc.]*

Plastics constitute a few percent of our solid wastes, but their tonnage increases each year. Reutilization of plastics usually requires a separation by type.

Ford Motor Company has successfully tested a process to recycle the plastic foam used in automobile upholstery. This plastic, polyurethane, can be obtained in a relatively clean form from junked cars, and Ford's process chemically converts this to compounds that can be sold to make more new polyurethane. Their data implied that the profit per car might be 35¢.

One interesting development has been to improve the biodegradability of plastics. Several companies have announced the development of "photodegradable" plastics which, when exposed to the sun, will slowly degrade to molecules that can be attacked by bacteria. The degradation begins 30 to 60 days after extensive exposure to sunlight (Figure 7.11).

Plastic bottles can be chemically cleaned to removed inks and markings, sanitized, and returned to use at a cost of about one third that of new containers, but this approach has never been promoted.

Metals

The recycling of metals is, in general, easier than the recycling of nearly any other item, and metals are so valuable that presently we recycle half of all copper and lead, one third of all aluminum, and one fifth of all iron consumed in the United States. As is true of most materials, the transportation costs of raw metal ores are lower than those of the secondary, that is recycled, metals. Rail rates for scrap steel, for example, were $4.60 per ton in 1973, whereas the rate for iron ore was only $1.84 per ton transported over the same distance.

"Tin" Cans. This most ubiquitous form of litter—"tin" cans—constitutes about two thirds of the total tonnage of metals in urban trash. The use of the word tin is inaccurate, a carryover dating from the early 1900s when cans were made of tin. Most of the cans in use today are made of iron, with a very thin tin coating to inhibit rusting. The "pull tab" cans are bimetallic; the sides are iron and the ends are aluminum. All-aluminum cans are used for some beverages. (A simple test of whether a can contains iron can be performed with a small magnet, which will attract iron and steel but not aluminum.)

The tin coating on iron cans hampers recycling. Used iron can be converted into steel in a blast furnace, but tin is an undesirable, difficult to remove impurity that weakens steel. Some companies are testing tin-free steel, using a plastic coating instead of tin as a rust inhibitor.

Tin cans are useful in chemically recovering copper from ore-processing solutions. Over 100,000 tons of cans, about 3% of all cans available, are used annually for this purpose in the southwestern United States.

Aluminum. Aluminum cans and other aluminum products are easily recycled. Nearly all aluminum manufacturers buy used aluminum cans at about 0.5₵ per can, which is about $10 per 100 lb. Other aluminum products bring from $10 to $30 per 100 lb. The EPA estimates that nearly one third of the aluminum cans sold are recycled.

Junked Cars. The number of junked automobiles in the environment is estimated to be over 20 million, with several million being added annually. In many cases the autos are simply abandoned, creating cleanup problems for cities; for example, over 70,000 cars are abandoned in the streets of New York each year. Most of the junked cars in the United States find their way into salvage yards which serve the useful purpose of making the parts available for reuse. After all recoverable parts are removed, the hulks may be reprocessed, dumped, or in a few cases used in some special application such as providing underwater habitats for fish.

The average automobile contains about 3300 lb of iron and steel, 74 lb of aluminum, 54 lb of copper, 50 lb of zinc castings, 24 ft² of laminated glass, and 29 lb of miscellaneous materials. The value of the recoverable materials is given in Table 7.7.

Reprocessing and recycling automobile bodies are hampered in part by the value of the materials; transportation costs alone may exceed the $56 total in Table 7.7. A principal limitation, however, is the presence of the metallic impurities, especially copper, in the auto body. Most steel mills are designed to extract iron from rocks and other nonmetallic substances, not to remove small percentages of metal impurities. The amount of copper present in automobiles has increased in recent years because of the increase in lights and in accessories

TABLE 7.7.   The value of materials in a junked automobile in 1970

Metal	Weight (lb)	Price	Value
No. 2 bundle iron	2614.0	$18.70 per ton	$24.44
Cast iron	429.3	$42.20 per ton	$ 9.06
Copper	31.9	$ 0.355 per lb	$11.34
Zinc, die castings	54.2	$ 0.0625 per lb	$ 3.39
Aluminum, cast, etc.	50.6	$ 0.124 per lb	$ 6.27
Lead			
Battery	20.0	$ 1.40 per battery	$ 1.40
Battery cable clamps	0.4	$ 0.11 per lb	$ 0.04
Total	3200.4		$55.94

Source: The Automobile Cycle: An Environmental and Resource Reclamation Problem. *U.S. Department of Health, Education, and Welfare, 1970.*

such as radios and tape players. Because steel is weakened if it contains more than 0.1% copper, only about 15% of the mix in a blast furnace can be made up of "as junked" auto bodies. As a result the price paid for the hulks has been depressed.

One means of avoiding this impurity bottleneck is to grind the automobile into myriads of small pieces in a giant shredder, then magnetically separate the iron shreds from the other metals. This affords a good separation of the iron, and means that there is no limit on the percentage of automobile scrap that can be loaded into the blast furnace with the raw ore. In this regard, shredding a junked car offers important advantages for recycling compared to merely processing the car by compressing it into a small bundle. (The large energy savings that can result from this procedure are discussed in the following paragraphs.) Shredders are expensive, however, and are presently located only in urban areas.

Energy and Recycling

The importance of energy in today's world has focused interest on the relative energy requirements of using natural resources as opposed to recycled materials. In many cases the actual energy cost of a product is unknown and may not be directly reflected in the price. Studies of the energy cost of products and goods are often done using a **systems** approach: calculating all the energy required to obtain, transport, and refine the raw materials; to manufacture, distribute, and market the product; and finally either to dispose of the product or recycle its components. By constructing such a cycle and evaluating each step, the energy cost of an item can be determined. This can be done not only for presently used technology but for alternate paths as well, perhaps including alternate methods of refining the ore or of increasing the amount of recycling in the system. For long-range planning, the system which uses the minimal energy would be preferred, but of course economic and resource considerations may alter the preference.

Aluminum is a prime example of the energy savings that can result from recycling. The total energy required to mine, process, and electrolytically refine 1 ton of aluminum is about 54,000 kw-hr taking into account all of the inefficiencies. This is 30 to 50 times more than the energy required to make a ton of aluminum from recycled materials! Other examples are not as extreme as aluminum, but energy savings of fivefold to tenfold resulting from manufacturing from recycled scrap metals rather than from raw ores are not unusual.

Energy Cost and Recycling of Automobiles

R. Stephen Berry and Margaret Fels, chemists at the University of Chicago, have studied the energy requirements of automobile manufacture. Their re-

TABLE 7.8.   Total (free) energy cost in the manufacture of one automobile

Item	Energy (kw-hr)
Manufacturing metallic materials	26,185
Manufacturing other materials	865
Fabrication of parts and assembly of auto	9,345
Transportation of materials	655
Transportation of assembled automobile	225
Total	37,275
Ideal thermodynamic requirement	1,035

Source: R. Stephen Berry and Margaret F. Fels, "The Energy Cost of Automobiles," Science and Public Affairs **29**, No. 10 (December 1973), 11.

sults yield the energy cost of the auto and provide an insight into the energy benefits derivable from the recycling of automobile hulks.

Berry and Fels used thermodynamics to analyze the materials cycle (mining, processing, fabrication, manufacturing, and disposal) for 1967 model automobiles. (They measured the requirements in thermodynamic potential or free energy, a concept we shall not develop; for our purposes, we may consider free energy as an approximate measure of the total energy.) As shown in Table 7.8, the total amount of energy required to make one automobile is equivalent to just over 37,000 kw-hr, of which 80% is due to the manufacture of the metals.

One point emphasized by the researchers is that the actual energy required in making a car is *35 times greater* than the minimum energy calculated to be necessary (1035 kw-hr) using thermodynamic analysis of the dominant tech-

TABLE 7.9.   Energy cost of various auto recycling and lifetime schemes

System	Energy Cost of One Automobile (kw-hr)	Energy Savings per Automobile (%)
Manufacture from raw materials, no recycling	37,275	—
15% of steel from recycled auto scrap	35,600	4.5
Shred and recycle one junked car for each new car	24,635	30.0
Annual cost—present (10 year lifetime)	3,728	—
Annual cost (triple lifetime of auto)	1,400	72.0

nology now in use. The inference to be drawn is that there is considerable room for improving the efficiency with which energy is used to manufacture automobiles.

Another important finding in this study is that significant energy savings can result from recycling auto hulks. As noted earlier, if the auto bodies are recycled without first separating the metals present, then they can constitute at most 15% of the furnace load for steel. Under this procedure, at most, 15% of the materials in a new automobile could come from recycled automobiles. If *all* new automobiles did in fact come from furnaces using 15% auto scrap, a *net* systems savings of 1600 kw-hr per auto would result (Table 7.9). By using a shredder, however, the metals can be separated as noted previously, and the new automobile could be made from completely recycled iron scrap (Figure 7.12). Berry and Fels demonstrated that the *maximum* system energy savings for recycling would arise if one car were scrapped in a shredder for each new automobile constructed: an energy savings of 30% would be achieved in the energy cost of an automobile (Table 7.9).

The required technology to operate under these maximum savings conditions is already in use. There are almost enough shredders in operation to permit the scrapping of one junked car for each new car manufactured, and

Figure 7.12. *By shredding these auto bodies and recycling the materials, the energy cost of an automobile can be reduced 30%. Even greater energy savings could be obtained if the useful lifetimes of the automobiles could be extended. [Photo courtesy of William D. McCormick.]*

there are enough cars already in salvage yards to provide a 20-year pool for increased automobile registrations. The bottlenecks lie elsewhere: with the collection and transportation of hulks and with the supply-demand-price situation in the scrap metals market.

If the 8 million automobiles manufactured annually in the United States were made from recycled hulks, the energy saved would be equivalent to the capacity of ten of the largest electrical power plants under construction.

Even more impressive energy savings could be achieved by extending the useful lives of automobiles. The energy required to extend the lifetime of an automobile is largely that required to produce replacement parts. To triple the lifetime, the total energy required to replace all estimated failing parts is only about 15% of the energy required to make the car. By tripling the lifetime of the car, from about 10 to 30 years, the total energy *savings* would be about 23,000 kw-hr per present vehicle lifetime (Table 7.9).

Energy and Recycling in the Beverage Industry

Bruce Hannon, Director of the University of Illinois' Center for Advanced Computation, has analyzed the systems energy of the beverage industry, and calculated the relative energy requirements of returnable versus nonreturnable containers and of recycled versus nonrecycled materials. Hannon compiled his data by analyzing actual energy use in all phases of the production, distribution, and disposal of beverage containers.

Before discussing Hannon's analysis, it is instructive to examine the trend toward nonreturnable containers that has occurred in the beverage industry during the past two decades. This trend was initiated by the introduction in the mid-1950s of the metal can to the beverage industry. Can manufacturers realized that, since returnable bottles averaged 40 refills, each returnable bottle represented a poential sale of 40 cans. They entered the market, and certain consumer tastes contributed to the can's success. Canned beverages achieved impressive success in the inner city. The generally small storage and cooling space available to the inner-city dweller favored purchases of one compact, cooled package per visit to the market. Furthermore, the 3¢ to 4¢ return deposit on bottles lacked the appeal to offset the convenience of nonreturnables, especially in view of the difficulty of returning the empties when shopping trips were often made by foot. Suburbanites with different life-styles were more prone to the use of returnables. Merchants also favored nonreturnable containers because of their convenience, and because they could convert the space previously required to store empty bottles into merchandise storage space. The success of the can encouraged glass manufacturers to develop the nonreturnable or one-way or "throwaway" bottle. Its success is shown in Table 7.10.

TABLE 7.10.   Market share of returnable bottles

Industry	1958	Percentage 1966	1976 (est.)
Soft drink	98	80	32
Beer	58	35	20

*Source: "The Role of Packaging in Solid Waste Management,"
Environmental Protection Agency, Washington, D.C., 1969.*

One loser in the battle has been the returnable bottle. Its decline has not been due to bottle fragility but rather to general affluence, to competition from other packagings, and to advertising, the mechanism for change in consumer habits.

Another loser in the battle, at least in economic aspects, has been the consumer. Beverages average 30% less per gallon when purchased in returnables, assuming the deposit is recovered. Furthermore, the increase in highway litter and solid waste volume created by nonreturnables has caused a significant increase in pickup and disposal costs, paid by the consumer through taxes or monthly billings. The resource and energy consumption per gallon also increased with the shift to nonreturnables.

Hannon studied the energy requirements of glass containers for four systems:

A. Returnable bottles refilled eight times, not recycled after breakage.
B. Returnable bottles, refilled eight times, 30% of the glass recycled after breakage.
C. Nonreturnable bottles, not recycled after use.
D. Nonreturnable bottles, 30% of glass recycled after use.

A summary of the energy costs of beverages in glass bottles is shown in Table 7.11 and Figure 7.13. In this table, a figure of eight refills, the average for Chicago, is assumed. The largest *energy* price is in the bottle manufacture. Note that the final energy tabulation is in energy per gallon of beverage, so that the energy required to make returnable bottles and handle their disposal is given assuming eight refills are possible, whereas the energy required to make and dispose of nonreturnables reflects the manufacture of sufficient bottles to hold 1 gal. Two important points are evident from Table 7.11:

1. A nonreturnable system requires more than *three times* as much energy per gallon as a returnable bottle beverage system. If the national average of 15 refills is used, nonreturnables require *4.5 times* as much energy per gallon!

TABLE 7.11. Energy expended (in Btu) for 1 gal of soft drink in 16-oz returnable or throwaway bottles

Operation	Returnable (8 Refills)	Throwaway
Raw material acquisition	990	5,195
Transportation of raw materials	124	650
Manufacture of container	7,738	40,624
Manufacture of cap (crown)	1,935	1,935
Transportation to bottler	361	1,895
Bottling	6,100	6,100
Transportation to retailer	1,880	1,235
Retailer and consumer	—	—
Waste collection	89	468
Separation, sorting, return for processing, 30% recycle	1,102	5,782
Total energy expended in Btus per gallon		
Recycled	19,970	62,035
Not recycled	19,220	58,100

TABLE 7.12. Energy ratios for various beverage container systems[a]

Container Type		Quantity	Beverage	Returnable Fills	Energy Ratios
Throwaway	Returnable				
Glass	glass	16 oz	soft drink	15	4.4
Can[b]	glass	12 oz	soft drink	15	2.9
Glass	glass	12 oz	beer	19	3.4
Can[b]	glass	12 oz	beer	19	3.8
Paper	glass	1/2 gal	milk	33	1.8
Plastic[c]	plastic	1/2 gal	milk	50	2.4

[a] The energy per unit beverage expended by a throwaway container system divided by the energy per unit beverage expended by a returnable container system. Discarded bottles and cans are not returned for recycling.
[b] Bimetallic can (iron with aluminum ends). All-aluminum cans require 33% more energy than the bimetal can system.
[c] High-density polyethylene.

2. Recycling the glass requires more energy than making the containers from raw material. That is, the recycling of glass cannot be justified if the minimal energy system is preferred.

Table 7.12 compares the energy requirement of other beverage containers with the minimal energy system, that is, the returnable bottle without recycling. Bimetallic soft drink cans require about 11% less energy per consumed

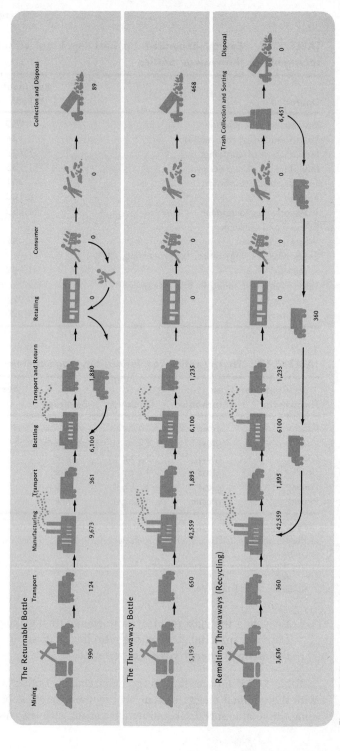

Figure 7.13. *Energy requirements for the various processes in the returnable and throwaway bottle cycles. [Originally appeared in "Bottles, Cans Energy," by Bruce Hannon, Environment, Vol. 14, No. 1. Copyright © Committee for Environmental Information, 1972. Reprinted with permission.]*

gallon than do nonreturnable glass containers, but still require nearly three times as much energy as returnable bottles. Aluminum cans, without recycling, are 33% more expensive in energy per gallon than bimetallic cans. Even paper containers require more energy per gallon than returnable glass bottles.

Hannon's studies also indicate that the cost per gallon of beverages in returnable containers was 30% less (after deposit recovery) than their cost in nonreturnable containers. A further result of the study is that certain segments of the beverage industry benefit from higher profits if nonreturnable containers are used, while other segments have followed the trend because they were forced to by adjacent industries or by unrestricted interindustry competition.

The implications of Hannon's study are clear. For minimal environmental impact and for maximum consumer economic benefit, a return to the use of returnable glass bottles should be encouraged. Because of the abundance of sand, the unfavorable energy requirement of recycling means that the effort, publicity, and money currently expended on glass recycling might be better spent in attempting to reverse the trend toward nonreturnable containers.

Integrated Utilization Schemes for Urban Trash

A major problem blocking the extensive utilization of municipal trash is illustrated by Table 7.13: the materials recovered from a ton of trash are worth less than $15 after separation and recovery. The most abundant material is

TABLE 7.13.   Potential values of resources recoverable from trash (1970)

Material	Value as Virgin Material in Bulk ($/ton)	Expected Value as Reclaimed from Trash ($/ton of material)	Value Potentially Obtainable ($/ton of trash)
Inorganic			
Scrap iron	20–40	15	1.20
Tin cans	20–40	17	0.90
Aluminum	540	200	2.00
Other metals			
(Cu, Zn)	—	380	1.00
Glass, ceramics	25	20	2.00
Organic			
Paper	90–125	18	5.00
Protein	—	230	2.30
Plastics	300	30¢/million Btu as fuel (none as plastic)	0.10
Other alternatives		Total	14.50
Compost	—	7.50	3.75
Fuel	—	30¢/million Btu	3.00

Source: Adapted from Robert R. Grinstead, "Bottlenecks," Environment 14 (April 1972), 2–13.

paper, which has a value of only $5 per ton of trash. The more valuable components are too dilute to afford major economic interest in reclamation on an individual basis. Realizing an economic benefit from recycling and reutilization of urban trash can apparently be achieved by large-scale integrated schemes designed to separate and process for recycling all the various materials present.

One of the most promising developments is that of the Black-Clawson Company's Hydraspoil/Fibreclaim™ process, which has been demonstrated in a pilot plant at Franklin, Ohio. The plant first pulps the input wastes with water in a giant machine that operates much like a kitchen blender. The wood and paper fibers float in the froth from which they are separated for resale to a fiberboard manufacturer (Figure 7.14). Glass is separated from the other wastes, crushed, and separated according to color by air jets directed by color-detecting light beams. Iron is removed magnetically, and the other metals are separated by a variety of mechanical and magnetic techniques.

Figure 7.14. Cellulose fiber that has been recovered from municipal trash. [Courtesy of Black-Clawson Company.]

Organic and food wastes are mixed with sewage sludge and incinerated. The plant can process 150 tons of refuse per day (the wastes from a city of 60,000) to yield 30 tons of paper fiber, 12 tons of glass, 12 tons of iron, and 2.5 tons of aluminum. Costs of running the plant were about $11/ton in the early 1970s. The company estimates that a 1000 ton/day plant costing $12 to $15 million would produce 370 tons of industrial materials per day worth $8600, at a daily cost of $2900.

The U. S. Bureau of Mines has described a process for separating incinerated wastes that sorts out iron, nonferrous metals, and glass (sorted as colored or clear) at a cost of $4 per ton to yield products worth $15 per ton.

In Baltimore a 1000 ton/day plant is being constructed under an EPA demonstration grant. Solid waste will be converted into fuel by pyrolysis, steam generated with the waste heat will be sold, and the glass and metals from the solid residue will be separated and sold.

Policies Affecting Resource Recovery

One important economic limitation on resource recovery from solid wastes is the disparate transportation costs which make ore shipment much cheaper than the shipment of secondary (recycled) materials. Another limitation is the depletion allowance given to encourage the utilization of virgin raw materials. Both of these were set by congressional action and both can be changed by Congress to reduce the emphasis on raw materials use; both changes, however, would be accompanied by other changes in addition to the anticipated shift toward the use of recycled materials instead of raw materials.

To encourage resource recovery and develop stable markets for recycled materials, it is essential for several large-scale, integrated recovery systems to begin operation. This would increase the level of confidence in the recycled materials market and stimulate a manufacture-use-recycle approach rather than the present manufacture-use-discard approach. These first steps will probably require stimulation in the form of government contracts and grants. Unhappily the budget for solid waste disposal research has been decreased rather drastically for the mid-1970s compared to the relatively high 1973 level.

It has often been suggested that the waste-sorting problem would be simplified if the householder separated trash into metals, glass, paper, and garbage. Such a policy, however, would require either altering the design of most garbage trucks or increasing the number of garbage collection trips.

Oregon and Vermont have taken legislative steps to reduce the use of nonreturnable containers. These actions have been taken in part to reduce the volume of solid waste, the costs of disposal, and the volume of unsightly litter, a particularly galling problem in states such as these which have magnificent scenery. Both states passed the laws over the vigorous opposition of the beverage and container industry and many retail organizations.

The Oregon law, which became effective in October 1972, outlaws "pull tab" beverage cans and requires a 5¢ deposit on most beverage containers, with a lower, 2¢, deposit for containers certified by the state as usable by several manufacturers. The lower rate is to encourage the uniform design of bottles, a development that would facilitate sorting and reuse. Deposits as such are not required, but are assumed to be added to the cost of the product. Before the law, 50% of the soft drink and beer containers in Oregon were returnable bottles; now 95% of the containers are returnable. Cans now account for less than 1% of beer and soft drink sales. Beverage container litter along Oregon highways has decreased 81% since the new law went into effect. Soft drink prices remained about the same; beer prices increased about 15%, due partly to inflation and partly to the law. Total beer and soft drink sales appeared unaffected by the legislation. Can manufacturers were the hardest hit economically, but overall the employment in the beverage industries was expected to increase under the law.

Vermont's law, which took effect in September 1973, requires a deposit of 5¢ on all beer, soft drink, and soda water containers.

Resource utilization could be improved by changes in our approach to the lifetime of goods. By extending the effective lifetimes of automobiles, appliances, and machinery, we would reduce the demand on raw materials.

Figure 7.15. In 1972 the chemical elements comprising the human body could be purchased for $3.50. [Drawing by S. Harris. Reproduced by permission.]

> **What's Along the Highway?**
> 59% of the items are paper, 16% are metal, 6% are plastics, 6% are glass and 13% are of miscellaneous composition.

Summary   Solid waste is our only growing resource, but our traditional approach has been to dispose of the wastes as expeditiously as possible. Ocean dumping, open dumps, and sanitary landfills, which account for over 90% of the waste disposal, do not encourage resource recovery or utilization.

Methods that increase the emphasis on recource recovery do exist. Inciner-ation, the combustion of wastes, can be used to reduce waste volume while simplifying utilization of the noncombustibles and providing electrical power generation in the process. Air pollution control is a special problem with incineration, however. Pyrolysis, heating wastes in the absence of air, converts the wastes into low-quality oils and solids and may be especially applicable to tires and plastics.

Composting converts organic materials into soil humus. While done by many householders and by many European cities, composting has never been widely successful on a commercial scale in the United States because of oper-ational problems. Aerobic decay and proper carbon-to-nitrogen ratios are necessary to ensure rapid, odor-free composting. Nitrogen robbing can cause poor crops the first year after compost humus is added to the soil.

Recycling and other resource recovery techniques offer the advantages of longer resource lifetimes, reduced pollution, and, in most cases, reduced en-ergy consumption. Tax incentives and cheap freight rates instituted years ago to encourage the use of raw materials create an economic barrier to the development of recycling. About 18% of the paper in the United States is recycled; most of this paper comes from relatively clean sources such as print-ers. Technical developments since World War II have in general improved the efficiency of using virgin materials to make paper but have tended to decrease the ease with which paper can be recycled.

Glass can be remelted and recycled easily, but it usually must be sorted by color before remelting, and this adds to the expenses of recycling. The abun-dance and low price of sand makes glass recycling a borderline economic operation unless volunteer labor is used.

About 30% to 50% of the common metals are recycled. The tin present as a thin rust inhibitor in "tin" cans (which are made principally of iron) hampers their recycling because it weakens the steel; the small amount of tin present cannot be easily removed. Copper and other metal "impurities" present in small amounts in automobile bodies hamper in a similar manner the recycling of junked automobiles. Bulk auto bodies can form only 15% of the mix in a blast furnace. Using an auto body shredder, however, allows relatively pure iron to be separated from the rest of the metals; this procedure allows almost 100% recycling of the iron in auto bodies. Manufacturing using recycled metals rather than raw ores affords a 30- to 50-fold reduction in the energy required in the case of aluminum, and a 5- to 10-fold reduction in the case of other metals.

A systems analysis shows that the energy required to manufacture an automobile is 35 times greater than the minimum energy necessary, as calculated from thermodynamics; this energy requirement indicates that the potential for efficiency improvement in the manufacture of automobiles is very large. Most of the energy demand comes from the manufacture of the metals, and by manufacturing automobiles completely from scrap auto bodies the energy cost per automobile can be reduced by 30%. By extending the life of the automobile, the energy requirement per vehicle is reduced even more.

Returnable bottles are the most energetically efficient form of beverage container; nonreturnable bottles cost more than three times as much energy per gallon. Recycling of broken glass increases slightly the energy cost per gallon of the containers; consequently, if only energy is considered, glass recycling cannot be justified.

Many of the economic limitations of recycling can be overcome if mass recource recovery of urban trash is undertaken. The Black-Clawson process, which separates fibers, metals, and glass, then incinerates the garbage with sewage sludge, appears profitable when done on a large scale.

Administrative and legislative policies can have a significant impact on solid waste problems. An Oregon law banning "pull tab" cans and placing a deposit on all nonreturnable beverage containers resulted in an 81% decrease in roadside litter with little effect upon beverage sales.

Review and Study Questions

1. What is the rank of urban solid waste in a list of wastes classified by annual tonnage in the United States?
2. What are the two largest components (by weight percentage) in collected municipal trash?
3. What factors should be considered in establishing ocean dumping sites?
4. Describe the operation of a sanitary landfill from the initial dumpings until the land area is filled and the operation is closed.
5. What are some disadvantages of incineration?
6. What substances can be composted?
7. Why is aerobic decay preferred in a compost pile?
8. How can nitrogen robbing caused by the addition of fresh compost humus to the soil be avoided?

9. What technological changes have made the recycling process for paper more difficult?

10. Why can't waste plastics be melted then remolded into new plastic products?

11. Why are metals recycled a greater percentage of the time than other materials?

12. Why is the shredding of junked cars superior to merely compressing them for reuse?

13. For most metals, how much energy can be saved by using recycled metal instead of ore as the starting product for metals manufacture?

14. What system affords the maximum energy savings in the manufacture and recycle approach to automobile construction?

15. What social factors contributed to the development of the nonreturnable bottle?

16. Rank the following in terms of increasing energy costs per gallon of beverage consumed: nonreturnable bottles, 30% recycled; cans (bimetallic); nonreturnable bottles, no recycle.

17. In which container are soft drinks cheaper per gallon consumed: nonreturnable bottles or returnable bottles?

Questions Requiring Outside Reading

18. What are the *net* costs of solid waste disposal if electrical power is generated in the operation? (See W. C. Kasper and R. R. Grinstead, Suggested Outside Reading.)

19. What technical limitations exists in trash separation by mechanical devices? (See R. R. Grinstead, Suggested Outside Reading.)

20. *Environment* was one of the first magazines to be printed almost exclusively on recycled paper. What problems were they warned to expect, and what problems actually materialized? (*Environment*, September 1971.)

21. What is the possibility of toxic metals in municipal trash being transferred to and taken up by plants if trash is composted? (See J. G. Meyer, Suggested Outside Reading.)

22. Animal manure can be converted into an oil suitable for fuel. How is this accomplished, what is the potential fuel value? (*Chemical and Engineering News*, November 17, 1969, 43; August 16, 1971, 43; May 29, 1072, 14.)

23. What problems have volunteer recycling programs caused for professional salvage companies? ("Recycling Backlash," *Wall Street Journal*, May 9, 1973, 1.)

24. How can the thermodynamic potential (free energy) be translated into policy? (See R. S. Berry and M. Fels, Suggested Outside Reading.)

25. Representatives of what industries formed and are active in the "Keep

America Beautiful" organization, which promotes antilitter campaigns? (See B. Hannon, Suggested Outside Reading.)

26. Which industries supported and which opposed the 1972 Oregon law that put a deposit on all returnables and outlawed pull tab cans? (J. McCaull, "Back to Glass," *Environment* 16, [January/February 1974], 6–11.

Suggested Outside Reading

BASCOM, WILLARD. "The Disposal of Waste in the Ocean," *Scientific American* **231** (August 1974), 16–25. Discusses ocean disposal and dumping. Recommendations are made for further study.

BERRY, R. S., and M. FELS. "The Energy Cost of Automobiles," *Science and Public Affairs (Bulletin of the Atomic Scientists)* (December 1973), 11. An excellent analysis of the free energy cost of automobile manufacture, the savings possible from scrap recycling, and the policies that might achieve those savings.

Chemical and Engineering News. "Cryogenic Process Recycles Tires," January 17, 1974, 21–22; "Economics are Bugaboo in Scrap Tire Recycling," August 14, 1972, 8–11. General articles on industrial applications of recycling, reusing, or recovering components of tires.

Cleaning Our Environment. The Chemical Basis for Action. The American Chemical Society, Washington, D. C., 1969. Includes a general discussion of the problems and successes in solid waste disposal and resource utilization techniques. The emphasis is on chemical aspects of these topics, but the approach is quite broad.

GOTAAS, H. B. *Composting.* World Health Organization, Geneva, 1956. A detailed look at the chemical, physical, and biological aspects of composting, especially as applied in underdeveloped nations.

GRINSTEAD, ROBERT R. "Bottlenecks," *Environment* 14 (April 1972), 2–13; "Machinery for Trash Mining," *Environment* 14 (May 1972), 34–42. A two-part series discussing general aspects of resource recovery. A very useful comparison of costs, credits, and net expenses for various resource recovery approaches is included.

HANNON, BRUCE M. "Bottles, Cans, Energy." *Environment* 14 (March 1972), 11–21. An analysis of the energy requirements of beverage containers. A brief history of the development of the nonreturnable beverage container is included, and the economics of returnable vs. nonreturnable containers is discussed.

KASPER, W. C. "Power from Trash," *Environment* 16 (March 1974), 34–38. A comparison of four solid waste refuse processing systems in terms of relative advantages and net costs.

MEYER, J. C. "Renewing the Soil," *Environment* 14 (March 1972), 22. An article on composting, its cost, its problems, its benefits and its potential. Commercial applications are a special focus.

WILCOX, DENNIS. "Fuel from City Trash," *Environment* 15 (September 1973), 36. A description of the trash processing power generation project of the Union Electric Co. and the City of St. Louis. Problems and costs are covered in some detail.

Pesticides

We have waged all-out chemical warfare against insects, and now we find the object of our attack stronger than ever. . . . Perhaps with the new ecological concern, people would be willing to eat some blemished apples or oranges or green vegetables with a few holes in the leaves if they knew that doing so would mean lower doses of insecticides released into the environment.

> F. R. Lawson, Former Head of the
> Biological Controls Laboratory,
> U.S. Department of Agriculture
> (*Environment*, May 1971)

It is becoming increasingly apparent that the benefits of using pesticides must be considered in the context of the present and potential risks of pesticide usage.

> Mrak Commission Report on Pesticides, 1969

Introduction

Man's existence involves competition with some organisms and dependence on many other organisms. Those organisms that compete with man for such necessities as food or such aesthetic pleasures as roses are termed pests. Some insects, rodents, weeds, worms, and fungi are included in this somewhat arbitrary classification. Less than 1% of all the other organisms of earth are classified as pests; the rest are beneficial, often essential, to man's success.

Agriculture involves special competition with certain insects. The recent history of agriculture is strongly influenced by man's efforts to overcome the effects of insect pests, which comprise about 3% of the total insect population. For centuries man had only his hands and the help of his domestic animals to combat pests: filling trenches with water to prevent pest migrations, plucking insects from plants, killing them with logs pulled behind horses, pulling weeds. As agriculture progressed, a variety of other techniques were used to prevent or minimize pest damage: timing planting and harvesting so that the growing period coincided as little as possible with periods when pests were abundant; removing crop residues and alternate host plants; rotating crops; using quarantines.

In recent years, especially, man has forsaken most of these and relied exclusively on the use of chemicals. Any chemical that causes the death of pests can be called a pesticide. Chemical pesticide use, as we shall see, is a mixed blessing, sometimes creating more problems than it solves.

371

For all his efforts, man has had little overall effect on insect populations; man's total impact is still much less than the control imposed by other insects. Unless held strongly in check, almost any insect species has the reproductive capacity to multiply so rapidly as to destroy the world, by physically covering it if nothing else. Other insects, the "beneficials," feed on the pests or compete with them for food or for habitat.

In a field with a diversity of plants, there will be a considerable diversity of insects, perhaps several hundred species. Maybe half of these are **phytophagus** insects; that is, they eat plant material. The others are **entomophagus,** meaning they eat other insects. For the farmer, the entomophagus insects are the beneficials, for they control the populations of the plant eaters, the pests. In the absence of disturbances the two groups exist in equilibrium. The diversity means that it is unlikely that any one phytophagus insect will be able to erupt to pest proportions; the limited amount of food and the abundance of predators work to maintain a natural control.

Because they are seldom natural plants, man's crops are especially vulnerable to attack by insects. Wild plants, on the other hand, exist in a balanced insect and plant community and have "earned their niche" (and ensured their continued survival) by fulfilling some essential role in the ecosystem, perhaps by providing a haven for beneficial insects. Man's crops are, to a large extent, intruders into this balanced system. An important factor that has made man's crops even more vulnerable to attack is the continuing trend toward monocultural farming: extensive planting (hundreds of acres) of a single crop. Under such conditions, phytophagus pests can flourish because of ample food; their control by other insects is limited by distance and by a lack of diversified plants.

Since World War II we have repeatedly attempted to use chemical pesticides to reduce the population of the pests. One danger inherent in this approach is that, if the pesticide has a greater effect on the beneficials than on the pests, the pests may reestablish themselves in even greater abundance. Many such unfortunate episodes have been recorded in the history of chemical pesticides. These problems are amplified if excessive amounts of pesticide are used, a procedure followed all too often in the past.

In looking at pesticides we must keep in mind that the goal of pest control should not be pest eradication but rather the reduction of the pest population to tolerable levels. Such a level is one where pest damage is not sufficient to justify the expense of further control efforts. Trying to reduce the populations below this level involves needless expense and environmental damage. No pest has, in fact, ever been totally eradicated with a chemical insecticide, and there is ample reason (discussed later) to assume it cannot happen. Even if possible, it would probably be a mistake to eradicate completely insect pests that are *native* to the area. (A justification for eradicating imported pests may be easier to find.) All organisms, including those we call pests, have important roles in a balanced ecosystem; their function may be food for other species, pollination of plants, assistance in the decay process, or some function of which we are as yet unaware.

Chemical Pesticides   The use of chemicals to control pests dates to the ancient Greeks, who used sulfur against insects and salt (sodium chloride) against weeds. Marco Polo brought pyrethrum, a naturally occurring pesticide isolated from plants, to Europe from the Far East.

A century ago farmers began using a variety of other naturally occurring organic compounds such as nicotine sulfate and creosote, as well as inorganic poisons such as Paris green (an arsenic compound used against the potato beetle), hydrogen cyanide, lead arsenate, copper sulfate, and a variety of mercury compounds. These chemical pesticides are referred to as "first generation pesticides." They were not especially effective, and the metal containing inorganic compounds in particular had several serious liabilities. Their poisonous nature represented a direct hazard to animals and man. Furthermore the metals were persistent and could undergo biological magnification in food chains.

In 1939 the discovery by a Swiss named Paul Mueller that the compound dichlorodiphenyltrichloroethane (mercifully called DDT) killed insects. This caused a revolution in pest control: the use of synthetic (man-made) chemical pesticides, many of which did not occur in nature. DDT was cheap, relatively nonpoisonous to mammals, and astonishingly more effective against almost all insects than any previously known control method. DDT had dramatic early success in World War II. It was used to kill mosquitoes on tropical islands and consequently saved millions from death due to malaria and yellow fever. It forestalled a threatened typhus epidemic in the Allied Army in Italy. And by increasing crop production, it eased the threat of starvation for many. During the war other synthetic pesticides were developed, many as an offshoot of German research on nerve gas. After the war enthusiastic supporters of DDT and other synthetic "second generation" pesticides predicted the complete destruction of all insect pests; Mueller received the Nobel Prize in 1948 for his work.

It is not surprising that few questioned the rapidly increasing use of pesticides in the 1940s and 1950s, the "Golden Age of Pesticides," even when their use was ineffective, as for example, when applied in a futile effort to control Dutch elm disease. But by the end of the 1950s, troubles began to appear. Insect resistance was widespread, and it became evident that DDT and related chemicals killed much more than insects. In 1962 biologist Rachel Carson published a best-selling book, *Silent Spring*, which detailed the harmful aspects of the pesticides. Her book began a period of environmental concern, and it received both lavish praise and vicious criticism. Ten years later, the side

bird pop. affected - pop. decline ; eggs didn't hatch

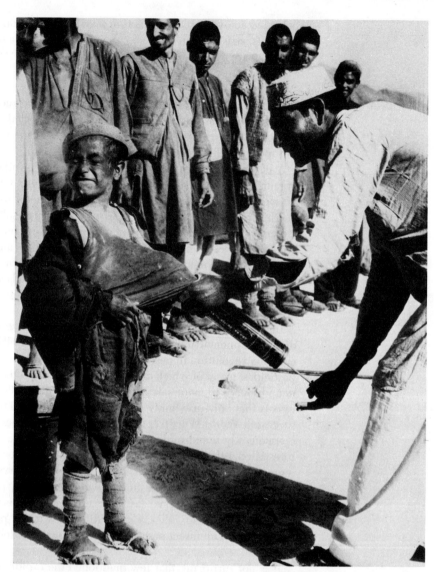

Figure 8.1. This young nomad boy is being dusted with DDT and talcum powder at a "delousing" station of a 1962 typhus control program in Pakistan. [UPI Photo, reprinted with permission.]

effects of DDT had caused enough harm and concern so that its use in the United States was banned beginning in 1973.

In this chapter we will study the environmental impact of pesticides and consider some alternatives to their use.

TABLE 8.1.   LD$_{50}$ values of some common drugs and chemical pesticides (oral doses in rats)

Compound	LD$_{50}$ (mg/kg)	Compound	LD$_{50}$ (mg/kg)
Hydrogen cyanide	4	2,4,5-T	100–300
Strychnine	5	copper sulfate	300
Arsenic acid	8	aspirin	1750
Morphine	1–50	methyl alcohol	500–5000
Sodium arsenite	10–50	ethyl alcohol	5000–15000
DDT	87–500		

Toxicity

Nearly all chemical pesticides are poisons, and an important characteristic is their toxicity. Toxicologists tabulate the strength of a poison by testing it on animals, then extrapolating the results to humans on the basis of body weight. Usually the toxicity is measured in terms of the dosage that kills 50% of the test animals; such a dosage, usually measured in milligrams of poison per kilogram of body weight, is called LD$_{50}$ (for *l*ethal *d*ose, *50%* die). It is necessary to test the poison for its activity if ingested, if brushed on the skin, or if inhaled. The LD$_{50}$ of some common drugs and pesticides is shown in Table 8.1

Toxicity in a pesticide measures its potential to kill organisms (usually test animals) upon contact or upon ingestion; the death must result from the immediate poisonous effects of the chemical. The chemical's action as a pesticide may or may not be due to the same mechanism that makes it toxic. Although toxicity is an important characteristic of pesticides, especially as regards their handling and application, they have other characteristics and side effects that are considerably more serious environmental concerns.

Characteristics of an Ideal Chemical Pesticide

From the standpoint of providing long-term, effective pest control with a minimum of environmental disruption, an ideal pesticide should possess two important characteristics. It should be

1. specific, affecting only the target pest; and
2. nonpersistent—that is, it should degrade in the environment in a few days or weeks.

In addition to these necessary properties, some helpful though not essential characteristics are low cost, low toxicity, and ease of use.

LD 50 (handwritten annotation)

Specificity. As noted earlier, the principal limitation on the population of an insect species is the control exerted by other insects, the beneficials. The best control will usually be achieved if only the pest is affected by the pesticide. Should the pesticide cause more damage to the beneficials than to the pests, the pest population may actually increase in the next generation. Many insect pests reproduce more rapidly than their predators, and therefore can adapt more rapidly.

None of the synthetic chemical second generation pesticides are specific; in fact they are broad-spectrum pesticides that affect scores of insect species as well as many other organisms. An additional problem created by a lack of specificity is that populations of certain species of insects can suddenly be elevated to pest status. For example, in Nova Scotia in the early 1930s there were devastating outbreaks of several apple pests, notably oystershell scale and the European red mite. It was found that normal predators and parasites of these insects had been almost exterminated as a side effect of the then new sulfur fungicides. Switching to other fungicides solved that problem. But in the 1940s and 1950s mite populations exploded wherever broad-spectrum chemical pesticides were used. Such problems led Nova Scotia's scientists to deemphasize the use of broad-spectrum pesticides where possible, developing natural control schemes instead. *central nervous system affected* (handwritten annotation)

Nonpersistence. Persistence measures how long chemical pesticide molecules remain in the environment before they are decomposed (degraded) or chemically or biologically deactivated. Degradation or deactivation may come from the action of water, constituents in the soil, sunlight, or of some biological organism such as bacteria. Persistence is usually measured as the time required for 90% of the molecules to be degraded or deactivated. The shorter the persistence, the greater the biodegradability of the chemical. Because the chemical molecules of a pesticide are mobile, capable of being transported by wind and water as well as by organisms, the mere fact that a pesticide cannot be found in soil where it was sprayed earlier does not necessarily imply that it has been degraded; it might have merely dispersed.

An ideal pesticide should be nonpersistent; that is, it should lose its activity in a few days or weeks. The persistence of pesticides varies widely. Some are nonpersistent; some, such as DDT, are persistent (lasting several years or decades); and some, such as metal compounds, are permanent. Persistence offers the possibility of accumulation and magnification in food chains (see Chapter 6), especially if the compound is much more soluble in fatty tissue than in water, as is DDT, for example. Food chain accumulation raises the possibility of long-term impacts, such as changes in reproductive and genetic systems. Extended persistence is also a contributor to another problem: the increase in insect resistance to the pesticide. Insects can adapt to changing conditions rapidly because of their rapid reproduction rate. Repeated repro-

non - gone within a week (handwritten annotation)
persistence - effects last days, years, forever (handwritten annotation)

duction in the presence of a slowly decreasing concentration of a persistent pesticide gives a genetic selection advantage to those mutant strains that are naturally resistant to the pesticide. Ultimately such strains become the dominant members of the species. The number of insect species immune to chemical pesticides has increased from 12 species in 1948 to over 200 by 1970. Persistence is not the only contributor to this development; the use of excessive amounts of pesticides also accelerates the growth of immune strains.

When insect resistance to a certain pesticide occurs, a different (usually more toxic) pesticide is usually used. Insects can develop an immunity to this second pesticide as well, or in fact, to several pesticides.

How Not to Do the Job. Peru has several coastal valleys that are isolated agricultural systems surrounded inland by deserts. One of these was sprayed in the late 1940s with DDT and two somewhat similar chemical pesticides, toxaphene and BHC. The initial success became a delayed disaster. After 4 years, cotton production had risen from 440 to 650 lb/acre, but insect resistance began to develop. By 1952, BHC could not control the cotton aphid. By 1954 toxaphene could not control leafworms. The fifth year the cotton yield dropped to only 350 lb/acre, nearly 100 lb less than before DDT was used. The 1955–1956 agricultural yields in the valley ranked with the lowest ever recorded. The pesticides had destroyed the beneficials and birds; the pests had developed resistance and were thriving.

The pest control policy was changed. Less-persistent pesticides were used, and their doses were kept small. Farming practices were altered. Restoration of natural controls in a few years improved production to new high levels.

Classification of Pesticides

A classification of chemical pesticides by similarities of chemical structure is given in Table 8.2. Three major classifications comprise most of the pesticides: organophosphates, carbamates, and chlorinated hydrocarbons.

Organophosphates

As a group, these pesticides with molecules containing phosphorus are relatively nonpersistent. The toxicity in this group varies widely. Some of the organophosphates are extremely toxic. The compound disyston is so toxic that children playing near recently sprayed gardens have been so severely poisoned that they were institutionalized for life. As a group, the organophosphates are in general considered more toxic than the other classifications and their use presents significant hazards for agricultural workers. Some specific examples of organophosphate chemical structures are given in Figure 8.2a.

The pesticidal action of the organophosphates lies in their ability to act as nerve toxins. A substance known as acetylcholine is responsible for the trans-

always contained phospherous
non-persistent

TABLE 8.2. Classification and properties of synthetic chemical pesticides

Classification	LD_{50} oral, rats (mg/kg)	Use (insecticide unless otherwise noted)	Relative Cost of Class	Relative Persistence of Class
Organophosphates				
Diazinon	66–600	mites		
Malathion	885–2800			
Parathion	3–30	mites	expensive	short
TEPP	0.5–2			
Carbamates				
Sevin (carbaryl)	307–986			
Baygon	95–175		expensive	short
Chlorinated Hydrocarbons				
Aldrin	39–60			
Benzenehexachloride (BHC)	600–1250			
Chlordane	283–590			
Dieldrin	40–100			
DDD	400–3400		cheap	long
DDT	87–500			
Endrin	3–45			
Lindane	76–200			
Mirex	235–702			
Toxaphene	40–283			
2,4-D	666	herbicide		
2,4,5-T	300	herbicide		

Source: Bulletin of Entomological Society of America **15** (1969), 85–135.

mission of nerve impulses from one nerve cell to another. After serving its function, the acetylcholine must be destroyed in order to enable the next nerve impulse to flow. The destruction is accomplished by the enzyme, acetylcholinesterase (ACHE). (Enzymes are proteins that regulate the rate of the bodily functions.) Organophosphate molecules inhibit the ability of ACHE to function. As a result, acetylcholine accumulates and activates a stream of unwanted nerve impulses, leading to death. The nature of their action ensures that the organophosphates are broad-spectrum pesticides.

Carbamates

Carbamates are best typified by the compound carbaryl, or sevin. The persistence of carbamates is similar to that of the organophosphates, and they are more toxic than the chlorinated hydrocarbons. Inhibition of ACHE in a man-

Common name	Structure	Uses and comments
Parathion	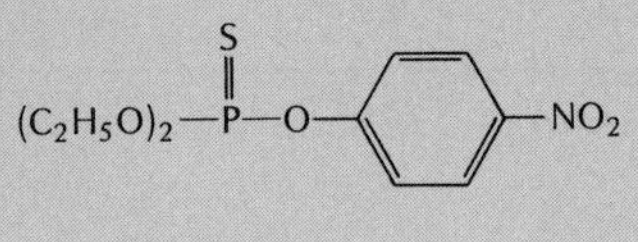 $(C_2H_5O)_2$—P—O—⟨benzene⟩—NO_2 (with S double-bonded to P)	*Most widely used of this class *Garlic odor *High toxicity to warm blooded animals *Broad spectrum insecticide *Originally used on potato beetle
Methyl parathion	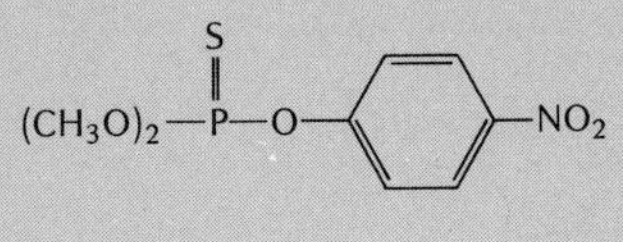 $(CH_3O)_2$—P—O—⟨benzene⟩—NO_2 (with S double-bonded to P)	*Less stable in storage *Shorter residue time *More toxic against aphids and beetles than parathion *Less toxic than parathion to warm blooded animals
Malathion	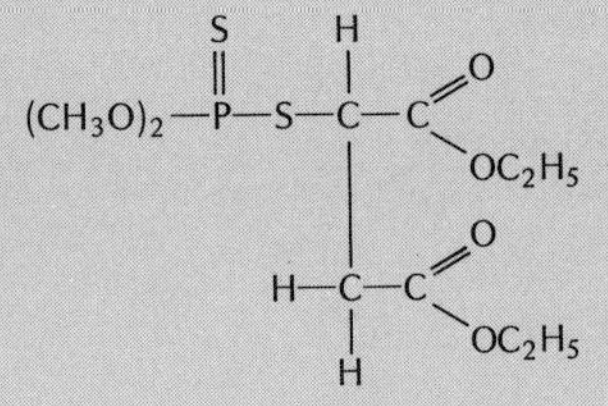 $(CH_3O)_2$—P—S—C—C (S double-bonded to P; H on first C; O and OC_2H_5 on C) H—C—C (O and OC_2H_5 on C; H below)	*Household, home garden, vegetable and fruit insect control *Control of mosquitoes, flies, lice
TEPP	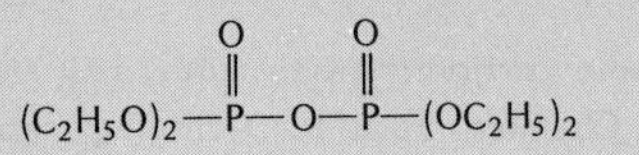 $(C_2H_5O)_2$—P—O—P—$(OC_2H_5)_2$ (O double-bonded to each P)	*Particularly rapid breakdown *Can be used on edible crops before harvest *Very toxic
DDVP	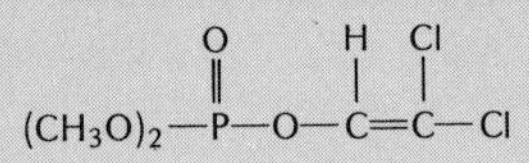 $(CH_3O)_2$—P—O—C=C—Cl (O double-bonded to P; H and Cl on the carbons)	*Baits and aerosols for rapid knockdown of flies, mosquitoes, moths *Fumigant for household pests *Active ingredient in "pest strips"

Figure 8.2a. *Some common organophosphate pesticides. All contain phosphorus (P).* [*All structures reprinted from* Cleaning Our Environment—The Chemical Basis for Action, *a report by the Subcommittee on Environmental Improvement, Committee on Chemistry and Public Affairs, American Chemical Society, 1969, pp. 199–200. Reprinted by permission of the copyright owner.*]

ner similar to that of the organophosphates accounts for carbamate biochemical activity. Some typical examples are given in Figure 8.2b.

Chlorinated Hydrocarbons

Chlorinated hydrocarbon insecticides are characterized by the presence of chlorine (Cl) substituted for hydrogen in certain hydrocarbons. Chlorine, like

persistent
largest group
cheap

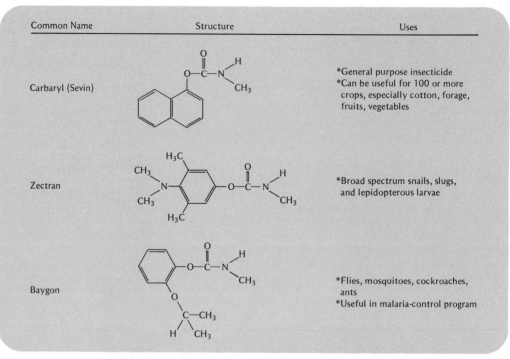

Common Name	Structure	Uses
Carbaryl (Sevin)		*General purpose insecticide *Can be useful for 100 or more crops, especially cotton, forage, fruits, vegetables
Zectran		*Broad spectrum snails, slugs, and lepidopterous larvae
Baygon		*Flies, mosquitoes, cockroaches, ants *Useful in malaria-control program

Figure 8.2b. *Some carbamate pesticides.* [*All structures reprinted from* Cleaning Our Environment: The Chemical Basis for Action, *a report by the Subcommittee on Environmental Improvement, Committee on Chemistry and Public Affairs, American Chemical Society, 1969, p. 200, reprinted with permission.*]

hydrogen, forms a single chemical bond. This class of pesticides includes DDT, 2,4,5-T, dieldrin, and aldrin—pesticides that have received limited bans on usage in the United States in recent years. Some structural examples are shown in Figure 8.3.

The chlorinated hydrocarbons do not occur in nature and consequently are not integrated into the biogeochemical cycles. Furthermore the carbon-chlorine bond is very strong; it does not react with water or components of the soil. As a result this class of pesticides is very persistent, lifetimes of 20 years or more in the environment having been demonstrated in careful tests. Toxicity to man is relatively low; death from DDT, for example, requires the ingestion of more than a pound at one time.

Chlorinated hydrocarbon insecticides act as central nerve toxins, inducing paralysis, convulsions, and, ultimately, death. They are nonselective broad-spectrum pesticides, capable of attacking any organism with a central nervous system.

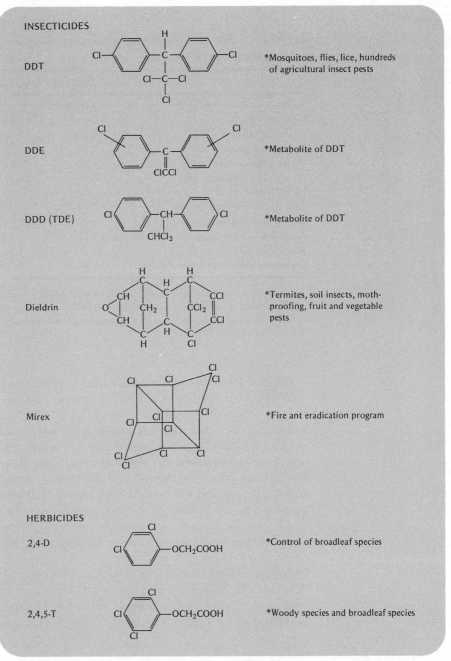

Figure 8.3. *Chlorinated hydrocarbon pesticides. Note that chlorine (Cl) is always present in the molecule. [All structures reprinted from Cleaning Our Environment: The Chemical Basis for Action, a report by the Subcommittee on Environmental Improvement, Committee on Chemistry and Public Affairs, American Chemical Society, 1969, p. 197–199, reprinted with permission.]*

TABLE 8.3. An example of food chain concentration of a persistent pesticide, DDT

	DDT Residues[a] (ppm)
Water	0.00005
Plankton	0.04
Silverside minnow	0.23
Sheephead minnow	0.94
Pickerel (predatory fish)	1.33
Needlefish (predatory fish)	2.07
Heron (feeds on small animals)	3.57
Tern (feeds on small animals)	3.91
Herring gull (scavenger)	6.00
Fish hawk (osprey) egg	13.80
Merganser (fish-eating duck)	22.80
Cormorant (feeds on larger fish)	26.40

[a] Parts per million (ppm) of total residues, DDT + DDD + DDE (all of which are toxic), on a wet weight, whole organism basis.

Source: Data from G. M. Woodwell, C. F. Wurster, and P. A. Isaacson, Science **156** (1967), 821–824.

Physical Properties of the Chlorinated Hydrocarbons. Chlorinated hydrocarbons have a number of physical properties that are important factors in their environmental impact. A dominant property is their persistence, which means that once used they will remain in the environment for decades.

As a group they have very limited solubility in water. DDT, for example, has one of the lowest solubilities ever measured, being almost completely insoluble. DDT and the other chlorinated hydrocarbons, however, do have a strong affinity for dust and soil particles and can be transmitted in this fashion by water and by air. In addition many of these compounds co-distill with water; that is, they pass into the air when the water evaporates.

The chlorinated hydrocarbons are, however, very soluble in fats, and this property combined with poor water solubility means that they can undergo biological magnification in food chains (see Chapter 6). An example is found in data on DDT concentrations in the food chain in an estuary on the eastern coast of the United States shown in Table 8.3. The species at the top of the chain have much higher levels; magnification of nearly 1000 times is common.

Environmental Impact of Chlorinated Hydrocarbons

Direct Kills

As with any poison, the use of chlorinated hydrocarbons results in immediate kills of many nonpest species that become exposed to a toxic concentration of the pesticide. For example, initial spraying of DDT in urban areas in the

1940s often caused a 90% mortality of robins. (This problem was compounded by the fact that earthworms accumulated substantial amounts of DDT in their body fat, thus providing an increased dietary intake for birds.) Fish mortality is also high when pesticides float onto water during spraying or are washed into rivers and lakes by rainfall.

The long-term (several years) effect of direct kills is usually relatively small assuming that the spraying is not universal, because in most cases species can repopulate the area.

Carcinogens

That most of the chlorinated hydrocarbons are carcinogens has been known for over 30 years. Their persistence and their accumulation in the food chain means that organisms will be exposed to these effects for decades. The long-term effect of this persistence and accumulation can only be speculated upon, but it is a cause for concern and has been a major factor in decisions to ban the use of some of the chlorinated hydrocarbons.

Enzyme Induction

One of the potentially most damaging aspects of chlorinated hydrocarbons is their ability to increase (induce) the action of certain enzymes and inhibit the action of other enzymes in living systems. In particular, many chlorinated hydrocarbons cause liver enzymes to metabolize steroid sex hormones such as estrogen (Chapter 2). This important biochemical property of these compounds was first noticed, almost by accident, in experiments on rats. The experiments were not related to pesticides, but they became a focus of study when it was noted that the spraying of rat cages with chlordane produced unusual results. This enzyme induction property of chlorinated hydrocarbons has manifested itself in reproductive problems in birds, mammals, and fish as discussed in the sections that follow.

DDT and Closely Related Insecticides

DDT is the most carefully studied pesticide, so it is instructive to consider this compound in detail. The widespread use of DDT for 30 years, its persistence, its physical properties, and its mobility have combined to create a truly ubiquitous pollutant. It is found in soils never sprayed, in birds and seals that never encounter continents, and in penguins in the Arctic where DDT was never used. It even comes down in the rain. DDT is passed upward into the food chain and is passed from generation to generation; when an organism dies, the DDT molecules are once more released into the environment. DDT does degrade slightly to form its metabolites, DDE and DDD (Figure 8.3). However, these metabolites are very stable. DDE in particular has many biological effects similar to those of DDT, and may in fact have caused some of the problems for which DDT is blamed.

As a firmly ensconced (at least for the next few decades) member of our ecosystem, it is suitable to ask, albeit somewhat tardily, what the effects of these compounds might be. The most serious effects that have appeared are those related to DDT's action on enzymes.

Reproductive Difficulties in Raptors

Beginning in the late 1940s many birds of prey—including the osprey, peregrine falcon, bald eagle, and brown pelican—underwent a sudden drop in population. The change was much too abrupt to have been a natural cycle. The peregrine falcon, for example, is known to have maintained a remarkably stable population for centuries; records of eyries that have been occupied more or less continuously by the peregrines go back in some cases to the Middle Ages. In Europe and North America, however, the peregrine populations began to decline beginning about 1950.

In all the affected species, the problem was not adult mortality but reproductive problems characterized by egg breakage and egg eating by parent birds, abnormally low numbers of young birds, abandonment of nests, and other obviously abnormal types of breeding behavior. An osprey colony in Connecticut that had 200 pairs of birds, averaging nearly 2.5 young birds per nest in 1938, had declined to 12 pairs, averaging slightly more than 1 young bird per nest in 1965. The brown pelican, the state bird of Louisiana, vanished from that state. A colony of 500 pairs of brown pelicans off the southern California coast produced exactly one bird in 1970.

The declines coincided geographically and in time with the introduction of chlorinated hydrocarbon insecticides. All of the species concerned existed near the top of the food chain and thus accumulated large amounts of DDT and DDE and other chlorinated hydrocarbons, many of which were present in the tissues and eggs of the birds. Thus chlorinated hydrocarbons were one of the suspected causes of the declines. But a suggested mechanism specifically relating these compounds to the declines did not become evident until the late 1960s.

In 1967 Derek Ratcliffe, a British ornithologist, examined preserved shells of the peregrine falcon. He found that the eggshell thickness and the calcium content were stable from 1900 to the 1940s. Then, beginning in the late 1940s, a sharp drop in eggshell thickness occurred, averaging 19% (Figure 8.4). Similar findings were subsequently found in the eggs of the other predatory birds whose populations were declining. (More than 34,000 collected and preserved eggshells were examined and measured to accumulate these results!) It became apparent that something must be wrong with the birds' calcium metabolism.

A surprisingly productive 1968 meeting in Rochester, New York, called to study the declining populations, finally assembled the evidence and outlined the mechanism responsible for the trouble. In the normal reproductive pro-

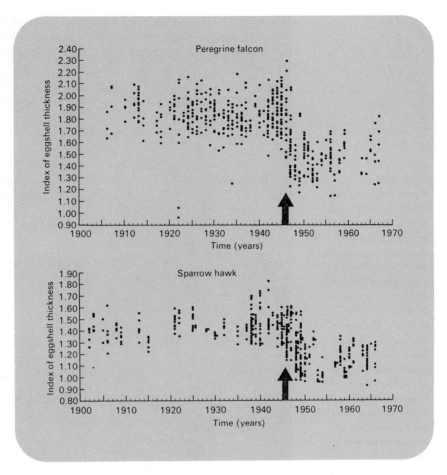

Figure 8.4. *Changes in thickness of eggshells of peregrine falcon and sparrowhawk in Britain. Arrows indicate first use of DDT. [From Population, Resources, Environment: Issues in Human Ecology, 2nd ed., by Paul R. Ehrlich and Anne H. Ehrlich, W. H. Freeman and Company. Copyright © 1972. Original data from D. Ratcliffe, Journal of Applied Ecology 7 (1970) pp. 67–115. Reprinted by permission.]*

cess, female birds increase the absorption of calcium from their diet, decrease their calcium excretion, and store excess calcium in their bone marrow. All of these processes are regulated by a steroid sex hormone, **estrogen.** In egg formation, the calcium is transferred from bone storage sites to the oviduct, where it becomes part of the eggshell. This transfer is controlled by the enzyme **carbonic anhydrase.** This crucial chain of events in the reproductive cycle can be interrupted by lowered estrogen levels and by inhibited carbonic anhydrase activity. Taking a cue from the recently discovered fact that chlorinated hydrocarbons caused enzyme induction in rats, the conferees focused on the possibility that enzyme induction could happen in birds. Later studies indicated that this reproductive problem had actually occurred.

The reproductive problems, it seems, are due in part to the fact that DDT increases the breakdown of estrogen by increasing the activity of certain

Figure 8.5. *Enzyme induction by DDT results in thin, fragile eggshells and few young birds. This eaglet and an egg (which never hatched) were photographed in April 1965 in a nest on the Muskegan River in Michigan. [Robert Harrington, Michigan Department of Conservation. Science,* **163,** *(February 7, 1969). Copyright 1969 by the American Association for the Advancement of Science. Used with permission.]*

enzymes in the liver. This alters the hormonal balance in the bird and changes reproductive behavior. Simultaneously the chlorinated hydrocarbon inhibits the function of the enzyme, carbonic anhydrase, which is required for normal calcium delivery. The result: thin-shelled eggs that break prematurely when the birds sit on them, and few, if any, young birds (Figure 8.5). The basic aspects of this mechanism have been verified in so many experiments that few scientists question that the effects are due to chlorinated hydrocarbons.

One of the first experiments was to see if chlorinated hydrocarbons could cause reproductive problems in birds that did not usually experience a large dietary intake of the compounds. Species low on the food chain, such as ringdoves, quail, and mallard ducks, subsist largely on grains and weed seeds and hence seldom ingest large amounts of pesticides. Experiments on these birds indicated that certain chlorinated hydrocarbons affect some species, whereas other chlorinated hydrocarbons are important to other species. For example, DDE decreased eggshell thickness in the ringdove, whereas dieldrin did not. In bobwhite quail 1 ppm of dieldrin in feed was effective in reducing hatching success, whereas it took 200 ppm of DDT to product the same effect. Groups of mallard ducks fed food containing 3 ppm DDE had shells 13% thinner; these were cracked or broken six times as often and produced less than half as many healthy chicks as did a control group of ducks whose diet had no DDE.

Another important aspect of the declining populations was that an artificial delay in breeding began to occur in many birds. Most birds do their breeding when food is most plentiful, thus giving the young an optimal chance for survival; the artificial delay reduced chances for reproductive success. From this point of view, it appears that dieldrin and some other chlorinated hydrocarbons are more of a threat to predatory birds than is DDT. Although eggshell thinning began in Golden Eagles in Scotland in 1952 when DDT was introduced, marked declines in breeding success did not occur until 1960 after the introduction of dieldrin. This delay is attributed to the lowered sex hormone levels caused by the enzyme induction action of the chlorinated hydrocarbons.

Reproductive Problems in Fish

DDT tends to concentrate in the egg yolks of fish and cause heavy mortality of fry. Widespread losses of as much as 50% of the salmon fry occurred in New Brunswick, Canada, for several years after DDT was sprayed in 1952. The Coho salmon, a top carnivore in Lake Michigan, experienced heavy fry mortality in the mid-1960s, a problem attributed to accumulated DDT residues in the egg yolks. In 1969 the Food and Drug Administration (FDA) banned the sale of certain shipments of Coho salmon because the tissues contained as much as 19 ppm of DDT, well above the allowed maximum of 5 ppm.

Other Problems

The action of the chlorinated hydrocarbons causes a variety of other effects in living organisms. In brook trout, nonlethal dosages of DDT result in decreased visual avoidance responses. In young salmon, temperature control mechanisms are upset by a few parts per billion of DDT. Oyster-shell growth has been reduced upon exposure to a few parts per billion of chlorinated hydrocarbons.

Reproductive problems observed in seals may be due to chlorinated hydrocarbons. Premature pupping in sea lions has been noted on the breeding islands off the Pacific and Gulf coasts of Baja California. In 1970 DDT and its metabolites, as well as polychlorinated biphenyls (next section) were eight times higher in the tissues of premature pupping females and their pups than in tissues of normal full-term females and their pups.

Studies have also reported that DDT reduces photosynthesis in phytoplankton, and that DDT in ducks may make them more susceptible to disease.

Polychlorinated Biphenyls (PCBs)

chlorinated hydrocarbons

man-made

Pesticides are not the only chlorinated hydrocarbons that are known causes of environmental disruption. Polychlorinated biphenyls (PCBs), a group of compounds in which chlorine replaces hydrogen in the biphenyl molecule, have become widely dispersed in the environment during the last 40 years. Figure 8.6 shows six chlorines substituted for hydrogens, but from one to ten chlorines may be present on any molecule. Commercial PCBs usually include a mixture of many types of chlorine-substituted biphenyl molecules.

PCBs are persistent compounds that exhibit many of the physical properties of chlorinated hydrocarbon pesticides. Because PCBs are colorless, fire-resistant fluids that are especially stable at high temperatures, they are used as heat transfer fluids in large electrical transformers and in heating and cooling systems. They have also been used, or have been present as accidental contaminants, in marine paints (PCBs serve as antifouling agents), lacquers, paper, plastics, and inks. Because of their chemical similarity to pesticides, PCBs are often created during the manufacture of pesticides and remain as accidental contaminants.

PCBs get into the environment through accidental leaks, through waste disposal, and through the weathering of materials containing them. The Monsanto Company, the sole manufacturer of PCBs in the United States, made more than 350,000 tons of these compounds in the 1960s, their output reaching a peak of 42,500 tons in 1970.

In the eastern United States, PCBs are about ten times more prevalent than DDT in the atmosphere, in the water, and in the fatty tissues of organisms. Levels of 3 ppm are often found in sediments near cities; mature trout in Cayuga Lake, near Ithaca, New York, have over 20 ppm. Ospreys may be more contaiminated than any other wildlife; near Long Island Sound osprey eggs contain PCB levels up to 2000 ppm. Even polar bears contain measurable amounts.

The effect of PCBs varies drastically with the particular molecular structure of the individual PCB and with the particular organism. Some studies indicate that PCB levels of 1 ppb may cause injury or death to many aquatic organisms, such as fish and crustaceans, whereas other studies indicate that the toxicity to fish is low but that invertebrates higher up the chain may suffer damage.

Experiments at the U.S. Department of Agriculture (USDA) indicated that adding 20 ppm of PCBs to hens' diets had no appreciable effects on the food consumed, on body weight, on mortality, on egg weight or on eggshell thickness; but the number of eggs *hatched* dropped drastically. Figure 8.6 shows

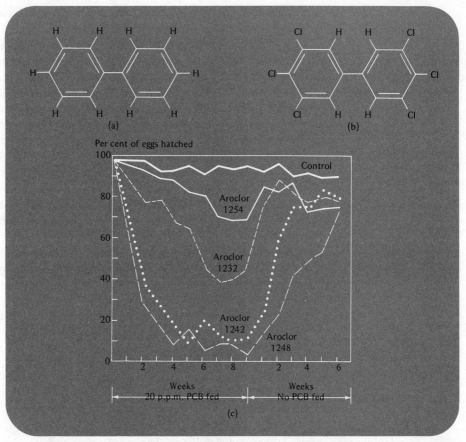

Figure 8.6. (a) The biphenyl molecule, an aromatic hydrocarbon. (b) A polychlorinated biphenyl is formed when chlorines replace hydrogens on the biphenyl molecule. (c) Egg hatchability in chickens declines as PCBs are added to the diet. Aroclor is the trade name for PCBs, and the numbers 1254, 1232, etc., refer to mixtures of molecules with differing numbers of chlorines substituted. [Source: U.S. Department of Agriculture in (c) illustration.]

that 9 weeks after the ingestion of PCBs began, the egg hatch dropped tenfold for certain types of PCBs. Perhaps more important is the fact that over one third of the dead embryos showed teratogenic abnormalities: small bodies, rotated ankles, shortened or crossed beaks. The chicks that did hatch appeared normal but did not grow as well as chicks from the control groups, which were not given PCBs. Removal of the PCBs from the feed brought a return to normal chick production. In 1971 accidental contamination of feed (fishmeal) with PCBs caused the FDA to confiscate over 100,000 chickens, several hundred thousand dozen eggs, and 50,000 lb of catfish because their PCB levels exceeded 5 ppm, the maximum level allowed in food.

High PCB levels are found in Great Lakes fish, especially the Coho salmon, a fact that is causing woe for mink ranchers who originally located in that area because of the availability of an abundant, cheap supply of fish. In the mid-1960s the ranchers noticed that mink reproduction was falling. The problem reached a peak in 1968 shortly after the Coho came onto the scene. Mink litters dropped from an average of four kits per litter to less than one-half kit per litter. At first the Coho itself was suspected, but in tests mink thrived on Coho imported from Oregon, whereas mink whose diets had included 30% Great Lakes Coho for 3 months whelped no kits at all. Controlled mink reproduction tests showed that DDT, dieldrin, and mercury added to the diets did not affect mink the way the Great Lakes fish did. But not even one kit survived when the mother's diet contained 5 ppm of PCBs. The Coho contained 15 ppm of PCBs, and the minks themselves contained 30 ppm after being on a diet of Great Lakes fishes. The mink ranchers must now feed ocean fish, poultry, or meat by-products, all of which are more expensive than Great Lakes fish. Also, the promising Great Lakes Coho salmon industry has been severely curtailed.

Although little can be done about present PCB levels in the food chain, newly instituted restrictions on the use of PCBs should result in a gradual decline in environmental contamination. The Monsanto Company voluntarily restricted sales in 1971 to those sources where the end uses of PCBs were known and could be controlled.

Herbicides and Defoliants

The chlorinated hydrocarbons 2,4,5-T and 2,4-D (Figure 8.3) are used as brush killers and defoliants. The persistence and environmental impact of these substances have received considerably less study than have the chlorinated hydrocarbon insecticides. Some evidence exists that these herbicides are less persistent than DDT, partly because the oxygen atom near the ring makes the aromatic ring more reactive and hence more likely to biodegrade.

Concern about the use of the herbicides arises from the ecological changes induced in the sprayed area and from their teratogenic nature. The widespread use of these defoliants by the United States in Vietnam and the reports of their effects, combined with the emotional impact of that conflict, have increased the attention given to the environmental effect of chlorinated hydrocarbons. A complication in assessing the effects was the presence in 2,4,5-T of a very highly toxic contaminant, dioxin or TCDD. Dioxin, almost inevitably created when 2,4,5-T is manufactured, kills embryos and causes birth defects (cleft palates) in guinea pigs. It is also teratogenic in mice, has an oral LD_{50} of only 0.0006 mg/kg of body weight in guinea pigs, and causes a skin disease, chloroacne, among workers in 2,4,5-T manufacturing plants. The presence of dioxin tended to cloud evidence of the impact of the defoliants. Subsequent experiments have established that 2,4,5-T and perhaps 2,4-D are also teratogenic.

In Vietnam the herbicides accounting for most of the spraying, Agent Orange and Agent White, both contained mixtures of 2,4-D and 2,4,5-T (Table 8.4). Agent Blue was used to destroy crops (rice) before harvesting.

More than one scientific panel has studied the safety of the herbicides themselves. Several other panels, including one appointed by the American Association for the Advancement of Science (1970) and one by the National Academy of Sciences (1974), studied the impact of herbicide use in Vietnam. Both of these Vietnam study panels agreed that widespread vegetation devastation had occurred as a result of the spraying. An area greater than the size of Connecticut was sprayed, and more than one third of the mangrove forests was destroyed. These forests were the major sources of wood and charcoal, the principal fuels used by the Vietnamese. The panels estimated that reforestation of the mangroves might take a century, if they in fact reestablished themselves at all over subsequent growth. About one fifth of the upland hardwood forests was destroyed and invasions of bamboo occurred. In the sprayed regions, the relatively open forest floor was replaced by dense undergrowth.

Both panels noted, albeit cautiously, that they could find no conclusive evidence that the defoliants caused an increase in birth defects in the population, although both stressed that reliable data were extremely limited and difficult to obtain. The National Academy of Sciences panel noted that stories of human illness among mountain tribesmen after sprayings were too prevalent to dismiss. Both reports caused heated public discussion. Charges of bias and a lack of objectivity on the part of the panelists, reminders that the herbicides were used to save lives, arguments that the data were too sparse to draw conclusions, and charges related to the emotionalism of the war itself were made by those on all sides of the issue.

The full extent of the environmental and medical damage of the herbicide use in Vietnam will not be known for many years. Should dioxin (TCDD) show any of the propensity for accumulating in food chains that chlorinated hydrocarbon pesticides exhibit, it will present a perhaps uniquely insidious threat to health.

The use of 2,4,5-T has been greatly restricted in the United States since 1971

TABLE 8.4.   Herbicide Application in Vietnam

Name	Defoliants	Amount Used (million gal)
Agent Orange	2,4-D and 2,4,5-T	11.3
Agent White	2,4-D and 2,4,5-T	5.3
Agent Blue	an arsenic compound	1.1

due to a decision by then EPA administrator William Ruckelshaus. Ruckelshaus' decision to prevent the use of 2,4,5-T on food crops was made despite a recommendation by an EPA advisory panel that no restrictions be made. In 1971 the military also discontinued use of the defoliants in Vietnam.

Risks to Man

Studies of the medical effects of chlorinated hydrocarbons on humans have been devoted largely to toxicity and to the possible creation of nervous disorders. Little data have been obtained on enzyme induction and inhibition, in part because of the difficulties inherent in enzyme research in humans.

DDT and its metabolites average about 10 ppm in human fatty tissue in the United States. They tend to be concentrated in human milk to a greater degree than in other mammals. The acceptable daily intake of DDT is presently 0.01 mg/day. An American adult usually consumes about one tenth of this, but a breast-fed infant receives about 0.02 mg/day. Although a cause for concern, the lack of relevant data about the effects of such dosages precludes recommendations concerning the curtailment of breast feeding for this reason.

Recent experiments on rats indicate that the combined effects of DDT ingestion plus a low protein diet create significantly greater problems (stunted growth, infant mortality, nervous disorders) than one would predict from merely adding together the individual effects of each dietary problem. If this is true for humans, the effects of pesticides on humans may first appear in the poorer nations, where low protein diets are common and where breast feeding is done until the child is 3 or 4 years of age.

One of the most complete compilations of the effects of pesticides on humans was done in 1969 by the Mrak Commission, a panel appointed by the Secretary of the Department of Health, Education, and Welfare. The Commission, headed by Emil Mrak, former chancellor of the University of California at Davis, studied all the relevant data on pesticides, received testimony from many experts, and reported their findings together with recommendations. The Commission was particularly concerned with the impact of the chlorinated hydrocarbons on humans, especially in view of the continuing use of the pesticides while their effects on humans were unknown. Some of the comments in the report are as follows:

> This phenomenon of enzyme induction has been extensively studied in animals and is discussed in detail in the report of the Panel on Interactions. Comparable enzyme induction in the human liver is brought about by many drugs and also by DDT. It is a sad comment on the dearth of knowledge of human physiology to point out that the threshold dose of DDT for induction of metabolizing enzymes in human liver is unknown . . . the field of pesticide toxicology exemplifies the absurdity of a situation in which 200 million Americans are undergoing lifelong exposure, yet our knowledge of what is happening to them is at best fragmentary and for the most part indirect and inferential.

The Mrak Commission made several recommendations, including

1. Eliminate within two years all uses of DDT and DDD in the United States excepting those uses essential to the preservation of human health or welfare and approved unanimously by the Secretaries of the Departments of Health, Education, and Welfare, Agriculture, and Interior; and
2. Restrict the usage of certain persistent pesticides in the United States to specific essential uses which create no known hazard to human health or to the quality of the environment and which are unanimously approved by the Secretaries of the Departments of Health, Education, and Welfare, Agriculture, and Interior. [The Commission specifically listed aldrin, dieldrin, endrin, heptachlor, chlordane, benzenehexachloride, lindane, and compounds containing arsenic, lead, or mercury.]

Legal Efforts to
Restrict the Use of
Chlorinated
Hydrocarbons

The DDT Trial

The direct impetus to ban DDT and other chlorinated hydrocarbons came not so much from the Mrak Commission's report as from the action of a citizen's organization called the Environmental Defense Fund (EDF). The EDF was founded to achieve environmental improvement in the courts by combining the best legal and scientific talent to pursue court cases. In 1968 this organization was asked to enter a lawsuit that was attempting to have DDT classified as a pollutant in Wisconsin and therefore subject to a ban under Wisconsin state law. The scientific coordinator of the EDF team was Dr. Charles F. Wurster, a chemist at the State University of New York at Stony Brook. A wide array of scientific talent, most of whom were volunteers, helped with the EDF's case. In opposition to the EDF was the Task Force for DDT, organized by the National Agricultural Chemicals Association; this group had the support of most agriculture-related businesses.

The case was heard before a federal examiner in Madison, Wisconsin, and the format included cross-examination by attorneys for each side, a procedure that gave witnesses an incentive to be well prepared and confident of the validity of their testimony—or not to testify at all. It also meant that the tactics of the lawyers were important, and clever strategies and histrionics abounded. The Madison case was remarkable for the great diversity of disciplines that played a role in its organization and its presentations. After nearly 6 months of testimony and 300 pages of transcript from 32 witnesses, DDT was ruled a pollutant and banned in Wisconsin. (About the same time, DDT was banned in Michigan, in part because of the troubles it caused in the Coho salmon.) A lengthy series of appeals followed, but they were overshadowed by efforts to ban DDT nationally.

The EPA, which had recently come into existence, was petitioned by the EDF to hold hearings on DDT. The EPA hearings lasted 7 months, heard 125 witnesses, and took 10,000 pages of testimony. The end result was a decision by EPA administrator William Ruckelshaus to ban virtually all uses of DDT in the United States beginning in 1973. DDT can be sold abroad and it can still be used in special emergency cases approved by the EPA Administrator; such a case occurred in 1974 when DDT was approved for use against the Tussock moth in forests in the northwestern United States.

Ruckelshaus' decision to ban DDT was contrary to the final recommendation of the EPA hearing examiner and his staff, and it was certainly not met with complete approval. Among the critics was Norman Borlaug (Chapter 2), who felt the decision was a prelude to banning all pesticides and who felt DDT was essential to crop production. For most United States crops, the DDT ban was of little practical consequence, since the development of resistance among insects had already precluded its use.

Aldrin and Dieldrin

These chlorinated hydrocarbon insecticides, aldrin and dieldrin are closely related. Aldrin slowly converts to dieldrin (Figure 8.3), which is one of the most persistent of all insecticides. Both are known from laboratory experiments to affect enzyme action, cause eggshell thinning, and kill fish. The EDF also petitioned the EPA to have hearings on aldrin and dieldrin; these began in 1973 and lasted for more than 1 year. In August 1974, before the hearings were over, EPA Administrator Russell Train announced that the future manufacture, sale, and use of aldrin and dieldrin for agricultural purposes had been suspended beginning in 1975. The compounds could still be used for domestic purposes, particularly for termite control. Administrator Train based his ruling on evidence presented by the EPA's Office of Hazardous Materials Control that the compounds caused liver tumors in mice, and as such constituted an "imminent hazard to the public."

Alternatives to Chemical Pesticides

Many promising approaches to insect control exist that offer the advantage of greatly reduced chemical pesticide applications, or, in some cases, no need at all for chemical pesticides. We shall now consider a few of these possible alternatives.

Cultivation Practices

As noted earlier, the trend to monocultural farming has given an advantage to phytophagus insect pests because it provides abundant food while reducing the impact of the beneficials. A return to multicultural farming is not likely, given the extent to which mechanization has become a part of farming. However, some changes can help. For example, the grape leafhopper, a pest of vineyards, can be controlled by an egg parasite that winters in blackberry bushes. Knowledgeable vineyard owners therefore maintain blackberry thickets near their grapes. Crop rotation and alterations in plowing and planting schedules to favor predators over pests would also help reduce pest populations.

Resistant Plants

In some cases, plant strains can be developed that are resistant to specific pests. During the 1930s wheat strains resistant to attack by the Hessian fly were developed, so that by the 1940s the impact of that pest was much less. With time, some of the strains show a tendency to lose their effectiveness in some regions, but the problem does not seem to be as serious as the development of resistance to chemical pesticides. When areas populated by resistant flies are discovered, another of the more than 20 varieties of resistant wheat are planted; this change is usually effective for 10 years.

Resistant plant techniques are costly and time consuming, however; development of a resistant strain often requires 10 years or more.

Sterile Male Techniques

For many years, Edward Knipling, Director of Entomological Research of the USDA Agricultural Research Service, has advocated the programmed release of sterile male insects as a means of reducing pest populations. Although sometimes difficult to carry out in practice, the basic technique is simple. Male insects are sterilized in such a way that their normal mating habits are not altered. Female mates of the sterile males lay unfertilized eggs, so their offspring do not develop. When a high ratio of sterilized to normal males is maintained, the population will decrease. This technique is especially successful in the case of insect species in which the female mates but once before laying eggs.

Knipling directed the development of the most famous sterile male program: the eradication of the screwworm fly in Florida and then its control in the southwestern United States. The screwworm lays eggs under the skin of cattle, and as the larvae hatch, they live on the cattle. This causes weight reduction in the cattle and in some cases death from infection.

The screwworm fly migrates northward through Mexico to the United States. The USDA keeps the insect suppressed by rearing and releasing 125 million sterile males each week along a 300 mile buffer zone between the United States and Mexico. Sterilization is accomplished by exposing the laboratory-reared males to a source of gamma irradiation (Chapter 4). The cost of the program is about $6 million a year; prior to the inception of this program in 1958, the annual cost of damage from the screwworm fly was estimated to be over $120 million annually. Plans are under way to move the barrier zone to the Isthmus of Panama to provide protection for Mexico as well.

Obstacles to the development of sterile male techniques include the requirement for economical, large-scale rearing techniques for the males and the difficulty of mass release. An additional complication is the dearth of detailed information on the habits of many pests.

The sterile male screwworm program experienced control difficulties in the mid-1970s. The number of verified screwworm cases jumped from 473 in 1971 to 90,000 in 1972. The increase might have been due to weather problems that permitted the flies to get past the barrier, but other, more far-reaching possibilities, may account for the difficulties:

1. The wild population may contain genetic types that no longer mate with released flies. If so, genetic selection would give these strains an enormous advantage.
2. Genetic selection in response to the pressures of domestic rearing may have reduced the ability of laboratory-grown flies to adapt to wild conditions.

Either or both of these may be occurring. One needed addition to the application of the sterile male technique is genetic monitoring of the insects to discover whether genetic changes are occurring.

Bacteria and Viruses

A variety of bacteria, viruses, fungi, and protozoa are pathogenic to insects and have tremendous potential for use as control agents. Two bacterial pesticides have been licensed, and one viral pesticide had nearly completed testing for licensing by 1975. Insect pathogens in general and viruses in particular are very specific. They occur naturally and viruses in particular are very specific. They occur naturally and are instrumental in controlling populations in any balanced ecosystem. On the basis of extensive testing, they appear to be completely harmless to plants, nontarget insects, animals, and man.

Research has been done on bacterial pesticides for decades. The first commercially available bacterial insecticide was a preparation containing spores of *Bacillus popilliae*, which causes milky spore disease in Japanese beetles and a few closely related beetle species. It has been in use since the 1940s with moderate success.

The only other bacterial insecticides are all preparations containing spores of *Bacillus thuringiensis,* or BT. It is effective against more than 100 species of caterpillars, many of which (cabbage looper, alfalfa caterpillar, and tomato hornworm) are serious pests. When BT is ingested, it paralyzes the gut of the caterpillar, preventing feeding. The bacteria then penetrate the gut wall, and death results. BT is effective only when the metabolic system in the insect is basic (pH 9 or higher), a characteristic possessed only by certain caterpillars. A new strain of the bacteria, HD-1, isolated in the late 1960s at the USDA laboratories in Brownsville, Texas, appears to be effective against the cotton bollworm and the gypsy moth, which were not susceptible to attack by previous strains. Several companies market the HD-1 strain, but unfortunately few companies seem to be devoting research to bacterial insecticides.

Viruses, in contrast, are receiving more research attention. Over 300 insect viruses are known, although not all of these appear to be potential candidates for insect control. Unlike bacteria, which are grown in bulk using fermentation, viruses used for insect control can only grow in the tissues of hosts. Thus production of a virus requires first the mass production of large numbers of disease-free host insects, an expensive and often tedious process. Consequently the costs of developing a viral pesticide may be greater than those required to develop a new chemical pesticide.

Few companies are interested in developing viral pesticides for minor pests because the high selectivity of the viral pesticide means that the return on developmental costs would be too small. Several companies are actively pursuing research on viruses active on major pests, however, and an example is the nuclear polyhedrosis virus that is specific to the larvae of the *Helioliths* moth. This virus has a potentially larger market because it attacks two major pests that cause more than $130 million in crop damage annually. One pest affected is the tobacco budworm. The common name of the other depends upon the crop attacked; it is known variously as the tomato fruitworm, corn earworm, or cotton bollworm. Although the disease caused by this nuclear polyhedrosis virus has been known for nearly a century, the virus itself was only isolated in 1961 in San Benito, Texas. That year also marked the development of an artificial diet making possible the raising of cotton bollworm larvae on a year-round basis.

The International Mining and Chemical Company, using the trade name VIRON/H™ for the virus, filed in 1968 to have the agent registered as a pesticide. Because this was the first viral pesticide, extensive testing was done over a period of 7 years. In field tests 1 qt of the virus controlled bollworms on 10 acres of cotton. When eaten by the worms, the viruses kill young worms within 2 days and old worms within 6 days. The organisms are highly infectious and once introduced spread rapidly, destroying large populations of insects. The viral pesticide's performance in controlling the bollworm has exceeded that of chemical pesticides in 80% to 90% of the field tests.

A problem with bacterial and viral pesticides might be the development of resistance in the target insects. In tests, no resistance to VIRON/H™ virus was indicated over 25 generations of insects. (Resistance to chemical pesticides usually shows up in 10 to 15 generations.) Another important consideration is the extent to which insect viruses can cross to forms of life other than insects. The evidence is considerable that this cannot happen in the case of the nuclear polyhedrosis virus, which seems entirely restricted to insects.

One indication of the safety of this virus to man is that when the cabbage looper succumbs to a virus, its body dissolves and sheds onto the leaf large quantities of the virus that are not killed by any of the preparative steps in the making of coleslaw. By mid-October, when the natural mortality from this virus is highest, the average bowl of coleslaw contains about 4 billion live viruses. It would probably be apparent by now if this virus were harmful to man.

Hormones

As in humans, growth in insects is regulated by hormones, and understanding the complex hormonal system of insects is at present an area of active chemical research. Because hormones can function as pesticides by upsetting normal metabolic processes, pest control is an important aspect of insect hormone research. At least three types of hormones in insects serve to regulate growth and metamorphosis from egg to larva to pupa to adult. **Juvenile hormones** and **molting hormones** are two of the major types of hormones involved, and these (and perhaps others) are thought to be regulated by one or more **brain hormones**.

In order to grow, insect larvae must molt. That is, they must shed their rigid cuticles and replace them with new ones. Both juvenile and molting hormones are crucial to growth, but during certain stages of development the secretions

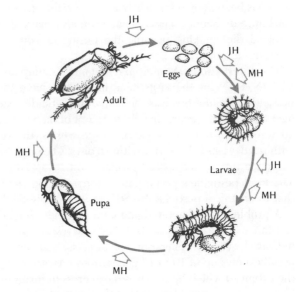

Figure 8.7. A very approximate representation of insect development and the hormones that must be present for successful development.

MH = molting hormone
JH = juvenile hormone

of one or both must temporarily cease. For example, juvenile hormone cannot be present in the insect eggs if they are to undergo normal embryonic development (Figure 8.7). It must be present in abundance, though (together with the molting hormone), for the proper larval stages to develop. Juvenile hormone secretion must cease for the insect to metamorphose into a mature adult, but later in the adult stage it is again necessary.

The key to juvenile hormone use as a pesticide is to keep its concentration high (perhaps from external sources) during any of the periods when it would normally be absent. For example, if eggs contact juvenile hormone they do not hatch. If larvae are continuously exposed to it, they either fail to metamorphose or they change into bizarre intermediate forms that do not reproduce (Figure 8.8). Because hormones are extremely specialized and because the development of resistance would probably require a drastic change in the biochemistry of the insects, Carroll Williams, a biology professor at Harvard, has referred to the hormones as "third generation pesticides." Their use requires their isolation and chemical structure characterization; neither of these is an easy task.

The history of attempts to isolate and determine the chemical structure of juvenile hormones goes back at least 40 years, but progress was hampered by the extremely small amounts of hormone produced. The isolation of a few milligrams from a ton or more of insects is common. Over the past decade, however, chemical technology has advanced to the point where millionths of a gram or less of a substance can be isolated from a few insects, then analyzed so that its chemical structure can be determined. Ideally, the development of a juvenile hormone pesticide would then involve synthesizing the compound in large lots for further testing. If the natural hormone lacked stability when applied in the environment, then a search for an **analog** would be undertaken. Analogs are chemically similar compounds that mimic the hormone's biochemical activity but might be more stable in the environment (and hopefully cheaper to make).

Juvenile hormones have now been successfully identified for several insects, the first being that of the Cecropia moth. This juvenile hormone (Figure 8.9), characterized by Herbert Röller and his colleagues at the University of Wisconsin, is so potent that 1 g will either kill or alter the development of 1 billion Cecropia larvae. The particular analog at the right in Figure 8.9 differs from the natural hormone by only two carbon atoms and is more stable, but it is only 0.02% as active on the insects.

The "Paper Factor" and Juvenile Hormone Mimics. Not all compounds showing hormonal activity are produced by insects. One unique development in the field involved a study of the European linden bug *(Pyrrhocoris apterus)* by Williams and Karel Sláma. A Czech biologist who had raised linden bugs for ten years in his laboratory in Prague, Sláma came to Harvard to work with

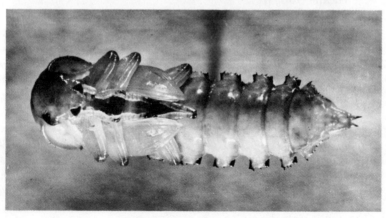

Figure 8.8. The pupa (left) of a yellow mealworm when continuously exposed to a synthetic juvenile hormone does not develop into a full adult (center). Instead, it develops an adult head and thorax, but not an adult abdomen (right). [Courtesy USDA]

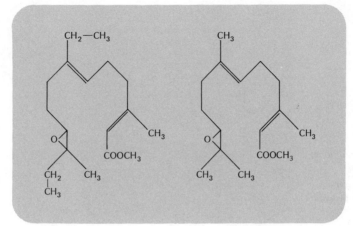

Figure 8.9. *The Cecropia juvenile hormone (left) and an analog mimic (right) which differs from the natural hormone by two carbon atoms (CH₂ groups).*

Williams. To their considerable mystification, linden bugs failed to undergo normal development when reared at Harvard. Instead of metamorphosing into sexually mature adults, they underwent an additional larval molt and died without completing metamorphosis. Such behavior had never before been noted in the tens of thousands of bugs Sláma had reared. The bugs were being exposed to an unknown source of juvenile hormone.

Clever detective work finally revealed the source as the fragment of paper towel put in the growing dishes for the insects to walk around on, and further investigation showed that American papers were much stronger sources of the juvenile hormone than were European papers. The initial report of Sláma and Williams solemnly noted that the *New York Times, The Wall Street Journal,* and *Science* were strong sources of the "paper factor," whereas *The Times* (London) and *Nature* (a British journal) were not! The essential "paper factor" turned out to be a natural product in certain evergreen trees (balsam fir), which are a major source of American paper pulp. When identified, the compound was found to have a chemical structure quite similar to that of the juvenile hormone; the "paper factor" was thus a juvenile hormone mimic. It was named **juvabione** (Figure 8.10).

Not the least interesting aspect of this work is the distinct possibility that this hormonal mimic in the balsam fir is present as a defense mechanism against certain insects. The evergreens are an ancient lot that were here before the insects. They are pollinated by wind and, unlike many other plants, do not seem to depend on insects for anything. The juvabione could have evolved in the tree's biochemical structure to combat some pest related to the linden bug, which is now extinct (for obvious reasons) or which has learned to avoid the balsam fir. Fir trees produce many other compounds somewhat similar chemi-

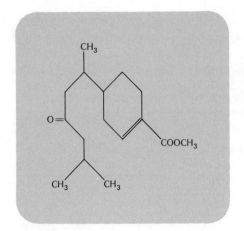

Figure 8.10. *Juvabione, the "paper factor" juvenile hormone mimic, isolated from the balsam fir tree.*

cally to juvabione. Williams has suggested that these might be a source of other species-specific pesticides.

Juvabione is very specific to linden bugs and a few close relatives. Unfortunately it is not effective on United States pests, although many important cotton pests are included in the family *Pyrrhocoridae.*

Juvabione and other hormonal mimics offer an unusual route to insect control. Sláma made the remarkable discovery that less than a millionth of a gram of a synthetic juvenile hormone mimic applied at any time to the body surface of female linden bugs sterilizes the insect for the rest of its life. Furthermore both male and female can tolerate without damage doses up to 10,000 times the quantity needed for sterilization, and neither sex was impaired in its desire or ability to mate. As a control scheme, males "painted" with 100 µg of the juvenile hormone mimic transmit enough material to females during mating to render the females sterile for life. This affords an unusual twist to the sterile male technique!

Since that discovery, research on insect hormone mimics has been feverish. Two companies applied to the EPA in 1973 to have hormones registered as insecticides. These initial applications are for hormones effective against some species of mosquitoes and flies. These will be subjected to an extensive testing program. Their toxicity is low; the LD_{50} values are typically 30 g/kg of body weight in rats, reflecting a mammalian toxicity about one thousandth that of most chemical pesticides.

Insect Attractants

Pheromones are chemicals secreted by an organism that affect the behavior of other individuals. An enormous variety of insect pheromones exist, communicating information on food, alarm, aggregation, and reproduction. Pheromones are thus chemical messengers, and the one most frequently considered for insect control is the sex attractant. It has long been known that females of many insect species depend on an attractant to lure males for mating; this is especially true for species in which the egg-laden female cannot fly. Efforts to control insects using attractants are directed toward using them to lure males to their death.

Before the 1960s, however, efforts to use this technique were tedious and expensive. Thousands of females had to be collected and a mixture of chemicals, including the attractant, was laboriously extracted from them. The extraction involved washing the crushed insects in a solvent, such as benzene, in the hope that the attractant would dissolve in the liquid. These natural extracts were expensive and were usually obtained only in amounts sufficient to conduct surveys to determine the extent of infestation. An alternative to this approach was to test the behavior of the insects to thousands of chemicals in the hope of locating, almost by chance, a chemical similar to the attractant. These approaches, though crude, worked for several species of fruit flies. One of the world's worst pests, the Mediterranean fruit fly, was eradicated from a million acres in Florida in 1957 at a cost of $11 million, which was much less than the annual damage it caused.

Isolating the Sex Attractant. A much more elegant method, however, is to use chemical technology available since the early 1960s to isolate the particular compound (or compounds) secreted by insects, determine the chemical structures, then synthesize the compounds or their close analogs for use. A USDA research team, headed by Morton Beroza, isolated and identified the first sex attractant, that of the gypsy moth, in 1969.

Gypsy moth. A native of Europe, the gypsy moth was imported to Massachusetts in 1869 for the purpose of producing silk for local industry. The effort failed, and unfortunately a few of the insects escaped. Because the gypsy moth has few natural enemies, it underwent a population explosion that continues today despite repeated efforts at control. The gypsy moth is a serious defoliator of trees, completely stripping hardwood forests. A single caterpillar eats about 1 ft^2/day of leaf surface. Opinion is somewhat divided as to the consequences of the stripping. The trees often survive the first stripping but may be more susceptible to other diseases. A second stripping produces considerably more damage, often killing most of the trees. The moth produces about one generation per year, and the population usually undergoes about a sevenfold increase every year. Many young caterpillars spin down on silken threads that break off and act as sails, allowing the wind to carry the insects off, sometimes for more than 5 miles, where they start new infestations. When laden with eggs, the female cannot fly and secretes a powerful sex attractant to lure males from distances as far away as ½ mile (Figure 8.11)!

Spraying for the gypsy moth was of course done, but its range continued to spread. In 1970 the pest defoliated 800,000 acres of forest in the Northeast; in

Figure 8.11. The female (top) and male (bottom) gypsy moths. The male's highly developed antennae help it detect the female sex attractant. (The male's wings are clipped behind it in a clothespin for mounting on a gas chromatograph.) [Courtesy U.S. Department of Agriculture. Photo at bottom by Larry Rana.]

1971 this rose to 1.9 million acres. Its geographic range has increased markedly in recent years due to the increase in camping and recreational travel; the pests can be transported hundreds of miles atop automobiles and campers. A principal concern is the impact of the pest should it spread to the forests of the South and Northwest.

The isolation and characterization of the sex attractant, attempted for 30 years by various experimenters, was achieved by the ingenious work of Beroza and coworkers. A gas chromatograph is a device used by chemists to separate thousands of volatile chemicals in a mixture. In this device, air blown across the sample is passed through a long tube packed with some inert material, usually a liquid or sometimes a powder (Figure 8.12). Because the various components move through the long tube at varying rates, the components are separated when they emerge several minutes or hours later from the other end. Beroza was able to achieve separation of the compounds in the natural extract by using a gas chromatograph, but the difficulty was to determine which of the hundreds of compounds isolated was the sex attractant. Realizing that male gypsy moths were the best detectors of the attractant, he attached 20 male moths to a board and passed the effluent over them (Figure 8.13). The insects went wild in response to the particular substance which was the attractant!

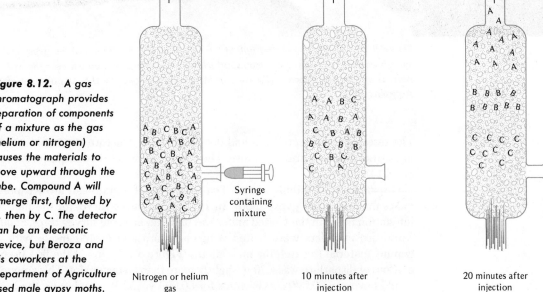

Figure 8.12. A gas chromatograph provides separation of components of a mixture as the gas (helium or nitrogen) causes the materials to move upward through the tube. Compound A will emerge first, followed by B, then by C. The detector can be an electronic device, but Beroza and his coworkers at the Department of Agriculture used male gypsy moths.

Figure 8.13. *A gas chromatograph of the type used in the isolation of the gypsy moth sex attractant. The male gypsy moths, attached by the wings to a board near the exit port, served as the detector. [Courtesy of Morton Beroza and the U.S. Department of Agriculture.]*

The researchers collected this and determined the structure from only a few micrograms of the natural sex lure. The compound was called disparlure (Figure 8.14).

Disparlure is incredibly active. Ten millionths of a gram can attract more males than a female gypsy moth. In 1971 an extensive survey of the moth's infestation range in the United States was undertaken using survey traps. Small cardboard cylinders were baited with disparlure and coated inside with a gummy material to catch the male moths (Figure 8.15). One of the first preparations of disparlure was a 30-g (slightly more than an ounce) lot. This amount would have run 60,000 survey traps for 50,000 years! By combining the lure with other compounds chosen to reduce its rate of "fragrance" loss, the lifetime of the traps could be made longer than 3 months.

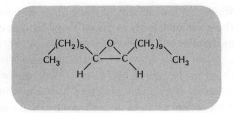

Figure 8.14. *Disparlure, the gypsy moth sex attractant.*

Two methods have been proposed to use disparlure for gypsy moth control. For low density infestations, devices similar to the survey traps could be used to kill the males. In tests this approach reduces moth populations by more than 90%. Beroza feels that the moths might even be eradicated under such conditions. For heavy insect concentrations control is much more difficult. One proposal is to sprinkle the entire area with disparlure in an effort to disrupt communications and reproduction completely. The costs of both approaches, even including the cost of aerial application, should be less than the cost of chemical pesticide use.

Figure 8.15. *Infestations of gypsy moths are detected by survey traps about 5 cm (2 in.) in diameter and 10 cm (4 in.) long. A solution of the attractant is applied to the wick of the trap which is then hung in a tree and later inspected for moths. [U.S. Department of Agriculture. Photo by Larry Rana.]*

Other insects. Sex attractants of more than 30 insects have been identified, including those for such pests as the cabbage looper, codling moth, corn borer, western pine beetle, and housefly. Not all insects rely on attractants as strongly as does the gypsy moth, however, so the technique is not universally suitable.

A test program began in 1972 to control the western pine beetle with pheromone-baited traps. This program uses a mixture of pheromones which cause synchronous invasion of target trees by thousands of insects; if the beetle fails to aggregate in this manner, they can neither feed nor reproduce. The results of this program have been somewhat mixed. Although 2.5 million beetles were trapped the first season, the number of infested trees on the 20 square mile plot increased from 1900 to 2100 trees. During the second season, tree mortality increased by another 200 trees, although the number of beetles captured nearly doubled. Thus both the beetle population and the number of trees infected increased in spite of the large number of beetles caught.

The Dutch elm disease, which has caused enormous devastation to elm trees in the United States over the last 50 years, is caused by a fungus that exists in a symbiotic relationship with the European elm bark beetle. The beetle, which can lay eggs only in dead or dying elm trees, picks up spores of the Dutch elm disease during its pupal stage and deposits them on healthy trees on which the adult beetle feeds. The fungus, by weakening the new trees, provides new sites for the beetle to mate and lay eggs. Recent research has shown that three pheromones are involved in beetle reproduction. Two come from the insects and, interestingly, one comes from the healthy elm tree itself. Efforts are now under way to use this information to control this disease.

The boll weevil, the most serious cotton pest, depends somewhat on sex attractants emitted by the male. Four separate attractants are released, and these have been undergoing field tests since 1969.

One recent, very interesting finding in the use of pheromones in insect control is that the insect's pheromone response can be blocked if it is exposed to small amounts of chemicals with a structure similar but not identical to that of the pheromone. Presumably these blocking chemicals clog or otherwise interfere with the insect's pheromone detection mechanism. Use of a pheromone blocking approach has the advantage that the adult insect lives its normal life cycle, providing whatever positive benefits it has for the environment, but is not "ordered" by the pheromone to cause damage. For example, no one would mind the presence of the boll weevil if it could be dissuaded from eating cotton. This approach is being developed for use on corn pests in the midwest.

Biological Controls

This form of insect control uses natural enemies such as predator insects and parasites. When biological control of a pest works, it is almost ideal. But achieving the control requires time, a thorough knowledge of the pest, and often a bit of luck.

A classic example of biological control is the use of the Vedalia beetle to subdue the population of the cottony cushion scale, a citrus pest in California. The scale appeared in California in 1868 and began to devastate the citrus crops. The Vedalia beetle was one of the control insects imported in 1887 from Australia because it fed almost exclusively on the scale. Three years later the scale was under control, and citrus shipments had tripled. Significantly troubles with the cottony cushion scale resurfaced in 1950 after the groves had been sprayed with DDT in an effort to kill thrips. The spray killed the Vedalia beetles, upsetting the balance and causing a cottony cushion scale crisis worse than that due to the thrips. Vedalia beetles were so reduced that a market developed for them, and they sold for as much as $2 each. DDT spraying was stopped and biological control by the beetle was reestablished.

Other such "permanent" biological controls are known. A major citrus pest in Israel, the Florida red scale, was controlled within 4 years after the importation of a parasite from Hong Kong in 1955. The olive scale in California has been controlled by the importation of two parasites, one from Persia and the other from Pakistan. In Mexico the citrus blackfly was completely controlled by four parasites imported from the Orient.

Parasitic Wasps. Not all biological control programs are established in the hope that "permanent" control can be established. Many projects instead involve the delivery to the field of large numbers of insectary-reared parasites that can overwhelm the existing pest population but that do not remain to provide future control. Several species of *Trichogramma* (or *Polistes*) wasps, a tiny wasp that can be produced by the millions, are sometimes used in this manner. *Polistes* wasps lay their eggs in the larvae of pests; as the wasp larvae develop, they feed on the host and kill it. Control of the cotton bollworm in certain areas of west Texas is achieved by the release of *Trichogramma*. The company responsible for the program achieves control for about one-third the cost of a chemical pesticide program. If *Trichogramma* is used, spraying in the area must be avoided.

Klamath Weed. Even some plants can be controlled using biological control. The Klamath weed appeared in California about 1900 and soon spread to 5 million acres of ranch land in California, Idaho, Oregon, Montana, and Washington. The weed displaces range grasses and secretes an oil which, when exposed to sunlight, can cause ulcerous skin conditions in cattle. In 1944 two natural enemies of the Klamath weed were imported. These insects, two species of chrysolina beetles, were imported only after extensive studies were made to determine that they would starve rather than eat such crops as sugar beets, flax, hemp, sweet potatoes, and cotton. Within 2 years no more importations were necessary. After 10 years, the weed had been eaten back to about 1% of its range in California. Chrysolina was not quite so successful in the other states, perhaps because it was not as well suited to their climates.

Future Prospects. Biological controls are often difficult to establish. In this century attempts have been made to introduce parasites and predators of about 80 pests into the United States. Of the 520 species imported, 115 have become established but only about 20 have provided significant control of destructive pests. Usually imported pests are most susceptible to this approach through the importation of their natural enemies.

For parasitic insects, one must also devise rearing techniques for the host in order to grow large numbers of parasites.

A special worry about importing organisms is of course that the import may become a pest. Usually this is less of a problem if the import is an entomophagus insect.

Integrated Pest Control

The difficulties associated with the use of pesticides plus recognition of the potential offered by alternate means of control have spurred a trend toward integrated pest control: utilizing all possible methods of control instead of an exclusive reliance on chemical pesticides. The emphasis of integrated control is to rely more on natural and biological controls, using pesticides only when absolutely necessary, then using them in only moderate amounts. The key to a successful integrated control scheme is an expert who can monitor insect populations, establish what level of infestation can be permitted, then prescribe corrective minimally disruptive controls should the pests exceed that level. This approach is not new, but it has certainly been neglected since the advent of synthetic chemical pesticides.

An example of the approach integrated control might take is afforded by a look at the research of F. R. Lawson, the recently retired head of the USDA's Biological Control Laboratories. In 1948 the USDA assigned Lawson to study the tobacco hornworm, a serious tobacco pest in the southeastern United States. Lawson's work pinpointed the natural weaknesses of the pest. First of all, the studies showed that the principal control on the insect was natural and that a biological control program would probably work: 30% of the eggs could be destroyed by the stilt bug, 60% could be killed by the *Trichogramma* wasp, and some of the remainder could be killed by other enemies. Lawson described an integrated control program, one step of which was erecting wooden boxes around the fields to serve as nesting places for *Trichogramma* to give them a further advantage. Pesticides were needed only when the hornworms exceeded a certain level, which from Lawson's work could be established by counting the number of caterpillars on 50 plants. Even then, he showed that, because the hornworm spent all its time on the upper quarter of the plant, only the tops of the plants needed to be sprayed. This would reduce the amount of pesticide applied per spraying to one fourth the amount usually used. But the program was never used; instead, widespread spraying was done on a regular basis.

Later, in the 1960s, Lawson and coworkers designed a method to reduce the population of hornworms 50% to 70% by an even simpler method. The farmer simply set up three mechanical light traps per square mile. The insects were attracted to the light at night and were killed by a mechanical device such as a fan. This plan was not used either. Spraying continued and DDT concentrations in tobacco increased until they exceeded by several times the levels permitted in food by the FDA. Ultimately the southeastern states greatly reduced the use of DDT on tobacco because European countries that imported 40% of the United States tobacco crop became worried about the resulting health hazards and consequently set low tolerances for chlorinated hydrocarbons in tobacco.

The present specialist in integrated pest control appears to be an entomologist in private practice who can provide independent advice not related to the sale of a chemical pesticide.

Robert van den Bosch, an entomologist at the University of California at Berkeley, has stated that, when integrated pest control has been used in California, it has decreased insecticide requirements 50% to 75% while increasing yields and profits to farmers.

Reasons for the Reliance on Chemical Insecticides

At the present time, we rely almost exclusively on chemical pesticides for insect control. Only a few of the alternatives just discussed are used. Many reasons account for this state of affairs.

The impact of the publicity DDT received in World War II was an important initial factor. Chemical pesticides were (and still are) easy to use, and they have an important psychological benefit: one can see the pests dying. Crop yields are improved if the pesticides are carefully selected and used. But agriculture is a very inexact and uncertain science. It is difficult to ascertain exactly what contribution pesticides make to increased crop production; better fertilizers, improved seeds, and increased mechanization also increase yields and, in addition, the weather is an enormous factor in the yield obtained in any one year. This means that it is very difficult for individual farmers to perform a "cost-benefit analysis" to determine if the increased cost of additional pesticides is justified by a sufficient improvement in production. Closely related to this is the fact that the farmer in the United States has traditionally been caught in a profit squeeze. A small percentage increase in yield may mean the difference between his survival or failure. He is virtually forced into using whatever technique his competitors (other farmers) are using.

Another crucial factor in our reliance on chemical pesticides is the role of the agricultural chemical companies. For most of his advice on pest control, the farmer relies on the salesman, a man who works for an agricultural chemical company and who is in the business of selling chemical pesticides.

The agricultural chemical companies are also closely involved in the determination of agricultural policy at the state and federal level. For decades there has been a close cooperation between these companies and entomology departments in colleges and universities. The companies have provided fellowships and sponsored research at the schools. Because the companies are interested principally in chemical pesticides, their funds are largely channeled into research on chemical pesticides. On economic grounds, the development of selective pesticides by private industry should not be expected. The return on the investment would be small, since in most cases the total sales would be small. As a result of the strong influence of agricultural chemical companies, graduates of most entomology departments have had little experience with nonchemical means of insect control. These same graduates become local, state, and federal officials who determine policy, including how federal research funding is used. As a result, the research support for nonchemical methods of control has been sadly neglected in the past.

Finally, increasingly stringent consumer demands contribute to an increased use of chemical pesticides. The requirement that only #1 grade fruit, that without insect holes or insect parts, can be sold on supermarket shelves gives an incentive to the farmer to produce fruit of that quality. To produce blemish-free fruit on a large scale requires that the crop be drenched with insecticides. This fruit quality standard is largely cosmetic; #2 fruit is consumed, insect parts and all, because it is allowed to be used in catsup, jams, and jellies.

Public Policy

Pesticides must be registered for use by the EPA. This registration is not to be granted unless scientific studies show that the product is both safe and effective. In the past, most of the safety aspects involved only toxicity, but there is an obvious need to include other effects of pesticides, such as their ability to interfere with enzyme action.

Should the pesticide be suspected of causing problems, the EPA has legal authority to cancel the registration or suspend the registration. Under **cancellation,** the use of the pesticide can continue while the effects of the pesticide are investigated, a process that can take years. Registration of DDT, for example, was canceled in January 1971, and the investigation took a year and a half. **Suspension** of pesticide registration is a much more definitive action; the pesticide can no longer be used.

Pesticide policy is determined largely by the USDA. A clear necessity in pesticide policy is that incentives must be forthcoming for the development of alternatives to synthetic chemical pesticides. The logical source for this incentive or for the advances themselves is the USDA and the federal government, since many alternatives are sufficiently selective that private enterprise is not interested in pursuing their development.

Summary   Ecological balance usually provides a limitation on the population of any particular species. The population of phytophagus insects, for example, is controlled by entomophagus insects and by competition with other phytophagus insects for food and for habitat. Nearly all of the synthetic chemical pesticides developed since 1940 may upset this balance because their broad spectrum action kills beneficial insects as well as pests. If the beneficials are decimated, pests may reestablish themselves in greater amounts, since their natural controls are depleted.

The ideal pesticide should be selective and nonpersistent. Persistence is undesirable because it lengthens the time a pesticide maintains effects in the environment, encourages the development of insect resistance, and provides the potential for biological magnification of the pesticide in the food chain. Of the pesticides presently used, chlorinated hydrocarbons are persistent, having lifetimes of several decades in the environment, whereas organophosphates and carbamates are relatively nonpersistent, having lifetimes of a few days or weeks.

The persistence of the chlorinated hydrocarbons, combined with their physical properties and their mobility, means that they are ubiquitous pollutants that accumulate in the food chain. A potentially serious long-range effect of the chlorinated hydrocarbons is their enzyme induction property. By causing liver enzymes to metabolize steroid sex hormones such as estrogen, and by interfering with the activities of certain other enzymes, chlorinated hydrocarbons cause reproductive problems in birds, mammals, and fish. These effects are also caused by chlorinated hydrocarbons such as the PCBs, which are not used as pesticides.

The herbicide 2,4,5-T is teratogenic; dioxin, a chlorinated hydrocarbon impurity created during the manufacture of 2,4,5-T, is teratogenic and very toxic. The widespread use of defoliants in Vietnam has created extensive ecosystem disruption and has resulted in potential health problems to humans.

Legal efforts to restrict the use of chlorinated hydrocarbons began with citizens' groups such as the Environmental Defense Fund which used scientific and legal experts in a successful effort to have DDT declared a pollutant and banned in Wisconsin. Uses of 2,4,5-T were greatly restricted in the United States (and in Vietnam) beginning in 1971. The EPA banned nearly all uses of DDT effective January 1, 1973, and nearly all agricultural uses of aldrin and dieldrin effective in 1975.

Many promising alternatives to insect control which reduce the amount of chemical pesticides needed or which eliminate entirely the need for pesticides are known, but little research has been devoted to these. In controlling pests, it is especially important to establish what level of pest infestation can be tolerated before expenses for pest control measures are economically justified. Alteration of cultivation practices can sometimes reduce pest impact, and crop strains resistant to the attack of certain pests can sometimes be developed. The release of laboratory-reared, sterilized male insects in such quantities that they greatly outnumber fertile males is used to control the screwworm fly in the southern United States.

Insect viruses and bacteria can be cultivated, then released to infect pests. These provide very specific control. The bacteria *Bacillus thurigiensis* is effective against some caterpillars, including the cabbage looper, alfalfa caterpillar, and tomato hornworm. Newer strains of this bacteria may be effective against the cotton bollworm and the gypsy moth. A viral pesticide for use against the cotton bollworm has been licensed for field trials.

Two alternatives to pesticides are areas of active chemical research: insect hormones and insect pheromones. A juvenile hormone is required for successful insect development, but its secretions must cease at certain critical intervals if metamorphosis to the adult stage is to occur. By keeping the level high at all times by using external applications of the hormone, insects' maturation can be stopped. The balsam fir possesses a juvenile hormone mimic (juvabione) which retards the growth of the linden bug. Application of other juvenile hormone mimics to the bodies of male linden bugs provides a means of insect control; a sufficient amount is transferred to the female during mating to make the insect sterile for life.

The use of pheromones, especially the sex attractant, provides a specific, safe means of controlling insects. A sex attractant, usually secreted by females to lure males for mating, can be used to lure males to their deaths; in the usual practice, it must be isolated, chemically characterized, and then synthesized by man. Sex attractants were used to eradicate the Mediterranean fruit fly from Florida in the 1950s and are now being developed for use against the gypsy moth and other pests. Other types of pheromones may be useful in controlling the western pine beetle and the Dutch elm disease.

Biological controls, such as predator and parasite insects, can provide ideal control when they can be devised. About 80 pests in the U.S. are controlled (with varying degrees of success) by imported biological controls. Parasitic wasps, such as *Trichogramma*, which can be reared by the millions, then released to lay their eggs in the larvae of other insects, are another form of biological control.

Future insect control programs will probably use integrated pest control. A specialist visits the field regularly, tabulates the insect populations, and recommends treatments (hopefully the least environmentally disruptive type) to reduce pest populations.

Our almost exclusive reliance on chemical pesticides is due to a number of factors. Our trend toward monocultural farming has increased the chances for pest survival. Low profit margins for farmers force them to follow the practices of their competitors. Agricultural chemical companies, seldom interested in nonchemical control, have influenced entomology departments, whose graduates often become the governmental decision makers on agricultural matters. Unrealistically high consumer standards for produce increase the need for heavy, frequent pesticide applications. Selective alternatives to chemical pesticides afford little financial return on developmental costs, hence private industry is seldom interested in such techniques. Thus it appears the development of these techniques requires a governmental role.

Review and Study Questions

1. What are entomophagus insects and what is their importance?
2. What are "second generation pesticides"?
3. What are the contributions, accomplishments, or importance of each of the following? Charles Wurster, F. R. Lawson, William Ruckelshaus, Carroll Williams, Rachel Carson, Wm. Beroza.
4. Define: LD_{50}, enzyme induction, acetylcholinesterase (ACHE), carbonic anhydrase, juvenile hormone mimic, teratogenic.
5. What two properties should be possessed by an ideal pesticide?
6. Explain how future pest populations can increase after the use of a pesticide.
7. Classify organophosphates, carbamates, and chlorinated hydrocarbon pesticides in terms of toxicity and persistence.
8. What is the evidence used to indict chlorinated hydrocarbons as the source of reproduction problems in raptors?
9. In what animals or birds are PCBs known to cause reproductive problems either in nature or in lab tests?
10. The presence of what impurity complicates an analysis of the effects of 2,4,5-T on living organisms?
11. What is the average concentration of DDT in human tissues (United States)?
12. Describe the 1969 DDT "trial" in Wisconsin. List those groups on either side of the issue.
13. How does a trend toward monocultural farming affect pest versus beneficial insect populations?
14. Describe the basic principles on which the sterile male insect control technique is based.
15. Name one far-reaching difficulty that may cause a reduction in the effectiveness of the sterile male technique.
16. For what insects is the first tested viral pesticide effective?
17. What is the "paper factor"?
18. What hormones exist in insects and what are their functions?
19. What are pheromones?
20. What two proposals have been made to use sex attractants to reduce gypsy moth concentrations?

21. What factor makes the development of a viral pesticide more complicated than the development of a bacterial pesticide?
22. What is integrated pest control?
23. How have agricultural chemical companies influenced our reliance on chemical pesticides?

Questions Requiring Outside Reading

24. Small invertebrates live in the soil and play a key role in its character. How is their ecological system affected by the use of pesticides? (C. A. Edwards, "Soil Pollutants and Soil Animals," *Scientific American* 220 [April 1969], 2–9.)
25. A monument to what insect, an important biological control, exists in Eureka, California? (See Kevin Shea, Suggested Outside Reading.)
26. The coyote is considered a pest in the western United States. Poisoning programs have been strongly promoted to reduce coyote populations. How effective has this been? What are some side effects? Does this program have any similarities to the use of broad spectrum chemical pesticides on insects? (Jack Olsen, "The Poisoning of the West," *Sports Illustrated,* March 8, 1971, 80–84; March 15, 1971, 36–40; March 22, 1971, 34–36; *Readers Digest* 99, August 1971, 69–74.)
27. Most rat poisons contain Warfarin, a slow-acting poison that causes rats to die of hemorrhaging because it is an anticoagulant. What problems have recently been encountered in its use? (*Newsweek,* July 10, 1972, 99; *Science* 176 [June 23, 1972], 1343–1344.)
28. What viruses other than the one discussed in this chapter are potential candidates for insect control? (Kevin Shea, "Infectious Cure," *Environment* 13 [January/February 1971], 43–45.)
29. Pyrethrum, a natural insecticide isolated from plants, is generally considered to be a relatively safe pesticide. In an aerosol spray, however, it may have serious side effects in humans. What are these? (Julian McCaull, "Mix With Care," *Environment* 13 [January/February 1971], 39–42.)
30. What political realities are involved in the control of the fire ant, a pest in the southern and southeastern U.S. with the chlorinated hydrocarbon Mirex? (Deborah Shapley, "Mirex and the Fire Ant: Decline in Fortunes of "Perfect" Pesticide," *Science* 172 [April 23, 1971], 358–360.)
31. Although the pesticide DDVP used in hanging insecticide strips is inhaled by millions annually, its effects on human physiology are unknown. How could it be registered despite this gap in knowledge? (The Committee for Environmental Information, "The Price of Convenience," *Environment* 12 [October 1970], 2–15, 44–48.)
32. In the mid 1960s, massive fish kills in the Mississippi River were caused by endrin, a chlorinated hydrocarbon insecticide that entered the river as

effluent from a manufacturing plant. What was involved in stopping the pollution? (Frank Graham, Jr., "The Mississippi Fish Kill" in *Disaster by Default,* by Frank Graham, Jr., M. Evans and Co., New York [1966].)

33. The boll weevil accounts for over one-third of the chemical pesticides used in the U.S. In recent years a proposal has been made to attempt to eradicate the boll weevil using all possible means of control. What would be involved in this enormous project? What would be the environmental impact? What are the chances for success? (Kevin Shea, "The Last Boll Weevil," *Environment* 16 [June 1974], 6–10. L.J. Carter, "Eradicating the Boll Weevil: Would It Be a No-Win War?" *Science* 183 [February 8, 1974], 494–499; R. van den Bosch, letter, *Science* 184 [April 12, 1974], 112.)

34. One means of reducing the amount of pesticide used annually is to package the pesticide in small pellets which provide a slow release over an extended period. How is this accomplished? What are the advantages and disadvantages of this approach? (*Chemical and Engineering News,* September 30, 1974, 20–22; August 12, 1974, 16–17.)

Suggested Outside Reading

BEROZA, MORTON. Insect Sex Attractants," *American Scientist* **59** (May/June 1971), 320–325. A general discussion of insect sex attractants with a special emphasis on the gypsy moth.

BEROZA, MORTON, and E. F. KNIPLING. "Gypsy Moth Control with the Sex Attractant Pheromone," *Science* (July 7, 1972), 19–27. A somewhat technical account of gypsy moth control techniques by those directing the project.

CARSON, RACHEL. *Silent Spring.* Houghton Mifflin Company, Boston, 1962. This book appeared as environmental concern in the U.S. was increasing. It was originally thought to be alarmist by many when published, but it may in fact understate the impact of pesticides. The book created a furor at the time.

Cleaning Our Environment: The Chemical Basis for Action. American chemical Society, Washington, 1969. Includes a general discussion of pesticides, their structures and their use. This book was prepared before the enzyme induction activity of chlorinated hydrocarbons became recognized.

GREER, F., C. G. IGNOFFO, and R. F. ANDERSON. "The First Viral Pesticide: Case History," *ChemTech,* June 1971, 342–347. An account of the development of the first viral pesticide by those instrumental in its development.

HENKIN, H., MARTIN MERTA, and JAMES STAPLES. *The Environment, The Establishment and The Law.* Houghton Mifflin Company, Boston, 1971. An engrossing description of the 1968 DDT hearings in Wisconsin. The authors focus on strategies, crucial testimony, and the personalities of those involved. The view of the authors is largely that of the anti-DDT forces.

Herbicides in Vietnam. The following discuss the pros and cons of the reports of two panels that assessed the impact of herbicide use in Vietnam: "Herbicides in Vietnam: AAAS Study Finds Widespread Devastation," *Science* 171 (January 8, 1971), 43–47; "war Herbicide Report Stirs Controversy," *Chemical & Engineering News,* March 11, 1974, 18–19.

HUFFAKER, CARL B. "Biological Control and a Remodeled Pest Control Technology," *Technology Review* 73 (June 1971), 30–37. An excellent review article.

MARX, JEAN L. "Insect Control: I. Use of Pheromones," *Science* 181 (August 24, 1973), 736–737; "II. Hormones and Viruses," *Science* 181 (August 31, 1973), 833–835. General review articles on the status of these methods.

McCAULL, JULIAN. "Know Your Enemy," *Environment* 13 (June 1971), 30–39. An interview with F. R. Lawson, retired director of the USDA's Biological Control of Insects Research Laboratory. Lawson discusses efforts at biological control, the role of chemical pesticides, and the influence of agricultural chemical companies on pesticide policy.

PEAKALL, D. B. "Pesticides and the Reproduction of Birds," *Scientific American* (April 1970), 73–74. An account of how chlorinated hydrocarbons affect breeding behavior and reproduction in birds.

RATCLIFFE, D. A. "Changes Attributable to Pesticides in Egg Breakage Frequency and Eggshell Thickness in Some British Birds," *Journal of Applied Ecology* 7 (1970), 67–115. A detailed discussion of the temporal and geographic correlations between eggshell strength and chlorinated hydrocarbon use. The bibliography contains references to papers which establish many of the effects of chlorinated hydrocarbons discussed in this chapter.

"Report of the Secretary's Commission on Pesticides and Their Relationship to Environmental Health, Parts I and II," U.S. Department of Health, Education and Welfare, December 1969. The report of the Mrak Commission. An authoritative look at pesticides and their impact on health. Loaded with references and data.

Science, 1967 to the present. Many technical articles on pesticides are published in this journal. Most of the enzyme induction experiments with chlorinated hydrocarbons were first published here, and conclusions were often debated in the "Letters" section of later issues.

SHEA, KEVIN P. "OLD WEAPONS ARE BEST," *Environment* 13 (June 1971), 40–49. A general discussion of the successes, problems, and failures of biological control.

WILLIAMS, CARROLL. "Third Generation Pesticides," *Scientific American* 217 (July 1967), 3–7. A discussion of insect hormones. Special emphasis is given to the linden bug and its juvenile hormone mimic, juvabione.

WURSTER, CHARLES F. "Aldrin and Dieldrin," *Environment* (October 1971), 33–45. An extended discussion of the health and environmental problems caused by these insecticides.

_____. "Chlorinated Hydrocarbon Insecticides and the World Ecosystem," *Biological Conservation* 1 (1969), 123–129. A summary of DDT's physical properties and environmental impact. This was part of an affidavit submitted by Wurster and the EDF to a Michigan Federal Court.

_____. "DDT Goes to Trial in Madison," *BioScience* 19 (September 1969), 809–813. Wurster's account of the DDT trial including the evidence submitted and the strategies used.

9 Legalities, Politics, and the Future

> But the time has come to inquire seriously what will happen when our
> forests are gone, when the coal, the iron, the oil, and the gas are exhausted,
> when the soil has been further impoverished and washed into the streams,
> polluting the rivers. . . .
> > Theodore Roosevelt
> > Conference on the Conservation of National Resources, 1908

> But it seems that the wind is setting East, and the withering of all woods
> may be drawing near.
> > J. R. R. Tolkien
> > *The Lord of the Rings,* copyright 1965 by J. R. R. Tolkien

Change and Altruism

In the preceding chapters we discussed a wide variety of environmental
problems and noted the potential for an even larger number of increasingly
severe problems. To solve present problems and eliminate future ones, changes
will obviously be required. Our purpose in this chapter is to examine some of
the vehicles for constructive change.

Although changes will inevitably occur as time progresses, desirable solutions to most of the problems mentioned will probably not occur spontaneously. Rather, they will have to be dictated and enforced by legal action or by
a governmental body. A number of factors contribute to such a conclusion.
Pollution costs can be viewed as external to the polluter; that is, as a cost to
others, because in disposing of residues of production and consumption, the
polluter uses essentially free resources, including air and water. The costs of
this pollution (if no pollution abatement is required) are borne collectively by
the rest of the populace. Thus, there is little incentive, except altruism, for
people to reduce their use of automobiles or for power plants, industries, and
manufacturing plants to reduce emissions. Economics is a hard taskmaster, and
an individual company or a plant that voluntarily attempts to reduce emissions
may increase operating costs to the point that it loses out to its (polluting)
competitors. Under a nonregulatory scheme, polluters have an economic advantage over nonpolluters who install expensive pollution controls. A coal
company that restricts strip mining to level ground, avoiding easily eroded
hillsides, and that follows an aggressive program of land reclamation and
resodding obviously incurs more costs than companies that pay little heed to
the problems, mining hillsides, and failing to refill the mined areas.

In a widely circulated essay, "The Tragedy of the Commons," Professor

421

Garrett Hardin has emphasized that, when an essentially free property is shared, the instinct to "get your own share" overcomes altruistic long-term goals. To see that altruism is a poor vehicle for change, consider the typical habits of two families:

An environmentally aware family will minimize its energy and resource consumption. Home heating and cooling requirements are minimized by insulation, by altering thermostat settings, by reducing artificial lighting, and by utilizing trees and breezes for cooling. The family automobile is chosen on the basis of fuel economy and the prospects for an extended (more than 10 years) performance lifetime. Engine tune-ups are done regularly to ensure good fuel economy. This family will plan meals to cook all food simultaneously using a minimum of fuel, wash dishes by hand, plan doing the laundry only when there is enough for a full machine load, and dry the laundry outdoors. Cloth diapers are used for the babies (at most two) in the family. They combine shopping and pleasure trips to reduce gasoline consumption, shop carefully to avoid over-packaged goods, and save cans, bottles, and papers for weekly trips to a recycling center.

An unenlightened family (unfortunately more common in United States society) follows few of these practices. Excessive energy and resource consumption are assumed to be part of the "good life." The house is brightly illuminated and kept at 72°F the year round; except for monthly complaints about the size of the utilities bills, little is done to minimize energy usage. Auto purchases (made every 2 years or so) are influenced principally by auto size (the larger the better), prestige, styling and accessories. Convenience foods are used often, each meal's dishes are done in a dishwasher, the clothes washer and dryer are used unnecessarily often. Disposable diapers, used on the babies in this family, are flushed down the toilet. Automobile trips are taken for convenience and pleasure, with the air conditioner going full blast in the summer, even on short trips.

What are the immediate advantages to the enlightened family? The answer is very few aside from lower utility and gasoline bills and some personal feelings of accomplishment for doing their part. Both families breathe foul air, encounter the same polluted lakes, and pay taxes at approximately the same rate. Public money is used to clean the sewers stuffed with paper diapers and to expand the utility system to handle excess consumption. The unenlightened family may in fact have more time for leisure activities, and little incentive exists for this unenlightened family to amend its ways. This illustrates why it is unrealistic to expect that goodwill alone will improve the environment. For altruism to work a large majority must be motivated.

Legal Remedies

The United States legal system is an interlocking system built upon precedent. This system is adapted in part from the traditions of common law (or civil law), inherited from England, and in part on statutes written by governmental bodies in the United States. Both components of our legal scheme are involved in seeking redress from environmental insult. The individual might have some impact in getting statutes passed (as discussed later in this chapter), but he may occasionally have a more immediate and heavier impact through court actions.

Damage suits and injunctions provide legal recourse against environmental degradation. In court actions one party, the **plaintiff,** alleges that he has been harmed by another party, the **defendant.** If it is judged that harm has indeed occurred, the defendant may be ordered to pay the plaintiff money to offset the damages. An **injunction,** a legal order requiring the defendant to cease his offending activities, is sometimes also requested and may be awarded by the court.

In order to be permitted by the court to bring suit, the plaintiff must have **standing.** This usually means that the plaintiff, if a private citizen, must have suffered more than the general public; if this is not so, the private citizen would not have standing before the court. In such cases only a public official could have standing.

Laws regulating the environment are very old, the first antismoke law having been passed in England in 1273.

Common Law

This body of law evolved to provide protection to English landowners, especially those whose rights to enjoy their property were directly affected. Many of the concepts in common law have required revision to handle technological advances such as motor vehicles, power plants, and airplanes.

Nuisance. Under common law, a nuisance exists when something or someone interferes with a landowner's right to enjoy and use his property without unreasonable interference. A nuisance may be private, that is, affecting an individual landowner more than other landowners, or it may be public. A public nuisance affects all landowners equally. For a public nuisance, only a public official, not a private landowner, can get standing to attempt court action. For a private nuisance, only the landowner or landowners directly affected can get court action.

A typical nuisance suit might be brought by one or more nearby affected landowners who allege that some form of pollution is interfering with their rights. The plaintiffs must show that the polluter is the principal (often the *only*) cause of the financial or irreversible physical harm incurred. This is often difficult; for example, showing that health damage is due to air pollution is particularly difficult if the plaintiff is a smoker. The suit may ask for damages plus an injunction. It often happens, however, that the court will award the damages but *not* the injunction. Such is especially likely when the court applies the "balancing of interests" test: do the financial benefits of having the defendant in business outweigh the damages the business creates? This test dates from the days the United States was an expanding industrial nation. Injunctions are much more likely to be granted when a public nuisance is involved.

Trespass. Trespass is an intentional and uninvited entry onto land. It need not affect enjoyment in any way and hence may require less proof than does a nuisance case. Unfortunately traditional interpretations have required that the entry be direct, that smoke, dust, or gases carried onto property by any intervening agency such as wind or water do not constitute trespass.

Water Laws. Common law provided much of the basis for our water laws. The states commonly use the **riparian doctrine** or the **prior appropriation concept** or a combination of both. The riparian doctrine holds that those with property along the waterway have the right to make reasonable use of the water, including in some cases using the water for waste disposal. On rare occasions, the **natural flow theory** is applied under this doctrine. This theory holds that the riparian owner is entitled to have water flow by him undiminished in either quantity or quality. Historically the natural flow theory has not been extensively used because the balancing of interests concept has favored industrial development.

The prior appropriation doctrine allows the first users of a source to maintain their use over later claimants, but does have some precedent for requiring maintenance of the original quality of the water.

Limitations. Because common law favors property owners, it cannot be easily extended to the public interest. A further limitation is that because action can be taken only after pollution has occurred, common law cannot be used to prevent environmental problems. Thus, common law does not provide a general solution to the problems; other legislation is necessary.

Statutory Law

Vast numbers of federal and state laws affect the environment. We shall investigate only a few of the more important ones.

Refuse Act of 1899. This act, one of the first specifically directed toward pollution control, was passed during the McKinley administration as the River and Harbors Act. It required industries, corporations, and municipalities to obtain a permit before discharging refuse into navigable bodies of water and their tributaries. Fines of $500 to $2500 were provided for each day of violation of the Act. An interesting aspect of this law is that half of the fine was to be paid as a reward to any citizens who had provided information leading to a conviction.

The law was largely ignored until 1970 when then President Nixon announced it was to be part of his environmental policy. Under his program, polluters would have to get a permit to be free from prosecution under the old law. By the time the program became effective, the EPA had responsibility for issuing the permits. A series of court decisions established the EPA's authority in such cases. Permit applications were slow to arrive, however, and several individuals received rewards under the 1899 act. In 1972, for example, the Hudson River Fishermen's Association was awarded $20,000 for information leading to the conviction of Anaconda Wire and Cable Company, which was dumping toxic discharges into the Hudson River. Few applications of the 1899 act are expected in the future because a 1972 law gives most of the enforcement powers to the EPA.

National Environmental Policy Act (NEPA)

And the Lord spake unto Moses, saying, "There is both good news and bad news. The good news is that plagues shall smite your Egyptian oppressors. The Nile shall be turned to blood, and frogs and locusts shall cover the fields, and gnats and flies shall infest the Pharaoh's people, and their cattle shall die and rot in the pastures, and hail and darkness shall visit punishment upon the land of Egypt! Then will I lead the children of Israel forth, parting the waters of the Red Sea so that they may cross, and thereafter strewing the desert with manna so that they may eat."

And Moses said, "O Lord, that's wonderful! But tell me, what's the bad news?"

And the Lord God replied, "It will be up to you, Moses, to write the environmental impact statement."

(From PLAYBOY Magazine; copyright 1975 by PLAYBOY)

The purpose of this 1969 act is to require government agencies to consider the environmental consequences of projects involving federal monies. The law has two major features. One established the three-man Council on Environmental Quality which is to advise the president on environmental matters and which is partly responsible for encouraging the government to comply with NEPA. The other feature is a broad statement of policy to the effect that the government should seek to enhance the environment by "all practical means." What lends muscle to NEPA is a provision that requires governmental agencies to prepare detailed statements of the environmental impact of any major activities they propose and to study in a similar manner all practical alternatives. This requirement for an "environmental impact statement" provided the act with an unanticipated approach for citizen involvement in environmental matters. Citizen suits to obtain injunctions against federal projects for which environmental impact statements had not been filed were upheld in court. This gave NEPA an importance it might never have had otherwise.

The most visible results of this provision have been setbacks to programs of the AEC, the Department of Interior, the Department of Transportation, and even the EPA. Hundreds of projects including dams, pesticide programs, highways, coastal projects, military installations, and public works projects were delayed, although usually temporarily, because environmental impact statements had not been filed in advance. The pipeline to transport oil from Alaska's north slope oil field to the southern Alaskan port of Valdez was delayed 2 years after the Department of the Interior's first impact statement (200 pages long) was declared inadequate by the courts. The final version accepted by the courts filled nine volumes and cost nearly $9 million to prepare.

The importance of NEPA may ultimately prove to be more that of a "consciousness raiser" than anything more substantive. Although impact statements must satisfy the courts and the Council on Environmental Quality, the final authority on the decisions to be made rests with the agency. While NEPA was intended to provide input to decision making, the environmental impact statement may in time become a new form of bureaucratic gamesmanship in which an agency's expertise is used to shape impact statements to fit preconceived decisions rather than the other way around. Be that as it may, the NEPA requirements do increase public and governmental awareness of environmental problems. NEPA procedures and the publicity surrounding them are given much credit for convincing then President Nixon to stop construction of the Cross Florida Barge Canal.

Standing to Sue

The question of standing has rendered the courts inaccessible for many environmental suits. For example, it would make no sense to allow the reader of a newspaper in New York to sue the Union Oil Company for the oil spill off Santa Barbara simply because the New Yorker was in sympathy with the Santa Barbara citizens and he felt compelled to act. He could, however, sue for damages if a boat he owned were moored in the channel and damaged by the spill; he would then be directly affected more than the public.

Since 1965, several important rulings affecting standing in environmental cases have occurred. In 1965 a court ruled in *Scenic Hudson Preservation Conference versus The Federal Power Commission* that the Conference, an organization of local citizens, had standing to sue. In the case, the Conference felt that a pumped water storage facility for electrical power generation should not be constructed at the proposed site (Storm King Mountain, N.Y.) because of that site's unique beauty and historical importance. The Conference ultimately won the suit. In 1969 a court ruled that the Sierra Club, a national organization long active in environmental affairs, had standing in a case in New York. However, in 1972, a California court ruled in a case contesting the proposed development of a ski area in a wilderness site that the Sierra Club had standing only to the extent that local members would be affected by the operation of the proposed area.

An increasingly important development in environmental law is the possibility of providing in statutes for a citizen's standing to sue in environmental matters. This would automatically provide standing in court cases. Such a clause is present in the Michigan Environmental Protection Act of 1970. Concerns that such a provision would clog the courts with needless and trivial suits have proved groundless. Few other states, however, have adopted such a statute.

Public Trust Concept. Many other proposals have been made for legal concepts that would offer a means of reducing environmental degradation. Professor Joseph Sax of the University of Michigan Law School, who was involved in the writing of the Michigan Environmental Protection Act of 1970, feels that ". . . of all the concepts of our law, only the public trust doctrine seems to have the breadth and substantive content which might make it useful as a tool of general application for citizens seeking to develop a comprehensive legal approach to resource management problems." Rivers, lakes, seashores, parks, and other public lands can be considered as being held in trust for the public and should not be diverted for more restrictive or special interest group use.

Limits of Court Powers

Legal action in environmental matters must work within the abilities of the courts, and these are in fact rather limited. Courts can establish procedures, such as requiring the filing of an environmental impact statement; they can arbitrate the location of enterprises and ensure licensing procedures. But courts should not be expected to decide complicated scientific matters such as determining what specific emission standards should apply or what amount of sewage can be handled by Lake Erie's bacterial system. A further limitation of court action is that usually a transgression must have occurred, or at least be well along in the planning process, before legal action is possible. Consequently court action may not be the best approach to ensure sound decisions.

Limitations of Regulatory Agencies and Commissions

It is customary in the United States and in many other governments to assign regulatory powers to agencies, commissions, and boards. Agency members are often federal employees, whereas commissions and boards are often composed of persons either appointed by some elected official (perhaps the president or a governor) or elected by the public or seated by virtue of their position as head of an agency. In many states, commissions regulate commercial activities, water use, resource development, and utility rates.

A principal charge of such regulatory bodies is to act in the public's interest, but a practical liability of such regulatory bodies is that with time, the body may be strongly influenced by, or in fact dominated by, the industries, businesses, or trades they are to regulate. Appointed members of the regulatory body may have been employed in a business in the area being regulated, for example, or retiring body members may take jobs with businesses they formerly regulated. Although this is not necessarily bad, the danger is that it can lead to a similarity in thought and points of view on the part of the regulatory body and of the regulated industry or trade. The results may be an unconscious favoring of the regulated activity over the public interest.

Self-policing commissions, formed by organizations or industries or businesses or trades, are particularly subject to a lack of objectivity in their task. The most glaring example of the failure of a commission to regulate an industry is the International Whaling Commission (IWC), which has in fact allowed that industry to commit virtual suicide while ignoring environmental concerns.

Whales

Although popular images may picture the whaling industry as being at its peak in the days of wooden sailing ships, the most intense whaling activities have occurred since 1930. Technological advances have made whaling an awesomely efficient industry in this century. Ship-based helicopters locate the whales and then guide killer boats to the site. The killer boats hunt by sonar, killing the quarry with an explosive harpoon head attached to a nylon line with an 18-ton test strength. The dead whale is inflated with compressed air so it will not sink, and a radio beacon is attached to the catch so that the towboat can find it and tow it to the factory ship, where it is rapidly processed.

What regulation has been imposed on whaling catches has been done by the IWC, a commission formed after World War II by the 17 nations (including the United States) involved in whaling. The IWC was to regulate catches and ensure compliance of the member nations. Actually, the IWC was given few inspection or enforcement powers. Unfortunately the IWC chose to regulate catches, not by the individual species of whale, but by the blue whale unit. The blue whale, the largest animal ever to live on Earth, was the prize catch because it yielded the most oil and meat. One blue whale unit is one blue whale or the equivalent in other species: two fin whales, two and one half humpback whales, or six sei whales. The IWC imposed a yearly limit of 16,000 blue whale units. Such a regulatory scheme unfortunately means that the largest whales will be killed first, then the next largest will be pursued, and so on.

The results of this whaling policy have been an overexploitation of whales and the hunting almost to extinction of the larger whales. Figure 9.1 compares technological changes and whale catches over the past 40 years. In 1933 nearly 29,000 whales were caught, producing 2.6 million barrels of oil. By 1966 twice as many whales were caught, but the amount of oil produced was only 1.5

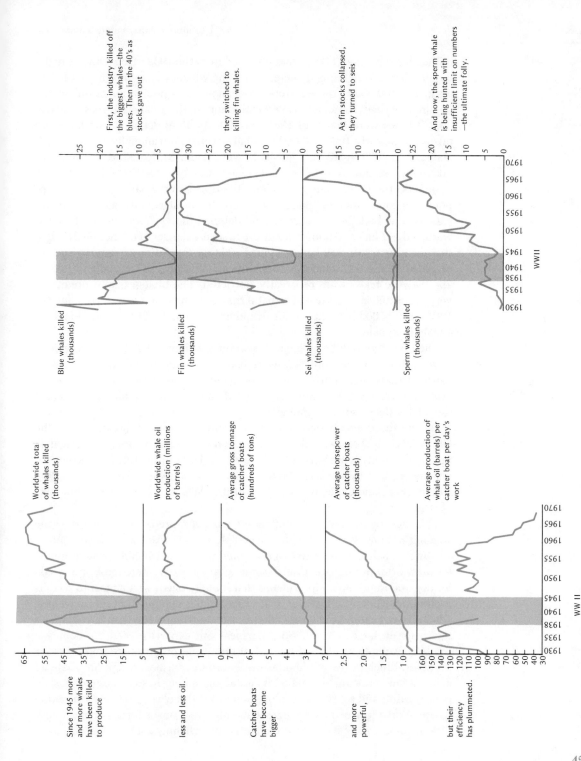

Figure 9.1. *The relative changes in the whaling industry and in whale catches. [From Population, Resources, Environment: Issues in Human Ecology, 2nd ed., by Paul R. Ehrlich and Anne H. Ehrlich, W. H. Freeman and Company. Copyright 1972. Original data from R. Payne, N.Y. Zoological Society Newsletter, November 1968. Reprinted with permission.]*

Worldwide total of whales killed (thousands)

Worldwide whale oil production (millions of barrels)

Average gross tonnage of catcher boats (hundreds of tons)

Average horsepower of catcher boats (thousands)

Average production of whale oil (barrels) per catcher boat per day's work

Since 1945 more and more whales have been killed to produce

less and less oil.

Catcher boats have become bigger

and more powerful,

but their efficiency has plummeted.

Blue whales killed (thousands)

Fin whales killed (thousands)

Sei whales killed (thousands)

Sperm whales killed (thousands)

First, the industry killed off the biggest whales—the blues. Then in the 40's as stocks gave out

they switched to killing fin whales.

As fin stocks collapsed, they turned to seis

And now, the sperm whale is being hunted with insufficient limit on numbers —the ultimate folly.

WW II

429

million barrels, about 60% of that obtained from the 1933 catch. Obviously the larger whales were being depleted. The blue whale, which once numbered an estimated 200,000 in the Antarctic alone, now has a population of only a few hundred, perhaps too small to avert extinction. In the 1930–1931 season nearly 30,000 blues were taken, in 1965 only 20. As blues declined, the fins were taken. In the early 1960s, 30,000 fin whales were taken annually, but the catch declined to 2000 by 1970 as fin stocks were depleted. No effort was made to take only a sustainable yield of the creatures.

In 1960 the IWC, concerned about the decreasing whale oil harvests, appointed a commission of biologists to study the problem and make recommendations. The biologists' report in 1963 detailed the exploitation problem and warned that blues and humpbacks were in severe danger of extinction. It urged that the blue whale unit be abandoned and that regulation be done by species of whale. It also recommended that blues and humpbacks be totally protected and that the take of fins be greatly restricted. The biologists also predicted, very accurately as it turned out, that if the blue whale unit were maintained in 1964, only 8500 blue whale units would be taken and that this would include 14,000 fin whales.

The IWC ignored the biologists' recommendations; 10,000 blue whale units were authorized, but only 8429 were taken including 13,900 fin whales, both figures astonishingly close to what the biologists predicted. Over the next 3 years, the biologists recommended that if the blue whale unit were used, it should be decreased, beginning at 4000, then declining to 2000 by 1967. Typically their recommendations were again ignored with Japan, Russia, the Netherlands, and Norway, the four nations involved in whaling, voting against the recommendations. Norway and the Netherlands ultimately dropped out of whaling so that by the late 1960s Japan and Russia were the major whaling nations (Peru and Chile, not members of the IWC, also do moderate amounts of whaling).

This picture of a self-regulated industry is not a pretty one. Since 1963 the whaling nations were aware that overexploitation meant an ultimate end to whaling. Whales, if harvested sanely, could provide a valuable protein source for the world. In practice, though some whale meat is eaten, most of it is sold for dog and mink food. Japan claims that it badly needs whale meat to feed its protein-poor people. However, half of its catch is composed of sperm whales, which are not eaten by anyone. Whale meat contributes less than 2% of Japan's total protein intake; in addition, Japanese fish exports exceed the domestic whale consumption. Most of the rest of the whale products are sold for other purposes. Much of the whale oil is made into margarine, and sperm whale oil makes a fine lubricant. Whale oil is also an ingredient in some cosmetics, shoe polish, paint, and soap. The exploitation of whales clearly reflects a situation where short-term economic gains triumph over longer-term benefits. Very little argument can be made for such intensive whaling as a basis for jobs, since only a few thousand are employed in the industry, and their jobs will vanish when the whales are extinct—which may be soon for the larger whales.

The loss of the larger whales may be important for several reasons in addition to the aesthetic and moral implications of exterminating such magnificent, gentle, and intelligent creatures. We know very little about whales, although we might benefit from detailed knowledge of how they maintain body temperatures in Arctic waters and how they navigate during migrations covering thousands of miles. Whales, like dolphins and bats, have an echo location system that serves as an excellent underwater radar system, but how the whales emit these sounds is unknown. Humpback whales, in particular, emit a great variety of underwater "calls" or "songs" that scientific analysis shows may last 30 minutes, then be repeated. Their purpose? No one knows.

We have little or no knowledge of how whales can survive the pressures of deep dives; sperm whales, for example, have been found entangled in cables 3700 ft below the surface where the pressure is over 2 tons/in.^{2}! Furthermore, whales apparently do not suffer from the bends during their rapid ascents to the surface. They presumably avoid the bends, which would kill a human diver coming up that rapidly from a mere fraction of the depth whales dive, by taking only a small amount of air and holding their breath for the duration of the dive. If this occurs, the question arises of how sperm whales can remain submerged for an hour and a half. A possible explanation is that they reduce their oxygen consumption by slowing their heartbeat and shutting down the flow of blood to all but the most important organs such as the heart and brain. Still another puzzle is how the whales, especially those that dive to great depths in search of food, regulate their buoyancy to compensate for the increasing density of water as they descend. Maintaining neutral buoyancy is presumably very important to a diving whale; otherwise, it would waste precious oxygen forcing its way down and back up to the surface.

The extinction of whales may prove to mean more than an irretrievable loss to our civilization. The principal food of most whales is *krill*, small shrimplike crustaceans found in abundance in the polar oceans. Krill feed on smaller animals (phytoplankton) and at times mass together in great shoals, measureable by the acre, in such concentrations that the color of the water changes. Whales skim through the water like great vacuum cleaners, straining out and consuming the krill. Before twentieth-century whaling so drastically reduced their numbers, the Antarctic whale herds alone ate an estimated 150 million tons/day of krill. They were an inestimable factor in regulating the krill-phytoplankton economy of the sea. What may happen to the seafood chain if the whale becomes extinct? The removal of such a potentially key element as the whale without careful consideration of the consequences might jeopardize the life of both man and whale.

In the early 1970s a moratorium on blue whale catches was finally achieved, but decisions on the others have lagged. In 1970 eight whale species were designated as endangered species by the U.S. Secretary of the Interior Walter Hickel and thus came under the protection of the 1969 Endangered Species Act. Under this Act, importation of whale products into the United States was to cease in December 1971. The 1972 U.N. Stockholm Conference on the Human Environment called for a ban on the killing of whales by a vote of 53 to 0; both the U.S. Senate and House of Representatives approved similar resolutions. But in 1973 and in 1974 the necessary three fourths majority of the IWC could not be found to vote for such a moratorium; there is, however, considerable doubt that Japan and Russia would obey such a moratorium even if it passed.

Political Action

Regardless of the impact of legal decisions and of public awareness campaigns, pushing for change in environmental and energy matters almost inevitably involves political action. It is in fact difficult to find a problem whose solution does not involve either direct political decisions by elected officials or politically influenced decisions by government agencies in response to public pressure. Direct political decisions are obviously required in such things as environmental legislation, federal research funding, and the utilization of natural resources. Zoning decisions in cities or suburbs, usually made by an appointed board, are subject to a final decision by a higher body such as a city council. Public utilities must raise money for energy systems by persuading the public to pass bond issues. Even those matters such as septic tank and sewage discharge permits or locations of sanitary landfills, usually handled routinely by bureaucratic agencies, can be strongly influenced by the opinions of elected officials.

A critical factor in nearly all political decisions is the interest and response of the public. An aggressive, involved stance by a portion of the community can have an important influence (for better or for worse) on the decisions. Public interest often precedes political action. The intense public concern with environmental problems in the late 1960s stimulated such environmental legislation as NEPA, the Clean Air Act of 1970 (Chapter 3), and the 1972 Amendments to the Federal Water Pollution Control Administration Act (Chapter 6). The cleanup of the Willamette River in Oregon (Chapter 6) involved enormous political action. Without public interest, legislative initiative in these areas would probably have been much less.

One prime example of the influence of public interest upon political decisions on environmental matters is offered by Project Sanguine, a proposed U.S. Navy submarine communications system. This project is a "last strike" capability, designed to transmit information to nuclear submarines after all other forms of communications are destroyed. It is a grid of underground cables, much like buried utilities cables, spaced a few miles apart. The entire assembly would cover several hundred square miles and cost over a billion dollars.

Only a few locations in the United States satisfied the geological requirements, the best site being in northern Wisconsin. Preliminary testing, done in a national forest in Wisconsin, resulted in strong public opposition. Critics cited the lack of information on the effects of low frequency electromagnetic radiation on ecosystems, questioned whether the system would in fact work, questioned the philosophy of adopting such a system, and worried that it might be a high priority target in the event of a nuclear attack. The public opposition grew despite extensive testing of environmental effects by the Navy, and political leaders in Wisconsin cooled toward the project. In particular, Senator Gaylord Nelson indicated some opposition.

The Navy then tentatively decided to try another location, a rural area in central Texas. Although the possible change was at first greeted warmly by elected officials in the area, public opposition soon changed the political climate. Public opposition to this site centered on environmental degradation and land use problems as well as on many of the objections raised in Wisconsin. One meeting in Llano, Texas, during the summer of 1972 attracted 1500 persons, an astonishing turnout for such a rural area. One United States senator from Texas who met with a small, highly motivated group of area landowners commented later that he "escaped town just ahead of the posse." Elected officials soon began to support the position that, since Wisconsin was apparently the best geologic site, perhaps the project should be located there. Eventually, Congress stopped, perhaps temporarily, funding for the project.

The activism of the late 1960s prompted many interested in environmental problems to pursue environmental improvement actively at the polls, attempting to get environmentally enlightened candidates elected to office. In 1972 Environmental Action, a private organization, was joined by a number of active environmental groups to push for the defeat of elected officials known to be insensitive to environmental problems. A "Dirty Dozen" list of House of Representatives members was compiled which listed the 12 key congressmen who represented blocks to the passage of environmental legislation., Environmental Action publicized the voting records of the "Dirty Dozen" and undertook a national fund raising campaign to provide funds for the campaigns of the opponents of the "Dirty Dozen". These activities were partly responsible for the defeat of Congressman Wayne Aspinall of Colorado. Aspinall was a powerful member of Congress who chaired the House Interior Committee and thus had enormous influence over the use of national forests and their resources, over the establishment of national parks and wilderness areas, and over federal environmental policy in general. Environmental Action felt he had favored special interests over environmental problems. Several others on the "Dirty Dozen" list also met defeat.

Many national and local organizations are actively engaged in lobbying on environmental matters and pursuing environmental change by political action at the polls. The Sierra Club, Wilderness Society, National Wildlife Federation, and Audubon Society have long-standing interests in environmental and energy legislation. The citizens lobby, Common Cause, has an active interest in these areas, and the League of Conservation Voters tabulates voting records of elected officials on environmental bills in the House and Senate and in many states. Several organizations active in environmental law are also active in Washington, lobbying for some bills and writing others. The National Resources Defense Council, formed by some of the members of the Scenic Hudson Preservation Conference, discussed earlier in this chapter, has a broad, continuing program in environmental law. The Environmental Defense Fund (Chapter 8), the Sierra Club Legal Defense Fund, and Ralph Nader's Center for Responsive Law are all active in environmental litigation.

Scientists can play an important role in providing input to such programs and in providing advice to elected officials. It is particularly important that public response be made with a firm knowledge of the facts, as known, about the problem. As noted in Chapter 1, science has a fundamental role in determining and presenting such information.

Land Use   An underlying issue in many of the topics of this book is the use of the land. As urbanization continues, it will be of increasing importance to make wise use of the land in order to minimize environmental disruption and maximize aesthetic and cultural benefits. Land use is important in environmental and energy problems. The location of power plants, highways, and sewage treatment plants affects air and water pollution and dictates the direction in which city expansion will occur. Coastal zones, especially important to aquatic ecosystems, are often hardest hit by man's intrusions: cities, industrial parks, refineries, and ports create disruption and pollution; draining and reclaiming marshes reduces the life support capacity of nearby ocean areas. Land use is important in agricultural production; it is often impossible to maintain agricultural operations at the edge of a city as land prices and taxes escalate. Keeping recreational and park lands represents a benefit to mental and physical health, but in the absence of planning the economic pressures to convert open space into apartments and commercial buildings is enormous.

Land use planning, an effort to regulate land use with the above (and other) problems in mind, is much more prevalent in European countries and other "older" nations than in the United States. In fact, land use planning has been alien to the freewheeling view of landownership in the United States. Urban pressure, however, has already resulted in at least partial regulation of land use. Zoning laws in many United States cities are intended to keep industrial and commercial activities separated from residences and to keep high-rise apartments separated from one-family residential areas.

Land use planning will undoubtedly be increasingly important in the United

States in the immediate future, and it is imperative that environmental and energy considerations be strong factors in those policies.

Over the past decade several completely new cities have been developed in this country. These "new towns" plan the layout of homes and businesses according to fresh ideas. Transportation by auto, for example, is separated from other modes of travel so that children can get to school using bike and footpaths that do not cross streets; adults can do most of their shopping without relying on the automobile. Careful use is made of the local topography to avoid erosion, scarred hillsides, and urban blight. Interconnecting open green spaces provide a pleasant diversion from more traditional urban scenes.

Land use planning is a complex, demanding task and many problems must be overcome before it can be effective.

A Look into the Future

What would be the end result of continuing indefinitely our present policies of resource consumption, population growth, and industrial growth? The Club of Rome, an organization of prominent European scientists, businessmen, and politicians, attempted to find the answer by funding the development of a computerized model of world dynamics. The Club chose Jay Forrester of MIT to head the project, and he and Dennis Meadows, a systems analyst, developed the model in 1972. A *perfect* model is obviously not achievable, but Forrester and Meadows attempted to approximate the world by assuming that five global variables—pollution, population, food production, industrialization, and the consumption of natural resources—could be used to describe the state of the world. Even this involves gross assumptions. Assigning one variable to pollution, for instance, means lumping together all types of pollution, being unable to distinguish between water pollution and air pollution, sulfur oxides and carbon monoxide, metals and chlorinated hydrocarbons.

Nevertheless, Forrester and Meadows worked out 100 causal relationships (or equations) between these five variables. These relationships express mathematically how pollution is related to industrial output and population growth, for example, and how consumption of natural resources changes as the other variables increase. The modeling procedure was to program the relationships on a computer, using data of past world history, and then altering the 100 relationships until the model accurately reproduced the past history of the five variables. Then, assuming the model was accurate, it could be used to predict future changes in the variables in response to various scenarios of industrial expansion, population growth, recycling of natural materials, and so on.

The result, if it represents even a crude approximation to world dynamics, has an urgent and alarming message; if current physical, economic, and social relationships continue unchanged, nonrenewable resources will be depleted, followed eventually by a soaring death rate and a rapid drop in world popula-

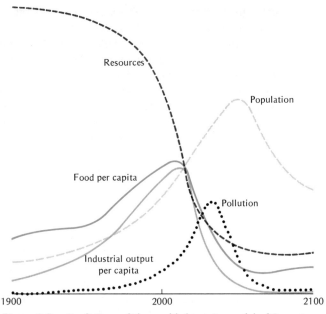

Figure 9.2. *Predictions of the world dynamics model of Forrester and Meadows. The variables are plotted on different vertical scales but are combined on the same graph to emphasize modes of behavior. Pollution is plotted as a multiple of the 1970 level, and resources are plotted as a fraction of 1900 reserves. [After* The Limits to Growth, *Universe Books, New York, 1972.]*

tion. The collapse is shown in Figure 9.2. An important aspect of the model is that even substantive changes in present policies, such as increasing the recycling of natural resources and increasing the reliance on nuclear fuels to save fossil fuels, have very little impact on the shape of the curves in Figure 9.2. They merely delay the collapse by a few decades. The only means by which collapse could be avoided in this model was to increase the doubling time of population and industrial growth to something like 1000 years. In short, the world must switch to nearly a steady state economy.

The model of Forrester and Meadows may be only an exercise in computer modeling. Much criticism has been leveled at the study, in particular that too much emphasis has been given to the results of such a crude model. Other critics have focused attention on more technical aspects of the programming used in the model. It is important to keep in mind that the model at best represents an extrapolation of present technology and present trends. It does not anticipate new technology, such as the utilization of solar energy or some other form of energy perhaps as yet unknown. One critic has modified the model to include a constantly advancing technology, and the results avert a collapse.

The model developed by Forrester and Meadows does not present an optimisic view of the future. If its predictions prove even approximately correct, it represents a tragic legacy of our civilization. It is small comfort, however, to seek consolation in the fact that the model is so crude relative to the real world that it may have no application whatsoever. Although the model predicts Hell on Earth in 50 years, hell already exists in Calcutta and numerous other large cities in the underdeveloped nations.

> What is civilization. . . . We sang songs that carried in their melodies
> all the sounds of nature—the running waters, the sighing of winds,
> and the calls of the animals. Teach these to your children that
> they may come to love nature as we love it.
>
> Grand Council Fire of American Indians
> from *I Have Spoken* by V. Armstrong

Summary

There is little evidence that good will and altruism will lead to desirable environmental change, but a number of legal and political actions may offer the means for constructive change.

Common law doctrines such as nuisance and trespass offer a means of redress in some instances, but in general they are limited because they favor only the immediately affected landowners. A variety of state and federal statutory laws apply to environmental problems. The National Environmental Policy Act of 1969 requires an environmental impact statement before federal funds can be used for large projects. In the long term, the major effect of this act may be to raise the consciousness of the public and government in environmental matters. Except for this act, courts have been largely inaccessible to citizens involved in environmental matters because standing is seldom granted. Recent court rulings have broadened the standing concept to include local and national organizations that have an interest in environmental problems. Declaring public lands and beaches to be a public trust, not to be used to benefit special interests, may offer a general means of reducing environmental degradation in those areas.

The use of commissions, boards, and governmental agencies as regulatory bodies has a potential glaring weakness: the slow development of a close association between the regulatory body and the activity to be regulated, so that the public's best interests may be slighted.

Organized efforts to improve the environment by political action, through the election of environmentally enlightened candidates and through lobbying for effective environmental legislation, may offer the best long-term solution to environmental problems. It is important that such efforts begin with all facts at hand, and providing these facts is an important role for scientists.

Review and Study Questions

1. Do you agree with the conclusion of the author that altruism and goodwill are not likely to bring about environmental improvement? State reasons for your answer.
2. Define or describe injunction, standing to sue, nuisance, natural flow theory, "Dirty Dozen," Project Sanguine, blue whale unit, Refuse Act of 1899.
3. What are two important aspects of NEPA?
4. Professor Joseph Sax is involved in what two ideas discussed in this chapter?
5. Discuss the chain of events leading to the decimation of the large whales. How might such international transgressions be prevented?
6. Discuss the goals of Environmental Action's "Dirty Dozen" campaign. What were the results?
7. Name three national organizations active in legal and political aspects of environmental change.
8. Give three examples of how land use has had an impact on you personally. Would land use planning have made significant changes in this impact?
9. What variables were used by Forrester and Meadows in their model of the world?
10. What have you done to reduce your energy consumption and environmental impact?

Questions Requiring Outside Reading.

11. What are the voting records on environmental matters of your representative and senators in Washington? (Contact the League of Conservation Voters, Suggested Outside Reading.)
12. How effective might special taxes and pollution residuals charges be in improving the environment? (A. M. Freeman, II, and R. H. Haverman, "Residuals Charges for Pollution Control: A Policy Evaluation," *Science* 117 [July 28, 1972], 322–329.)
13. The EPA's authority to issue permits to polluters was established in the courts by what legal decisions? ("Water Polluters Tangle with the Law—and Lose," *Business Week,* October 9, 1971, 64–65.)
14., What is the impact of adding to the Forrester-Meadows model a constantly improving technology? (Robert Boyd, "World Dynamics: A Note," *Science* 117 [August 11, 1972], 516–519.)

15. How were the calls of the humpback whale recorded and analyzed? (See McNulty, Payne, and McVea, Suggested Outside Reading.)

Suggested Outside Reading

Audubon **77** (January 1975). An entire issue devoted to dolphins and whales, including the history of whaling and the efforts to achieve a moratorium.

CARTER, L. J. "Environmental Law. I and II," *Science* **179** I (March 23, 1973), 1205–1209; II (March 30, 1973), 1310–1312. Discusses this emerging field, its problems, its successes, and its principal figures.

————. "Land Use Law. I and II," *Science* **182** I (November 16, 1973), 691–697; II (November 30, 1973), 902–908. An excellent series describing the problems inherent in writing land use laws. Florida is singled out for close study.

GILLETTE, ROBERT. "The Limits to Growth: Hard Sell for a Computer View of Doomsday," *Science* **175** (March 10, 1972), 1088–1091. A reportorial report on the Forrester and Meadows model and the publicity that surrounded its publication.

LEVI, DONALD R., and DALE COLYER. "Legal Remedies for Pollution Abatement," *Science* **175** (March 10, 1972), 1085–1087. A general article discussing common law and statutory law concepts applicable to pollution control.

McHARG, IAN. *Design With Nature*. Natural History Press, Garden City, New York, 1969. An effective scheme of land use planning developed and used by the author, a noted expert. Environmental impact is a special feature of the plan.

McNULTY, FAITH. *The Great Whales*. Doubleday & Company, Inc., Garden City, New York, 1974; and "Lord of the Fish," *New Yorker,* August 6, 1973, 38–67. A look at whales, whaling, and the IWC. Included is a description of trips with Roger Payne to record whale songs.

McVAY, SCOTT. "The Last of the Great Whales," *Scientific American,* August 1960, 13–21. A look at the larger whales, the whaling industry, and the overexploitation of these creatures.

MEADOWS, D. H., D. L. MEADOWS, J. RANDERS, and W. W. BEHRENS, II. *The Limits to Growth,* Potomac Associates—Universe Books, Inc., N.Y., 1972. The popularized version of the Forrester and Meadows model of world dynamics.

PAYNE, ROGER S., and SCOTT McVAY. "Songs of the Humpback Whales," *Science* **173** (August 13, 1971), 585–597. A scientific analysis of the songs is presented together with general information on whales. Both authors have been active in attempting to achieve a moratorium on hunting the large whales.

SAX, JOSEPH L. *Defending the Environment*. Alfred A. Knopf, Inc., New York,

1971. A strategy for citizen action written by a man eminent in environmental law. In this book he explains at length the need for and the role of citizen-initiated litigation. Written for a general audience.

Some organizations active in environmental change by legal and political means. All publish regular newsletters and have other information available:

Animal Welfare Institute, P.O. Box 3650, Washington, D.C. 20007

Committee for Environmental Information, 438 N. Skinker Blvd., St. Louis, Missouri 63130. Publishers of *Environment* Magazine.

Environmental Action, Room 200, 2000 P Street, N.W., Washington, D.C. 20036

Environmental Defense Fund, 162 Old Town Road, East Setauket, New York 11733

Federation of American Scientists, 203 C Street, N.E., Washington, D.C. 20002

Friends of the Earth, 30 E. 42nd Street, New York, N.Y. 10017

League of Conservation Voters, 324 C Street, S. E., Washington, D.C. 20003

National Audubon Society, 1130 Fifth Avenue, New York, N.Y. 10028

National Resources Defense Council, 15 W. 44th Street, New York, N.Y. 10036

National Wildlife Federation, 1412 16th Street, N.W., Washington, D.C. 20036

Scientists Institute for Public Information, 30 E. 69th Street, New York, N.Y. 10021

Sierra Club, 1050 Mills Tower, San Francisco, California 94104

Society for Protective Animal Legislation, P.O. Box 3719, Georgetown Station, Washington, D.C. 20007

The Wilderness Society, 729 Fifteenth Street, N.W., Washington, D.C. 20005

Zero Population Growth, Inc., 4080 Fabian Way, Palo Alto, California 94303

INDEX*

*Abbreviations used in index: *t.* (table); *f.* (figure).